# Stochastic Models

CONTEMPORARY MATHEMATICS

336

APORTACIONES MATEMÁTICAS
SOCIEDAD MATEMÁTICA MEXICANA

# Stochastic Models

Seventh Symposium on
Probability and Stochastic Processes
June 23 – 28, 2002
Mexico City, Mexico

José M. González-Barrios
Jorge A. León
Ana Meda
Editors

American Mathematical Society
Providence, Rhode Island

This volume contains the proceedings of the Seventh Symposium on Probability and Stochastic Processes that was held in Mexico City, Mexico, in the Antiguo Colegio de San Ildefonso, National University of Mexico (UNAM), from June 23–28, 2002.

2000 *Mathematics Subject Classification.* Primary 60E15, 60F10, 60G15, 60G50, 60H05, 60H10, 60H15, 60J60, 91B26, 91B30.

---

**Library of Congress Cataloging-in-Publication Data**

Symposium on Probability and Stochastic Processes (7th : 2002 : Mexico City, Mexico)
Stochastic models : Seventh Symposium on Probability and Stochastic Processes, June 23–28, 2002, Mexico City, Mexico / José M. González-Barrios, Jorge A. León, Ana Meda, editors.
p. cm. — (Contemporary mathematics, ISSN 0271-4132 ; 336) (Aportaciones matemáticas)
Includes bibliographical references.
ISBN 0-8218-3466-5 (alk. paper)
1. Stochastic analysis—Congresses. I. González-Barrios, José M., 1957–. II. León, J. A. (León Vázquez, Jorge A.), 1960–. III. Meda, Ana, 1965–. IV. Title. V. Series. VI. Contemporary mathematics (American Mathematical Society) ; v. 336.

QA274.2 .S96 2002
519.2′2—dc22    2003062763

---



♾ The paper used in this book is acid-free and falls within the guidelines
established to ensure permanence and durability.
Visit the AMS home page at http://www.ams.org/

10 9 8 7 6 5 4 3 2 1    08 07 06 05 04 03

# Contents

# Preface

The Probability and Stochastic Processes Symposia have been held in different cities of Mexico since 1988, every two years. The main purpose of these events is to promote Probability and Stochastic Processes among researchers and students interested in these and related fields.

The Seventh Symposium on Probability and Stochastic Processes was held in Mexico City in the Antiguo Colegio de San Ildefonso, National University of Mexico (UNAM) from June 23rd to June 28th, 2002.

This volume "Stochastic Models," Proceedings of the Seventh Symposium on Probability and Stochastic Processes, contains the lecture notes of three courses and a collection of refereed original research papers by participants in this event.

Besides the three courses, the Seventh Symposium included five invited talks and several short communications. The courses, which were lectured by David Nualart, Esa Nummelin and Winfried Stute, surveyed topics on fractional Brownian motion, credit risk, economic equilibrium and financial markets, and present some of the most recent advances in these fields. The five invited conferences were given by Ma. Emilia Caballero, Begoña Fernández, Jin Ma, Philip Protter and Jaime San Martín.

The research papers in this volume cover topics related to the courses, as well as several other topics of current interest in Probability and Stochastic Processes and their applications.

We are deeply grateful to Karim Anaya, José Luis Enríquez and Anabel Lagos for helping us with the editorial work of these proceedings.

Finally we acknowledge the financial support of Conacyt (National Science and Technology Council) through grants 32705-E and 37130-E.

José M. González-Barrios
Jorge A. León
Ana Meda

# Lecture Notes

Contemporary Mathematics
Volume **336**, 2003

# Stochastic Integration with Respect to Fractional Brownian Motion and Applications

David Nualart

ABSTRACT. Fractional Brownian motion (fBm) is a Gaussian stochastic process $B = \{B_t, t \geq 0\}$ with zero mean and covariance function given by $E(B_t B_s) = \frac{1}{2}\left(s^{2H} + t^{2H} - |t-s|^{2H}\right)$, where $0 < H < 1$ is the Hurst parameter. This process has stationary increments, self-similarity and long-range dependence properties. These properties make fBm a suitable driving noise in different applications like mathematical finance and network traffic analysis. In order to develop these applications, one needs to construct a stochastic calculus with respect to fBm. In the particular case $H = \frac{1}{2}$, the process $B$ is an ordinary Brownian motion but for $H \neq \frac{1}{2}$ it is not a semimartingale and we cannot use the classical Itô calculus. The objective of these notes is to present some recent advances in the stochastic calculus with respect to fractional Brownian motion (fBm) and their applications.

## 1. Fractional Brownian motion

A centered Gaussian process $B = \{B_t, t \geq 0\}$ is called *fractional Brownian motion* (fBm) of Hurst parameter $H \in (0,1)$ if it has the covariance function

$$R_H(t,s) = E(B_t B_s) = \frac{1}{2}\left(s^{2H} + t^{2H} - |t-s|^{2H}\right). \tag{1.1}$$

This process was first introduced by Kolmogorov [**K**] and studied by Mandelbrot and Van Ness in [**M-N**], where a stochastic integral representation in terms of a standard Brownian motion was established.

Fractional Brownian motion has the following *self-similar* property: For any constant $a > 0$, the processes $\left\{a^{-H} B_{at}, t \geq 0\right\}$ and $\{B_t, t \geq 0\}$ have the same distribution. This property is an immediate consequence of the fact that the covariance function (1.1) is homogeneous of order $2H$.

From (1.1) we can deduce the following expression for the variance of the increment of the process in an interval $[s,t]$:

$$E\left(|B_t - B_s|^2\right) = |t-s|^{2H}. \tag{1.2}$$

---

2000 *Mathematics Subject Classification.* Primary: 60H05; Secondary: 60H07, 60H10, 60G18.

*Key words and phrases.* Fractional Brownian motion. Stochastic integral. Malliavin calculus. Itô's formula. Stochastic differential equations driven by fractional Brownian motion.

This implies that fBm has *stationary increments.*

By Kolmogorov's continuity criterion and (1.2) we deduce that fBm has a version with continuous trajectories. Moreover, by Garsia-Rodemich-Rumsey inequality, we can deduce the following modulus of continuity for the trajectories of fBm: For all $\varepsilon > 0$ and $T > 0$, there exists a nonnegative random variable $G_{\varepsilon,T}$ such that $E\left(|G_{\varepsilon,T}|^p\right) < \infty$ for all $p \geq 1$, and

$$|B_t - B_s| \leq G_{\varepsilon,T}|t-s|^{H-\varepsilon},$$

for all $s, t \in [0, T]$. In other words, the parameter $H$ controls the regularity of the trajectories, which are Hölder continuous of order $H - \varepsilon$, for any $\varepsilon > 0$.

For $H = \frac{1}{2}$, the covariance can be written as $R_{\frac{1}{2}}(t,s) = t \wedge s$, and the process $B$ is an ordinary Brownian motion. In this case the increments of the process in disjoint intervals are independent. However, for $H \neq \frac{1}{2}$, the increments are not independent. The covariance between two increments $B_{t+h} - B_t$ and $B_{s+h} - B_s$, where $s + h \leq t$, and $t - s = nh$ is

$$\begin{aligned} \rho_H(n) &= \frac{1}{2}h^{2H}\left((n+1)^{2H} + (n-1)^{2H} - 2n^{2H}\right) \\ &\approx h^{2H}H(2H-1)n^{2H-2} \to 0 \end{aligned}$$

as $n$ tends to infinity. Therefore:

i) If $H > \frac{1}{2}$, $\rho_H(n) > 0$ and $\sum_{n=1}^{\infty} \rho_H(n) = \infty$.
ii) If $H < \frac{1}{2}$, $\rho_H(n) < 0$ and $\sum_{n=1}^{\infty} |\rho_H(n)| < \infty$.

In case i) two increments of the form $B_{t+h} - B_t$ and $B_{t+2h} - B_{t+h}$, are positively correlated and the process presents an aggregation behavior. In case ii) these increments are negatively correlated and we say that there is intermittency.

**1.1. Moving average representation.** Mandelbrot and Van Ness obtained in [**M-N**] the following integral representation of fBm in terms of a Wiener process on the whole real line (see also Samorodnitsky and Taqqu [**S-T**]):

$$B_t = \frac{1}{C_1(H)} \int_{\mathbb{R}} \left[\left((t-s)^+\right)^{H-\frac{1}{2}} - \left((-s)^+\right)^{H-\frac{1}{2}}\right] dW_s,$$

where $\{W(A), A \text{ Borel subset of } \mathbb{R}\}$ is a Brownian measure on $\mathbb{R}$ and

$$C_1(H) = \left(\int_0^\infty \left((1+s)^{H-\frac{1}{2}} - s^{H-\frac{1}{2}}\right)^2 ds + \frac{1}{2H}\right)^{\frac{1}{2}}.$$

PROOF. Set $f_t(s) = \left((t-s)^+\right)^{H-\frac{1}{2}} - \left((-s)^+\right)^{H-\frac{1}{2}}$, $s \in \mathbb{R}$, $t \geq 0$. Notice that $\int_{\mathbb{R}} f_t(s)^2 ds < \infty$. In fact, if $H \neq \frac{1}{2}$, as $s$ tends to $-\infty$, $f_t(s)$ behaves as $(-s)^{H-\frac{3}{2}}$ which is square integrable at infinity. For $t \geq 0$ set

$$X_t = \int_{\mathbb{R}} \left[\left((t-s)^+\right)^{H-\frac{1}{2}} - \left((-s)^+\right)^{H-\frac{1}{2}}\right] dW_s.$$

We have

$$\begin{aligned} E(X_t^2) &= \int_{\mathbb{R}} \left[ \left((t-s)^+\right)^{H-\frac{1}{2}} - \left((-s)^+\right)^{H-\frac{1}{2}} \right]^2 ds \\ &= t^{2H} \int_{\mathbb{R}} \left[ \left((1-u)^+\right)^{H-\frac{1}{2}} - \left((-u)^+\right)^{H-\frac{1}{2}} \right]^2 du \\ &= t^{2H} \left( \int_{-\infty}^0 \left[ (1-u)^{H-\frac{1}{2}} - (-u)^{H-\frac{1}{2}} \right]^2 du + \int_0^1 (1-u)^{2H-1} du \right) \\ &= C_1(H)^2 t^{2H}. \qquad (1.3) \end{aligned}$$

Similarly, for any $s < t$ we obtain

$$\begin{aligned} E(|X_t - X_s|^2) &= \int_{\mathbb{R}} \left[ \left((t-u)^+\right)^{H-\frac{1}{2}} - \left((s-u)^+\right)^{H-\frac{1}{2}} \right]^2 du \\ &= \int_{\mathbb{R}} \left[ \left((t-s-u)^+\right)^{H-\frac{1}{2}} - \left((-u)^+\right)^{H-\frac{1}{2}} \right]^2 du \\ &= C_1(H)^2 |t-s|^{2H}. \qquad (1.4) \end{aligned}$$

From (1.3) and (1.4) we deduce that the centered Gaussian process $\{X_t, t \geq 0\}$ has the covariance $R_H$ of a fBm with Hurst parameter $H$. □

Notice that the above integral representation implies that the function $R_H$ defined in (1.1) is a covariance function, that is, it is symmetric and nonnegative definite.

It is also possible to establish the following spectral representation of fBm (see Samorodnitsky and Taqqu [**S-T**]):

$$B_t = \frac{1}{C_2(H)} \int_{\mathbb{R}} \frac{e^{its} - 1}{is} |s|^{\frac{1}{2}-H} d\widetilde{W}_s,$$

where $\widetilde{W} = W^1 + iW^2$ is a complex Gaussian measure on $\mathbb{R}$ such that $W^1(A) = W^1(-A)$, $W^2(A) = -W^2(A)$, and $E(W^1(A)^2) = E(W^2(A)^2) = \frac{|A|}{2}$, and

$$C_2(H) = \left( \frac{\pi}{H\Gamma(2H) \sin H\pi} \right)^{\frac{1}{2}}.$$

**1.2. Hurst's statistical phenomenon of self-similarity.** Hurst developed in [**Hur**] a statistical analysis of the yearly water run-offs of Nile river. Suppose that $x_1, \ldots, x_n$ are the values of $n$ successive yearly water run-offs. Denote by $X_n = \sum_{k=1}^n x_k$ the cumulative values. Then, $X_k - \frac{k}{n} X_n$ is the deviation of the cumulative value $X_k$ corresponding to $k$ successive years from the empirical mean as calculated using data for $n$ years. Consider the range of the amplitude of this deviation:

$$\mathcal{R}_n = \max_{1 \leq k \leq n} \left( X_k - \frac{k}{n} X_n \right) - \min_{1 \leq k \leq n} \left( X_k - \frac{k}{n} X_n \right)$$

and the empirical mean deviation

$$\mathcal{S}_n = \sqrt{\frac{1}{n} \sum_{k=1}^n \left( x_k - \frac{X_n}{n} \right)^2}.$$

Based on the records of observations of Nile flows in 622-1469, Hurst discovered that $\frac{\mathcal{R}_n}{\mathcal{S}_n}$ behaves as $cn^H$, where $H = 0.7$. On the other hand, the partial sums

$x_1 + \cdots + x_n$ have approximately the same distribution as $n^H x_1$, where again $H$ is a parameter larger than $\frac{1}{2}$.

These facts lead to the conclusion that one cannot assume that $x_1, \ldots, x_n$ are values of a sequence of independent and identically distributed random variables. Some alternative models are required in order to explain the empirical facts. One possibility is to assume that $x_1, \ldots, x_n$ are values of the increments of a fractional Brownian motion. Motivated by these empirical observations, Mandelbrot has given the name of Hurst parameter to the parameter $H$ of fBm.

**1.3. Non semimartingale property.** We have seen that for $H \neq \frac{1}{2}$ fBm does not have independent increments. In this subsection we will show that for $H \neq \frac{1}{2}$ fBm is not a semimartingale. A proof in the case $H > \frac{1}{2}$ can be found in [**Li**] (see also Example 4.9.2 in Liptser and Shiryaev [**L-S**]). We will present here the proof given by Rogers in [**Ro**] for any $H \neq \frac{1}{2}$. The main arguments of this proof are as follows. For $p > 0$ set

$$Y_{n,p} = n^{pH-1} \sum_{j=1}^{n} \left| B_{j/n} - B_{(j-1)/n} \right|^p .$$

By the self-similar property of fBm, the sequence $\{Y_{n,p}, n \geq 1\}$ has the same distribution as $\{\widetilde{Y}_{n,p}, n \geq 1\}$, where

$$\widetilde{Y}_{n,p} = n^{-1} \sum_{j=1}^{n} \left| B_j - B_{j-1} \right|^p .$$

The stationary sequence $\{B_j - B_{j-1}, j \geq 1\}$ is mixing. Hence, by the Ergodic Theorem $\widetilde{Y}_{n,p}$ converges almost surely and in $L^1$ to $E(|B_1|^p)$ as $n$ tends to infinity. As a consequence, $Y_{n,p}$ converges in probability as $n$ tends to infinity to $E(|B_1|^p)$. Therefore,

$$V_{n,p} = \sum_{j=1}^{n} \left| B_{j/n} - B_{(j-1)/n} \right|^p$$

converges in probability to zero as $n$ tends to infinity if $pH > 1$, and to infinity if $pH < 1$. Consider the following two cases:

i) If $H < \frac{1}{2}$, we can choose $p > 2$ such that $pH < 1$, and we obtain that the $p$-variation of fBm (defined as the limit in probability $\lim_{n \to \infty} V_{n,p}$) is infinite. Hence, the quadratic variation ($p = 2$) is also infinity.

ii) If $H < \frac{1}{2}$, we can choose $p$ such that $\frac{1}{H} < p < 2$. Then the $p$-variation is zero, and, as a consequence, the quadratic variation is also zero. On the other hand, if we choose $p$ such that $1 < p < \frac{1}{H}$ we deduce that the total variation is infinite.

Therefore, we have proved that for $H \neq \frac{1}{2}$ fBm cannot be a semimartingale.

In a recent paper [**Che1**] Cheridito has introduced the notion of *weak semimartingale* as a stochastic process $\{X_t, t \geq 0\}$ such that for each $T > 0$, the set of random variables

$$\left\{ \sum_{j=1}^{n} f_j (B_{t_j} - B_{t_{j-1}}), n \geq 1, 0 \leq t_0 < \cdots < t_n \leq T, \right.$$
$$\left. |f_j| \leq 1, f_j \text{ is } \mathcal{F}_{t_{j-1}}^X\text{-measurable} \right\}$$

is bounded in $L^0$, where for each $t \geq 0$, $\mathcal{F}_t^X$ is the $\sigma$-field generated by the random variables $\{X_s, 0 \leq s \leq t\}$. It is important to remark that this $\sigma$-field is not completed with the null sets. Then, in [**Che1**] it is proved that fBm is not a weak semimartingale if $H \neq \frac{1}{2}$.

Let us mention the following surprising result also proved in [**Che1**]. Suppose that $\{B_t, t \geq 0\}$ is a fBm with Hurst parameter $H \in (0,1)$, and $\{W_t, t \geq 0\}$ is an ordinary Brownian motion. Assume they are independent. Set

$$M_t^H = B_t + W_t.$$

Then $\{M_t, t \geq 0\}$ is not a weak semimartingale if $H \in (0, \frac{1}{2}) \cup (\frac{1}{2}, \frac{3}{4}]$, and it is a semimartingale, equivalent in law to Brownian motion on any finite time interval $[0,T]$, if $H \in (\frac{3}{4}, 1)$.

**1.4. Fractional integrals and derivatives.** In this subsection we will recall the basic definitions and properties of the fractional calculus. For a detailed presentation of these notions we refer to [**S-K-M**].

Let $a, b \in \mathbb{R}$, $a < b$. Let $f \in L^1(a,b)$ and $\alpha > 0$. The left and right-sided *fractional integrals* of $f$ of order $\alpha$ are defined for almost all $x \in (a,b)$ by

$$I_{a+}^{\alpha} f(x) = \frac{1}{\Gamma(\alpha)} \int_a^x (x-y)^{\alpha-1} f(y)\, dy \tag{1.5}$$

and

$$I_{b-}^{\alpha} f(x) = \frac{1}{\Gamma(\alpha)} \int_x^b (y-x)^{\alpha-1} f(y)\, dy, \tag{1.6}$$

respectively. Let $I_{a+}^{\alpha}(L^p)$ (resp. $I_{b-}^{\alpha}(L^p)$) the image of $L^p(a,b)$ by the operator $I_{a+}^{\alpha}$ (resp. $I_{b-}^{\alpha}$).

If $f \in I_{a+}^{\alpha}(L^p)$ (resp. $f \in I_{b-}^{\alpha}(L^p)$) and $0 < \alpha < 1$ then the left and right-sided *fractional derivatives* are defined by

$$D_{a+}^{\alpha} f(x) = \frac{1}{\Gamma(1-\alpha)} \left( \frac{f(x)}{(x-a)^{\alpha}} + \alpha \int_a^x \frac{f(x)-f(y)}{(x-y)^{\alpha+1}} dy \right), \tag{1.7}$$

and

$$D_{b-}^{\alpha} f(x) = \frac{1}{\Gamma(1-\alpha)} \left( \frac{f(x)}{(b-x)^{\alpha}} + \alpha \int_x^b \frac{f(x)-f(y)}{(y-x)^{\alpha+1}} dy \right) \tag{1.8}$$

for almost all $x \in (a,b)$ (the convergence of the integrals at the singularity $y = x$ holds point-wise for almost all $x \in (a,b)$ if $p = 1$ and moreover in $L^p$-sense if $1 < p < \infty$).

Recall the following properties of these operators:

- If $\alpha < \dfrac{1}{p}$ and $q = \dfrac{p}{1-\alpha p}$ then
$$I_{a+}^{\alpha}(L^p) = I_{b-}^{\alpha}(L^p) \subset L^q(a,b).$$
- If $\alpha > \dfrac{1}{p}$ then
$$I_{a+}^{\alpha}(L^p) \cup I_{b-}^{\alpha}(L^p) \subset C^{\alpha - \frac{1}{p}}(a,b),$$
where $C^{\alpha-\frac{1}{p}}(a,b)$ denotes the space of $\left(\alpha - \frac{1}{p}\right)$-Hölder continuous functions of order $\alpha - \frac{1}{p}$ in the interval $[a,b]$.

The following inversion formulas hold:

$$I_{a+}^{\alpha}\left(D_{a+}^{\alpha}f\right)=f$$

for all $f\in I_{a+}^{\alpha}\left(L^{p}\right)$, and

$$D_{a+}^{\alpha}\left(I_{a+}^{\alpha}f\right)=f$$

for all $f\in L^{1}\left(a,b\right)$. Similar inversion formulas hold for the operators $I_{b-}^{\alpha}$ and $D_{b-}^{\alpha}$.

We will make use of the following *integration by parts formula*:

$$\int_{a}^{b}\left(D_{a+}^{\alpha}f\right)(s)g(s)ds=\int_{a}^{b}f(s)\left(D_{b-}^{\alpha}g\right)(s)ds, \tag{1.9}$$

for any $f\in I_{a+}^{\alpha}\left(L^{p}\right)$, $g\in I_{b-}^{\alpha}\left(L^{q}\right)$, $\frac{1}{p}+\frac{1}{q}=1$.

**1.5. Representation of fBm on an interval.** Fix a time interval $[0,T]$. Consider a fBm $\{B_t, t\in[0,T]\}$ with Hurst parameter $H\in(0,1)$. We denote by $\mathcal{E}$ the set of step functions on $[0,T]$. Let $\mathcal{H}$ be the Hilbert space defined as the closure of $\mathcal{E}$ with respect to the scalar product

$$\left\langle \mathbf{1}_{[0,t]},\mathbf{1}_{[0,s]}\right\rangle_{\mathcal{H}}=R_H(t,s).$$

The mapping $\mathbf{1}_{[0,t]}\longrightarrow B_t$ can be extended to an isometry between $\mathcal{H}$ and the Gaussian space $H_1(B)$ associated with $B$. We will denote this isometry by $\varphi\longrightarrow B(\varphi)$.

In this subsection we will establish the representation of fBm as a Volterra process, following the lines of [**A-N**] (case $H>\frac{1}{2}$) and [**A-M-N2**] (general case).

1.5.1. *Case* $H>\frac{1}{2}$. It is easy to see that the covariance of fBm can be written as

$$R_H(t,s)=\alpha_H\int_0^t\int_0^s|r-u|^{2H-2}dudr, \tag{1.10}$$

where $\alpha_H=H(2H-1)$. Formula (1.10) implies that

$$\langle\varphi,\psi\rangle_{\mathcal{H}}=\alpha_H\int_0^T\int_0^T|r-u|^{2H-2}\varphi_r\psi_u dudr \tag{1.11}$$

for any pair of step functions $\varphi$ and $\psi$ in $\mathcal{E}$.

We can write

$$\begin{aligned}|r-u|^{2H-2} &= \frac{(ru)^{H-\frac{1}{2}}}{\beta(2-2H,H-\frac{1}{2})}\\ &\quad\times\int_0^{r\wedge u}v^{1-2H}(r-v)^{H-\frac{3}{2}}(u-v)^{H-\frac{3}{2}}dv,\end{aligned} \tag{1.12}$$

where $\beta$ denotes the Beta function. Let us show Equation (1.12). Suppose $r > u$. By means of the change of variables $z = \frac{r-v}{u-v}$ and $x = \frac{r}{uz}$, we obtain

$$\begin{aligned}
&\int_0^u v^{1-2H}(r-v)^{H-\frac{3}{2}}(u-v)^{H-\frac{3}{2}}dv \\
&= (r-u)^{2H-2}\int_{\frac{r}{u}}^{\infty}(zu-r)^{1-2H}z^{H-\frac{3}{2}}dz \\
&= (ru)^{\frac{1}{2}-H}(r-u)^{2H-2}\int_0^1(1-x)^{1-2H}x^{H-\frac{3}{2}}dx \\
&= \beta(2-2H, H-\frac{1}{2})(ru)^{\frac{1}{2}-H}(r-u)^{2H-2}.
\end{aligned}$$

Consider the square integrable kernel

$$K_H(t,s) = c_H s^{\frac{1}{2}-H}\int_s^t (u-s)^{H-\frac{3}{2}}u^{H-\frac{1}{2}}du, \tag{1.13}$$

where $c_H = \left[\frac{H(2H-1)}{\beta(2-2H,H-\frac{1}{2})}\right]^{1/2}$ and $t > s$.

Taking into account formulas (1.10) and (1.12) we deduce that this kernel verifies

$$\begin{aligned}
\int_0^{t\wedge s} K_H(t,u)K_H(s,u)du &= c_H^2\int_0^{t\wedge s}\left(\int_u^t(y-u)^{H-\frac{3}{2}}y^{H-\frac{1}{2}}dy\right) \\
&\quad\times\left(\int_u^s(z-u)^{H-\frac{3}{2}}z^{H-\frac{1}{2}}dz\right)u^{1-2H}du \\
&= c_H^2\beta(2-2H,H-\frac{1}{2})\int_0^t\int_0^s|y-z|^{2H-2}dzdy \\
&= R_H(t,s). \qquad (1.14)
\end{aligned}$$

Formula (1.14) implies that the kernel $R_H$ is nonnegative definite and provides an explicit representation for its square root as an operator.

From (1.13) we get

$$\frac{\partial K_H}{\partial t}(t,s) = c_H\left(\frac{t}{s}\right)^{H-\frac{1}{2}}(t-s)^{H-\frac{3}{2}}. \tag{1.15}$$

Consider the linear operator $K_H^*$ from $\mathcal{E}$ to $L^2(0,T)$ defined by

$$(K_H^*\varphi)(s) = \int_s^T\varphi(t)\frac{\partial K_H}{\partial t}(t,s)dt. \tag{1.16}$$

Notice that

$$\left(K_H^*\mathbf{1}_{[0,t]}\right)(s) = K_H(t,s)\mathbf{1}_{[0,t]}(s). \tag{1.17}$$

The operator $K_H^*$ is an isometry between $\mathcal{E}$ and $L^2(0,T)$ that can be extended to the Hilbert space $\mathcal{H}$. In fact, for any $s,t\in[0,T]$ we have using (1.17) and (1.14)

$$\begin{aligned}
\left\langle K_H^*\mathbf{1}_{[0,t]}, K_H^*\mathbf{1}_{[0,s]}\right\rangle_{L^2([0,T])} &= \left\langle K_H(t,\cdot)\mathbf{1}_{[0,t]}, K_H(s,\cdot)\mathbf{1}_{[0,s]}\right\rangle_{L^2(0,T)} \\
&= \int_0^{t\wedge s}K_H(t,u)K_H(s,u)du \\
&= R_H(t,s) = \left\langle\mathbf{1}_{[0,t]},\mathbf{1}_{[0,s]}\right\rangle_{\mathcal{H}}.
\end{aligned}$$

The operator $K_H^*$ can be expressed in terms of fractional integrals:

$$(K_H^*\varphi)(s) = c_H\Gamma(H-\frac{1}{2})s^{\frac{1}{2}-H}(I_{T-}^{H-\frac{1}{2}}u^{H-\frac{1}{2}}\varphi(u))(s). \tag{1.18}$$

This is an immediate consequence of formulas (1.15), (1.16) and (1.6).

For any $a \in [0,T]$, the indicator function $\mathbf{1}_{[0,a]}$ belongs to the image of $K_H^*$ and applying the rules of the fractional calculus yields

$$(K_H^*)^{-1}(\mathbf{1}_{[0,a]}) = \frac{1}{c_H\Gamma(H-\frac{1}{2})}s^{\frac{1}{2}-H}\left(D_{a-}^{H-\frac{1}{2}}u^{H-\frac{1}{2}}\right)(s)\mathbf{1}_{[0,a]}(s). \tag{1.19}$$

Consider the process $W = \{W_t, t \in [0,T]\}$ defined by

$$W_t = B((K_H^*)^{-1}(\mathbf{1}_{[0,t]})). \tag{1.20}$$

Then $W$ is a Wiener process, and the process $B$ has the integral representation

$$B_t = \int_0^t K_H(t,s)dW_s. \tag{1.21}$$

Indeed, for any $s,t \in [0,T]$ we have

$$\begin{aligned} E(W_tW_s) &= E\left(B((K_H^*)^{-1}(\mathbf{1}_{[0,t]}))B((K_H^*)^{-1}(\mathbf{1}_{[0,s]}))\right) \\ &= \left\langle (K_H^*)^{-1}(\mathbf{1}_{[0,t]}), (K_H^*)^{-1}(\mathbf{1}_{[0,s]})\right\rangle_{\mathcal{H}} \\ &= \left\langle \mathbf{1}_{[0,t]}, \mathbf{1}_{[0,s]}\right\rangle_{L^2(0,T)} = s \wedge t. \end{aligned}$$

Moreover, for any $\varphi \in \mathcal{H}$ we have

$$B(\varphi) = \int_0^T (K_H^*\varphi)(t)dW_t.$$

Notice that from (1.19), the Wiener process $W$ is adapted to the filtration generated by the fBm $B$ and (1.20) and (1.21) imply that both processes generate the same filtration. Furthermore, the Wiener process $W$ that provides the integral representation (1.21) is unique. Indeed, this follows from the fact that the image of the operator $K_H^*$ is $L^2([0,T])$, because this image contains the indicator functions.

The elements of the Hilbert space $\mathcal{H}$ may not be functions but distributions of negative order (see Pipiras and Taqqu [**Pi-Ta1**], [**Pi-Ta2**]). In fact, from (1.18) it follows that $\mathcal{H}$ coincides with the space of distributions $f$ such that $s^{\frac{1}{2}-H}I_{0+}^{H-\frac{1}{2}}(f(u)u^{H-\frac{1}{2}})(s)$ is a square integrable function.

We can find a linear space of functions contained in $\mathcal{H}$ in the following way. Let $|\mathcal{H}|$ be the linear space of measurable functions $\varphi$ on $[0,T]$ such that

$$\|\varphi\|_{|\mathcal{H}|}^2 = \alpha_H\int_0^T\int_0^T |\varphi_r||\varphi_u||r-u|^{2H-2}drdu < \infty. \tag{1.22}$$

It is not difficult to show that $|\mathcal{H}|$ is a Banach space with the norm $\|\cdot\|_{|\mathcal{H}|}$ and $\mathcal{E}$ is dense in $|\mathcal{H}|$. On the other hand, it has been shown in [**Pi-Ta2**] that the space $|\mathcal{H}|$ equipped with the inner product $\langle\varphi,\psi\rangle_{\mathcal{H}}$ is not complete and it is isometric to a subspace of $\mathcal{H}$.

The following estimate has been proved in [**M-M-V**]

$$\|\varphi\|_{|\mathcal{H}|} \le b_H\|\varphi\|_{L^{\frac{1}{H}}([0,T])}, \tag{1.23}$$

for some constant $b_H > 0$.

PROOF OF (1.23). Using Hölder's inequality with exponent $q = \frac{1}{H}$ in (1.22) we get

$$\|\varphi\|_{|\mathcal{H}|}^2 \le \alpha_H \left(\int_0^T |\varphi_r|^{\frac{1}{H}} dr\right)^H \left(\int_0^T \left(\int_0^T |\varphi_u| (r-u)^{2H-2} du\right)^{\frac{1}{1-H}} dr\right)^{1-H}.$$

The second factor in the above expression, up to a multiplicative constant, it is equal to the $\frac{1}{1-H}$ norm of the left-sided fractional integral $I_{0+}^{2H-1}|\varphi|$. Finally is suffices to apply the Hardy-Littlewood inequality (see [**St**, Theorem 1, p. 119])

$$\left\| I_{0+}^{\alpha} f \right\|_{L^q(0,\infty)} \le c_{H,p} \|f\|_{L^p(0,\infty)}, \tag{1.24}$$

where $0 < \alpha < 1$, $1 < p < q < \infty$ satisfy $\frac{1}{q} = \frac{1}{p} - \alpha$, with the particular values $\alpha = 2H - 1$, $q = \frac{1}{1-H}$, and $p = \frac{1}{H}$. □

As a consequence

$$L^2(0,T) \subset L^{\frac{1}{H}}(0,T) \subset |\mathcal{H}| \subset \mathcal{H}.$$

The inclusion $L^2(0,T) \subset |\mathcal{H}|$ can be proved by a direct argument:

$$\begin{aligned} \int_0^T \int_0^T |\varphi_r| |\varphi_u| |r-u|^{2H-2} drdu &\le \int_0^T \int_0^T |\varphi_u|^2 |r-u|^{2H-2} drdu \\ &\le \frac{T^{2H-1}}{H-\frac{1}{2}} \int_0^T |\varphi_u|^2 du. \end{aligned}$$

This means that the Wiener-type integral $\int_0^T \varphi(t)dB_t$ (which is equal to $B(\varphi)$, by definition) can be defined for functions $\varphi \in |\mathcal{H}|$, and

$$\int_0^T \varphi(t)dB_t = \int_0^T (K_H^* \varphi)(t)dW_t. \tag{1.25}$$

1.5.2. *Case* $H < \frac{1}{2}$. We claim that the kernel

$$K_H(t,s) = c_H \left[ \left(\frac{t}{s}\right)^{H-\frac{1}{2}} (t-s)^{H-\frac{1}{2}} - (H-\frac{1}{2}) s^{\frac{1}{2}-H} \int_s^t u^{H-\frac{3}{2}} (u-s)^{H-\frac{1}{2}} du \right]$$

where $c_H = \sqrt{\frac{2H}{(1-2H)\beta(1-2H, H+1/2)}}$, satisfies

$$R_H(t,s) = \int_0^{t\wedge s} K_H(t,u) K_H(s,u) du. \tag{1.26}$$

To verify this relation is not so easy as in the case $H > \frac{1}{2}$. In the references [**D-U**] and [**Pi-Ta2**] this property is proved using the analyticity of both members as functions of the parameter $H$. We will give here a direct proof using the ideas of [**N-V-V**]. Notice first that

$$\frac{\partial K_H}{\partial t}(t,s) = c_H (H - \frac{1}{2}) \left(\frac{t}{s}\right)^{H-\frac{1}{2}} (t-s)^{H-\frac{3}{2}}. \tag{1.27}$$

PROOF OF (1.26). Consider first the diagonal case $s = t$. Set $\phi(s) = \int_0^s K_H(s,u)^2 du$. We have

$$\begin{aligned}\phi(s) &= c_H^2 \Bigg[ \int_0^s (\frac{s}{u})^{2H-1} (s-u)^{2H-1} du \\ &-(2H-1) \int_0^s s^{H-\frac{1}{2}} u^{1-2H} (s-u)^{H-\frac{1}{2}} \left( \int_u^s v^{H-\frac{3}{2}} (v-u)^{H-\frac{1}{2}} dv \right) du \\ &+(H-\frac{1}{2})^2 \int_0^s u^{1-2H} \left( \int_u^s v^{H-\frac{3}{2}} (v-u)^{H-\frac{1}{2}} dv \right)^2 du \Bigg].\end{aligned}$$

Making the change of variables $u = sx$ in the first integral and using Fubini's theorem yields

$$\begin{aligned}\phi(s) &= c_H^2 \Big[ s^{2H} \beta(2-2H, 2H) \\ &-(2H-1) s^{H-\frac{1}{2}} \int_0^s v^{H-\frac{3}{2}} \left( \int_0^v u^{1-2H} (s-u)^{H-\frac{1}{2}} (v-u)^{H-\frac{1}{2}} du \right) dv \\ &+2(H-\frac{1}{2})^2 \int_0^s \int_0^v \int_0^w u^{1-2H} (v-u)^{H-\frac{1}{2}} (w-u)^{H-\frac{1}{2}} \\ &\times w^{H-\frac{3}{2}} v^{H-\frac{3}{2}} du dw dv \Big].\end{aligned}$$

Now we make the change of variable $u = vx$, $v = sy$ for the second term and $u = wx$, $w = vy$ for the third term and we obtain

$$\begin{aligned}\phi(s) &= c_H^2 s^{2H} \Bigg[ \beta(2-2H, 2H) - (2H-1)(\frac{1}{4H} + \frac{1}{2}) \\ &\times \int_0^1 \int_0^1 x^{1-2H} (1-xy)^{H-\frac{1}{2}} (1-x)^{H-\frac{1}{2}} dx dy \Bigg] \\ &= s^{2H}.\end{aligned}$$

Suppose now that $s < t$. Differentiating Equation (1.26) with respect to $t$, we are aimed to show that

$$H(t^{2H-1} - (t-s)^{2H-1}) = \int_0^s \frac{\partial K_H}{\partial t}(t,u) K_H(s,u) du. \tag{1.28}$$

Set $\phi(t,s) = \int_0^s \frac{\partial K_H}{\partial t}(t,u) K_H(s,u) du$. Using (1.27) yields

$$\begin{aligned}\phi(t,s) &= c_H^2 (H-\frac{1}{2}) \int_0^s \left(\frac{t}{u}\right)^{H-\frac{1}{2}} (t-u)^{H-\frac{3}{2}} \left(\frac{s}{u}\right)^{H-\frac{1}{2}} (s-u)^{H-\frac{1}{2}} du \\ &-c_H^2 (H-\frac{1}{2})^2 \int_0^s \left(\frac{t}{u}\right)^{H-\frac{1}{2}} (t-u)^{H-\frac{3}{2}} u^{\frac{1}{2}-H} \\ &\times \left( \int_u^s v^{H-\frac{3}{2}} (v-u)^{H-\frac{1}{2}} dv \right) du.\end{aligned}$$

Making the change of variables $u = sx$ in the first integral and $u = vx$ in the second one we obtain

$$\begin{aligned}\phi(t,s) &= c_H^2 (H-\frac{1}{2}) (ts)^{H-\frac{1}{2}} \gamma(\frac{t}{s}) \\ &-c_H^2 (H-\frac{1}{2})^2 t^{H-\frac{1}{2}} \int_0^s v^{H-\frac{3}{2}} \gamma(\frac{t}{v}) dv,\end{aligned}$$

where $\gamma(y)=\int_0^1 x^{1-2H}(y-x)^{H-\frac{3}{2}}(1-x)^{H-\frac{1}{2}}dx$ for $y>1$. Then, (1.28) is equivalent to

$$c_H^2\left[(H-\frac{1}{2})s^{H-\frac{1}{2}}\gamma(\frac{t}{s})\ -(H-\frac{1}{2})^2\int_0^s v^{H-\frac{3}{2}}\gamma(\frac{t}{v})dv\right]$$
$$= H(t^{H-\frac{1}{2}}-t^{\frac{1}{2}-H}(t-s)^{2H-1}). \tag{1.29}$$

Differentiating the left-hand side of equation (1.29) with respect to $t$ yields

$$c_H^2(H-\frac{3}{2})\left[(H-\frac{1}{2})s^{H-\frac{3}{2}}\delta(\frac{t}{s})-(H-\frac{1}{2})^2\int_0^s v^{H-\frac{5}{2}}\delta(\frac{t}{v})dv\right]:=\mu(t,s), \tag{1.30}$$

where, for $y>1$,

$$\delta(y)=\int_0^1 x^{1-2H}(y-x)^{H-\frac{5}{2}}(1-x)^{H-\frac{1}{2}}dx.$$

By means of the change of variables $z=\frac{y(1-x)}{y-x}$ we obtain

$$\delta(y)=\beta(2-2H,H+\frac{1}{2})y^{-H-\frac{1}{2}}(y-1)^{2H-2}. \tag{1.31}$$

Finally, substituting (1.31) into (1.30) yields

$$\begin{aligned}\mu(t,s) &= c_H^2\beta(2-2H,H+\frac{1}{2})(H-\frac{3}{2})(H-\frac{1}{2})\\ &\quad\times t^{-H-\frac{1}{2}}s(t-s)^{2H-2}+\frac{1}{2}t^{-H-\frac{1}{2}}((t-s)^{2H-1}-t^{2H-1})\\ &= H(1-2H)\left(t^{-H-\frac{1}{2}}s(t-s)^{2H-2}+\frac{1}{2}(t-s)^{2H-1}t^{-H-\frac{1}{2}}-\frac{1}{2}t^{H-\frac{3}{2}}\right).\end{aligned}$$

This last expression coincides with the derivative with respect to $t$ of the right-hand side of (1.29). This completes the proof of the equality (1.26). □

The kernel $K_H$ can also be expressed in terms of fractional derivatives:

$$K_H(t,s)=c_H\Gamma(H+\frac{1}{2})s^{\frac{1}{2}-H}\left(D_{t-}^{\frac{1}{2}-H}u^{H-\frac{1}{2}}\right)(s). \tag{1.32}$$

Consider the linear operator $K_H^*$ from $\mathcal{E}$ to $L^2(0,T)$ defined by

$$(K_H^*\varphi)(s)=K_H(T,s)\varphi(s)+\int_s^T(\varphi(t)-\varphi(s))\frac{\partial K_H}{\partial r}(t,s)dt. \tag{1.33}$$

Notice that

$$\left(K_H^*\mathbf{1}_{[0,t]}\right)(s)=K_H(t,s)\mathbf{1}_{[0,t]}(s). \tag{1.34}$$

From (1.26) and (1.34) we deduce as in the case $H>\frac{1}{2}$ that the operator $K_H^*$ is an isometry between $\mathcal{E}$ and $L^2(0,T)$ that can be extended to the Hilbert space $\mathcal{H}$.

The operator $K_H^*$ can be expressed in terms of fractional derivatives:

$$(K_H^*\varphi)(s)=d_H s^{\frac{1}{2}-H}(D_{T-}^{\frac{1}{2}-H}u^{H-\frac{1}{2}}\varphi(u))(s), \tag{1.35}$$

where $d_H=c_H\Gamma(H+\frac{1}{2})$. This is an immediate consequence of (1.33) and the equality

$$\left(D_{t-}^{\frac{1}{2}-H}u^{H-\frac{1}{2}}\right)(s)\mathbf{1}_{[0,t]}(s)=\left(D_{T-}^{\frac{1}{2}-H}u^{H-\frac{1}{2}}\mathbf{1}_{[0,t]}(u)\right)(s).$$

As a consequence,

$$C^\gamma([0,T]) \subset \mathcal{H} \subset L^2([0,T]),$$

if $\gamma > \frac{1}{2} - H$.

Using the alternative expression for the kernel $K_H$ given by

$$K_H(t,s) = c_H(t-s)^{H-\frac{1}{2}} + s^{H-\frac{1}{2}} F_1(\frac{t}{s}), \tag{1.36}$$

where

$$F_1(z) = c_H(\frac{1}{2} - H) \int_0^{z-1} \theta^{H-\frac{3}{2}}(1-(\theta+1)^{H-\frac{1}{2}})d\theta,$$

one can show that $\mathcal{H} = I_{T-}^{\frac{1}{2}-H}(L^2)$ (see **[D-U]** and Proposition 6 of **[A-M-N2]**). In fact, from (1.33) and (1.36) we obtain, for any function $\varphi$ in $I_{T-}^{\frac{1}{2}-H}(L^2)$

$$\begin{aligned}
\left(K_H^*\varphi\right)(s) &= c_H\left[(T-s)^{H-\frac{1}{2}}\varphi(s) + (H-\frac{1}{2})\int_s^T (\varphi(r)-\varphi(s))(r-s)^{H-\frac{3}{2}}dr\right] \\
&\quad + s^{H-\frac{3}{2}}\int_s^T \varphi(r) F_1'\left(\frac{r}{s}\right)dr \\
&= c_H\Gamma(\frac{1}{2}+H) D_{T-}^{\frac{1}{2}-H}\varphi(s) + \Lambda\varphi(s),
\end{aligned}$$

where the operator

$$\Lambda\varphi(s) = c_H(\frac{1}{2}-H)\int_s^T \varphi(r)(r-s)^{H-\frac{3}{2}}\left(1-\left(\frac{r}{s}\right)^{H-\frac{1}{2}}\right)dr$$

is bounded in $L^2$.

On the other hand, (1.35) implies that

$$\mathcal{H} = \{f : \exists\phi \in L^2(0,T) : f(s) = d_H^{-1} s^{\frac{1}{2}-H}(I_{T-}^{\frac{1}{2}-H} u^{H-\frac{1}{2}}\phi(u))(s)\},$$

with the inner product

$$\langle f, g\rangle_{\mathcal{H}} = \int_0^T \phi(s)\psi(s)ds,$$

if

$$f(s) = d_H^{-1} s^{\frac{1}{2}-H}(I_{T-}^{\frac{1}{2}-H} u^{H-\frac{1}{2}}\phi(u))(s)$$

and

$$g(s) = d_H^{-1} s^{\frac{1}{2}-H}(I_{T-}^{\frac{1}{2}-H} u^{H-\frac{1}{2}}\psi(u))(s).$$

Consider process $W = \{W_t, t \in [0,T]\}$ defined by

$$W_t = B((K_H^*)^{-1}(\mathbf{1}_{[0,t]})).$$

As in the case $H > \frac{1}{2}$, we can show that $W$ is a Wiener process, and the process $B$ has the integral representation

$$B_t = \int_0^t K_H(t,s)dW_s.$$

Therefore, in this case the Wiener-type integral $\int_0^T \varphi(t)dB_t$ can be defined for functions $\varphi \in I_{T-}^{\frac{1}{2}-H}(L^2)$, and (1.25) holds.

Define the left and right-sided fractional derivative operators on the whole real line for $0 < \alpha < 1$ by

$$D_-^\alpha f(s) := \frac{\alpha}{\Gamma(1-\alpha)} \int_0^\infty \frac{f(s) - f(s+u)}{u^{1+\alpha}} du$$

and

$$D_+^\alpha f(s) := \frac{\alpha}{\Gamma(1-\alpha)} \int_0^\infty \frac{f(s) - f(s-u)}{u^{1+\alpha}} du,$$

$s \in \mathbb{R}$, respectively. Then, the scalar product in $\mathcal{H}$ has the following simple expression

$$\langle f, g \rangle_{\mathcal{H}} = e_H^2 \left\langle D_-^{\frac{1}{2}-H} f, D_+^{\frac{1}{2}-H} g \right\rangle_{L^2(\mathbb{R})}, \tag{1.37}$$

where $e_H = C_1(H)^{-1}\Gamma(H + \frac{1}{2})$, $f, g \in \mathcal{H}$, and by convention $f(s) = g(s) = 0$ if $s \notin [0, T]$.

In [**A-M-N2**] these results have been generalized to Gaussian Volterra processes of the form

$$X_t = \int_0^t K(t,s) dW_s,$$

where $\{W_t, t \geq 0\}$ is a Wiener process and $K(t,s)$ is a square integrable kernel. Two different types of kernels can be considered, which correspond to the cases $H < \frac{1}{2}$ and $H > \frac{1}{2}$:

i) *Singular case*: $K(\cdot, s)$ has bounded variation on any interval $(u, T]$, $u > s$, but $\int_s^T |K|(dt, s) = \infty$ for every $s$.

ii) *Regular case*: The kernel satisfies $\int_s^T |K|((s,T], s)^2 ds < \infty$ for each $s$.

## 2. Stochastic calculus of variations with respect to fBm

Let $B = \{B_t, t \in [0,T]\}$ be a fBm with Hurst parameter $H \in (0,1)$. Let $\mathcal{S}$ be the set of smooth and cylindrical random variables of the form

$$F = f(B(\phi_1), \ldots, B(\phi_n)), \tag{2.1}$$

where $n \geq 1$, $f \in C_b^\infty(\mathbb{R}^n)$ ($f$ and all its partial derivatives are bounded), and $\phi_i \in \mathcal{H}$.

The *derivative operator* $D$ of a smooth and cylindrical random variable $F$ of the form (2.1) is defined as the $\mathcal{H}$-valued random variable

$$DF = \sum_{i=1}^n \frac{\partial f}{\partial x_i}(B(\phi_1), \ldots, B(\phi_n))\phi_i.$$

The derivative operator $D$ is then a closable operator from $L^p(\Omega)$ into $L^p(\Omega; \mathcal{H})$ for any $p \geq 1$. For any integer $k \geq 1$ we denote by $D^k$ the iteration of the derivative operator. For any $p \geq 1$ the Sobolev space $\mathbb{D}^{k,p}$ is the closure of $\mathcal{S}$ with respect to the norm

$$\|F\|_{k,p}^p = E\left(|F|^p\right) + \sum_{j=1}^k E\left(\left\|D^j F\right\|_{\mathcal{H}^{\otimes j}}^p\right).$$

In a similar way, given a Hilbert space $V$ we denote by $\mathbb{D}^{k,p}(V)$ the corresponding Sobolev space of $V$-valued random variables.

The *divergence operator* $\delta$ is the adjoint of the derivative operator. We say that a random variable in $L^2(\Omega;\mathcal{H})$ belongs to the domain of the divergence operator, denoted by Dom $\delta$, if

$$|E(\langle DF,u\rangle_{\mathcal{H}})| \leq c_u \|F\|_{L^2(\Omega)}$$

for any $F \in \mathcal{S}$. In this case $\delta(u)$ is defined by the duality relationship

$$E(F\delta(u)) = E(\langle DF,u\rangle_{\mathcal{H}}), \tag{2.2}$$

for any $F \in \mathbb{D}^{1,2}$.

The following are two basic properties of the divergence operator:

i) $\mathbb{D}^{1,2}(\mathcal{H}) \subset$Dom $\delta$ and for any $u \in \mathbb{D}^{1,2}(\mathcal{H})$

$$E\left(\delta(u)^2\right) = E\left(\|u\|_{\mathcal{H}}^2\right) + E\left(\langle Du,(Du)^*\rangle_{\mathcal{H}\otimes\mathcal{H}}\right), \tag{2.3}$$

where $(Du)^*$ is the adjoint of $(Du)$ in the Hilbert space $\mathcal{H}\otimes\mathcal{H}$.

ii) For any $F$ in $\mathbb{D}^{1,2}$ and any $u$ in the domain of $\delta$ such that $Fu$ and $F\delta(u)+\langle DF,u\rangle_{\mathcal{H}}$ are square integrable, then $Fu$ is in the domain of $\delta$ and

$$\delta(Fu) = F\delta(u) - \langle DF,u\rangle_{\mathcal{H}}. \tag{2.4}$$

**2.1. Transfer principle.** Recall that the operator $K_H^*$ is an isometry between $\mathcal{H}$ and a closed subspace of $L^2(0,T)$. Moreover, $W_t = B((K_H^*)^{-1}(\mathbf{1}_{[0,t]}))$ is a Wiener process such that

$$B_t = \int_0^t K_H(t,s)dW_s,$$

and for any $\varphi \in \mathcal{H}$ we have $B(\varphi) = W(K_H^*\varphi)$.

A similar relation holds for the derivative and divergence operators with respect to the processes $B$ and $W$. That is:

(i) For any $F \in \mathbb{D}_W^{1,2} = \mathbb{D}^{1,2}$

$$K_H^* DF = D^W F,$$

where $D^W$ denotes the derivative operator with respect to the process $W$, and $\mathbb{D}_W^{1,2}$ the corresponding Sobolev space.

(ii) Dom$\delta$ $= (K_H^*)^{-1}$ (Dom$\delta_W$), and for any $\mathcal{H}$-valued random variable $u$ in Dom $\delta$ we have $\delta(u) = \delta_W(K_H^* u)$, where $\delta_W$ denotes the divergence operator with respect to the process $W$.

Suppose $H > \frac{1}{2}$. We denote by $|\mathcal{H}|\otimes|\mathcal{H}|$ the space of measurable functions $\varphi$ on $[0,T]^2$ such that

$$\|\varphi\|_{|\mathcal{H}|\otimes|\mathcal{H}|}^2 = \alpha_H^2 \int_{[0,T]^4} |\varphi_{r,\theta}|\,|\varphi_{u,\eta}|\,|r-u|^{2H-2}\,|\theta-\eta|^{2H-2}\,drdud\theta d\eta < \infty.$$

Then, $|\mathcal{H}|\otimes|\mathcal{H}|$ is a Banach space with respect to the norm $\|\cdot\|_{|\mathcal{H}|\otimes|\mathcal{H}|}$. Furthermore, equipped with the inner product

$$\langle\varphi,\psi\rangle_{\mathcal{H}\otimes\mathcal{H}} = \alpha_H^2 \int_{[0,T]^4} \varphi_{r,\theta}\psi_{u,\eta}\,|r-u|^{2H-2}\,|\theta-\eta|^{2H-2}\,drdud\theta d\eta$$

the space $|\mathcal{H}|\otimes|\mathcal{H}|$ is isometric to a subspace of $\mathcal{H}\otimes\mathcal{H}$. A slight extension of the inequality (1.23) yields

$$\|\varphi\|_{|\mathcal{H}|\otimes|\mathcal{H}|} \leq b_H \|\varphi\|_{L^{\frac{1}{H}}([0,T]^2)}. \tag{2.5}$$

For any $p > 1$ we denote by $\mathbb{D}^{1,p}(|\mathcal{H}|)$ the subspace of $\mathbb{D}^{1,p}(\mathcal{H})$ formed by the elements $u$ such that $u \in |\mathcal{H}|$ a.s., $Du \in |\mathcal{H}| \otimes |\mathcal{H}|$ a.s., and

$$E\left(\|u\|_{|\mathcal{H}|}^{p}\right) + E\left(\|Du\|_{|\mathcal{H}|\otimes|\mathcal{H}|}^{p}\right) < \infty.$$

## 3. Stochastic integrals with respect to fractional Brownian motion

In the case of an ordinary Brownian motion, the adapted processes in $L^2([0,T]\times\Omega)$ belong to the domain of the divergence operator, and on this set the divergence operator coincides with Itô's stochastic integral. Actually, the divergence operator coincides with an extension of Itô's stochastic integral introduced by Skorohod in [**Sk**]. In this context a natural question is to ask in which sense the divergence operator with respect to a fractional Brownian motion $B$ can be interpreted as a stochastic integral. Note that the divergence operator provides an isometry between the Hilbert Space $\mathcal{H}$ associated with the fBm $B$ and the Gaussian space $H_1(B)$, and gives rise to a notion of stochastic integral in the space of deterministic functions $|\mathcal{H}|$ included in $\mathcal{H}$ (case $H > \frac{1}{2}$) or in the space $I_{T-}^{\frac{1}{2}-H}(L^2)$ (case $H < \frac{1}{2}$).

Different approaches have been used in the literature in order to define stochastic integrals with respect to fBm. Lin [**Li**] and Dai and Heyde [**D-H**] have defined a stochastic integral $\int_0^T \phi_s dB_s$ as limit in $L^2$ of Riemann sums in the case $H > \frac{1}{2}$. This integral does not satisfy the property $E(\int_0^T \phi_s dB_s) = 0$ and it gives rise to change of variable formulae of Stratonovich type. A new type of integral with zero mean defined by means of Wick products was introduced by Duncan, Hu and Pasik-Duncan in [**D-H-P**], assuming $H > \frac{1}{2}$. This integral turns out to coincide with the divergence operator.

A construction of stochastic integrals with respect to fBm with parameter $H \in (0,1)$ by a regularization technique was developed by Carmona, Coutin and Montseny in [**C-C-M**]. The integral is defined as the limit of approximating integrals with respect to semimartingales obtained by smoothing the singularity of the kernel $K_H(t,s)$. The techniques of Malliavin Calculus are used in order to establish the existence of the integrals. The ideas of Carmona and Coutin were further developed by Alòs, Mazet and Nualart in the case $0 < H < \frac{1}{2}$ in [**A-M-N1**].

The interpretation of the divergence operator as a stochastic integral has been first studied by Decreusefont and Üstünel in [**D-U**]. A stochastic calculus for the divergence process has been developed by Alòs, Mazet and Nualart in [**A-M-N2**].

In this section we will discuss the relation between the divergence operator and the path-wise stochastic integral with respect to fBm with parameter $H \in (0,1)$ defined as the limit of the integrals with respect to a regularization of fBm by the convolution with a constant function. The results will be based on the papers [**A-N**] (case $H > \frac{1}{2}$), [**A-L-N**] and [**C-N**] (case $H < \frac{1}{2}$).

The following definition of the *symmetric stochastic integral* was introduced by Russo and Vallois in [**R-V**]. By convention we will assume that all processes and functions vanish outside the interval $[0,T]$.

DEFINITION 3.1. *Let $u = \{u_t, t \in [0,T]\}$ be a stochastic process with integrable trajectories. The symmetric integral of $u$ with respect to the fBm $B$ is defined as* *the limit in probability as $\varepsilon$ tends to zero of*

$$(2\varepsilon)^{-1}\int_0^T u_s(B_{s+\varepsilon}-B_{s-\varepsilon})ds,$$

*provided this limit exists, and it is denoted by* $\int_0^T u_t dB_t$.

**3.1. The divergence integral in the case $\mathbf{H > \frac{1}{2}}$.** The following proposition gives sufficient conditions for the existence of the symmetric integral, and provides a representation of the divergence operator as a stochastic integral (see [**A-N**]).

PROPOSITION 3.2. *Let $u=\{u_t, t\in[0,T]\}$ be a stochastic process in the space $\mathbb{D}^{1,2}(|\mathcal{H}|)$. Suppose also that a.s.*

$$\int_0^T\int_0^T |D_s u_t|\,|t-s|^{2H-2}\,dsdt<\infty. \tag{3.1}$$

*Then the symmetric integral exists and we have*

$$\int_0^T u_t dB_t=\delta(u)+\alpha_H\int_0^T\int_0^T D_s u_t\,|t-s|^{2H-2}\,dsdt. \tag{3.2}$$

REMARK 3.3. Under the assumptions of the Proposition 3.2 the integral $\int_0^T u_t dB_t$ also coincides with the forward and backward integrals.

REMARK 3.4. A sufficient condition for (3.1) is

$$\int_0^T\left(\int_s^T |D_s u_t|^p\,dt\right)^{1/p} ds<\infty$$

for some $p>\frac{1}{2H-1}$.

SKETCH OF THE PROOF. Approximate $u$ by

$$u_t^\varepsilon=(2\varepsilon)^{-1}\int_{t-\varepsilon}^{t+\varepsilon} u_s ds.$$

We have

$$\|u^\varepsilon\|^2_{\mathbb{D}^{1,2}(|\mathcal{H}|)}\le d_H\,\|u\|^2_{\mathbb{D}^{1,2}(|\mathcal{H}|)},$$

for some positive constant $d_H$. Using (2.4) we obtain

$$(2\varepsilon)^{-1}\int_0^T u_s(B_{s+\varepsilon}-B_{s-\varepsilon})ds=\delta(u^\varepsilon)+(2\varepsilon)^{-1}\int_0^T \left\langle Du_s,\mathbf{1}_{[s-\varepsilon,s+\varepsilon]}\right\rangle_{\mathcal{H}} ds.$$

Finally, take the limit as $\varepsilon$ tends to zero. □

3.1.1. *Estimates for the divergence integral.* Suppose that $u=\{u_t,t\in[0,T]\}$ is a stochastic process in the space $\mathbb{D}^{1,2}(|\mathcal{H}|)$ such that condition (3.1) holds. Then, for any $t\in[0,T]$ the process $u\mathbf{1}_{[0,t]}$ also belongs to $\mathbb{D}^{1,2}(|\mathcal{H}|)$ and satisfies (3.1). Hence, by Proposition 3.2 we can define the indefinite integral $\int_0^t u_s dB_s=\int_0^T u_s\mathbf{1}_{[0,t]}(s)dB_s$ and the following decomposition holds

$$\int_0^t u_s dB_s=\delta(u\mathbf{1}_{[0,t]})+\alpha_H\int_0^t\int_0^T D_r u_s\,|s-r|^{2H-2}\,drds.$$

The second summand in this expression is a process with absolutely continuous paths. Therefore, in order to deduce $L^p$ estimates and to study continuity properties of $\int_0^t u_s dB_s$ we can reduce our analysis to the process $\delta(u\mathbf{1}_{[0,t]})$. In this section we will establish $L^p$ maximal estimates for this divergence process. We will make use of the notation

$$\int_0^t u_s \delta B_s = \delta\left(u\mathbf{1}_{[0,t]}\right).$$

By Meyer's inequalities (see for example [**N**]), if $p > 1$, a process $u \in \mathbb{D}^{1,p}(|\mathcal{H}|)$ belongs to the domain of the divergence in $L^p(\Omega)$, and we have

$$E\left(|\delta(u)|^p\right) \leq C_{H,p}\left(\|E(u)\|_{|\mathcal{H}|}^p + E\left(\|Du\|_{|\mathcal{H}|\otimes|\mathcal{H}|}^p\right)\right).$$

As a consequence, applying (2.5) we obtain

$$E\left(|\delta(u)|^p\right) \leq C_{H,p}\left(\|E(u)\|_{L^{1/H}([0,T])}^p + E\left(\|Du\|_{L^{1/H}([0,T]^2)}^p\right)\right).$$

Let $pH > 1$. Denote by $\mathbb{L}_H^{1,p}$ the space of processes $u \in \mathbb{D}^{1,2}(|\mathcal{H}|)$ such that

$$\|u\|_{p,1} := \left[\int_0^T E(|u_s|^p)ds + E\left(\int_0^T\left(\int_0^T |D_r u_s|^{\frac{1}{H}} dr\right)^{pH} ds\right)\right]^{\frac{1}{p}} < \infty.$$

Using Meyer's inequality and a convolution argument the following maximal $L^p$ inequality for the divergence integral has been established in [**A-N**]:

$$E\left(\sup_{t\in[0,T]}\left|\int_0^t u_s\delta B_s\right|^p\right) \leq C\,\|u\|_{p,1}^p,$$

where the constant $C > 0$ depends on $p$, $H$ and $T$.

Assume $pH > 1$ and suppose that $u \in \mathbb{L}_H^{1,p}$. Set $X_t = \int_0^t u_s \delta B_s$. Then, the process $X_t$ has a version with continuous trajectories and for all $\gamma < H - \frac{1}{p}$ there exists a random variable $C_\gamma$ such that

$$|X_t - X_s| \leq C_\gamma\,|t-s|^\gamma.$$

This result is also proved in [**A-N**]. As a consequence, for a process $u \in \cap_{p>1}\mathbb{L}_H^{1,p}$, the indefinite integral process $X = \left\{\int_0^t u_s\delta B_s,\ t\in[0,T]\right\}$ is $\gamma$-Hölder continuous for all $\gamma < H$. If we assume also that hypothesis (3.1) holds, we deduce analogous continuity results for the symmetric integral process $\int_0^t u_s dB_s$.

3.1.2. *Itô's formula for the divergence integral.* Suppose that $f, g : [0,T] \longrightarrow \mathbb{R}$ are Hölder continuous functions of orders $\alpha$ and $\beta$ with $\alpha + \beta > 1$. Young [**Y**] proved that the Riemann-Stieltjes integral $\int_0^T f dg$ exists. As a consequence, if $F$ is a function of class $C^2$, and $H > \frac{1}{2}$, the path-wise Riemann-Stieltjes integral $\int_0^t F'(B_s)dB_s$ exists for each $t \in [0,T]$. Moreover the following change of variables formula holds:

$$F(B_t) = F(0) + \int_0^t F'(B_s)dB_s. \tag{3.3}$$

In fact, it suffices to show that the second order term

$$R_\pi := \sum_{i=1}^n F''(X_i)(B_{t_i} - B_{t_{i-1}})^2$$

converges to zero almost surely as the norm of the partition $\pi = \{0 = t_0 < t_1 < \cdots < t_n = t\}$ tends to zero, where $X_i$ is an intermediate value between $B_{t_i}$ and $B_{t_{i-1}}$. This follows immediately from the estimate

$$|R_\pi| \le C_\varepsilon \|F''\|_\infty \sum_{i=1}^n |t_i - t_{i-1}|^{2H-\varepsilon}.$$

Moreover, the Riemann-Stieltjes path-wise integral $\int_0^t F'(B_s)dB_s$ coincides with the symmetric integral in the Russo-Vallois sense introduced in Definition 3.1.

Suppose that $F$ is a function of class $C^2(\mathbb{R})$ such that

$$\max\{|F(x)|, |F'(x)|, |F''(x)|\} \le ce^{\lambda x^2}, \tag{3.4}$$

where $c$ and $\lambda$ are positive constants such that $\lambda < \frac{1}{4T^{2H}}$. This condition implies

$$E\left(\sup_{0\le t\le T} |F(B_t)|^p\right) \le c^p E\left(e^{p\lambda \sup_{0\le t\le T}|B_t|^2}\right) < \infty$$

for all $p < \frac{T^{-2H}}{2\lambda}$. In particular, we can take $p = 2$. The same property holds for $F'$ and $F''$.

Then, if $F$ satisfies the growth condition (3.4), the process $F'(B_t)$ belongs to the space $\mathbb{D}^{1,2}(|\mathcal{H}|)$ and (3.1) holds. As a consequence, from Proposition 3.2 we obtain

$$\begin{aligned} \int_0^t F'(B_s)dB_s &= \int_0^t F'(B_s)\delta B_s + H(2H-1)\int_0^t\int_0^s F''(B_s)(s-r)^{2H-2}drds \\ &= \int_0^t F'(B_s)\delta B_s + H\int_0^t F''(B_s)s^{2H-1}ds. \end{aligned} \tag{3.5}$$

Therefore, putting together (3.3) and (3.5) we deduce the following Itô's formula for the divergence process

$$F(B_t) = F(0) + \int_0^t F'(B_s)\delta B_s + H\int_0^t F''(B_s)s^{2H-1}ds. \tag{3.6}$$

The divergence operator has the following local property:

Lemma 3.5. *Let $u$ be an element of $\mathbb{D}^{1,2}(\mathcal{H})$. If $u = 0$ a.s. on a set $A \in \mathcal{F}$, then $\delta(u) = 0$ a.s. on $A$.*

Given a set $L$ of $\mathcal{H}$-valued random variables we will denote by $L_{loc}$ the set of $\mathcal{H}$-valued random variables $u$ such that there exists a sequence $\{(\Omega^n, u^n)\}, n \ge 1\} \subset \mathcal{F} \times L$ with the following properties:

i) $\Omega^n \uparrow \Omega$ a.s.

ii) $u = u^n$, a.e. on $[0,T] \times \Omega_n$.

We then say that $\{(\Omega^n, u^n)\}$ localizes $u$ in $L$. If $u \in \mathbb{D}^{1,2}_{loc}(\mathcal{H})$ by Lemma 3.5 we can define without ambiguity $\delta(u)$ by setting

$$\delta(u)|_{\Omega^n} = \delta(u^n)|_{\Omega^n}$$

for each $n \ge 1$, where $\{(\Omega^n, u^n)\}$ is a localizing sequence for $u$ in $\mathbb{D}^{1,2}(\mathcal{H})$.

We state the following general version of Itô's formula proved in [**A-N**]:

THEOREM 3.6. *Let $F$ be a function of class $C^2(\mathbb{R})$. Assume that $u = \{u_t, t \in [0,T]\}$ is a process in the space $\mathbb{D}^{2,2}_{loc}(|\mathcal{H}|)$ such that the indefinite integral $X_t = \int_0^t u_s \delta B_s$ is a.s. continuous. Assume that $\|u\|_2$ belongs to $\mathcal{H}$. Then for each $t \in [0,T]$ the following formula holds*

$$
\begin{aligned}
F(X_t) &= F(0) + \int_0^t F'(X_s) u_s \delta B_s \\
&\quad + \alpha_H \int_0^t F''(X_s) u_s \left( \int_0^T |s-\sigma|^{2H-2} \left( \int_0^s D_\sigma u_\theta \delta B_\theta \right) d\sigma \right) ds \\
&\quad + \alpha_H \int_0^t F''(X_s) u_s \left( \int_0^s u_\theta \, (s-\theta)^{2H-2} \, d\theta \right) ds. \qquad (3.7)
\end{aligned}
$$

REMARK 3.7. If the process $u$ is adapted, then the third summand in the right-hand side of (3.7) can be written as

$$
\alpha_H \int_0^t F''(X_s) u_s \left( \int_0^s \left( \int_0^\theta |s-\sigma|^{2H-2} D_\sigma u_\theta d\sigma \right) \delta B_\theta \right) ds.
$$

REMARK 3.8. $\frac{2H-1}{s^{2H-1}}(s-\theta)^{2H-2}\mathbf{1}_{[0,s]}(\theta)$ is an approximation of the identity as $H$ tends to $\frac{1}{2}$. Therefore, taking the limit as $H$ converges to $\frac{1}{2}$ in Equation (3.7) we recover the usual Itô's formula for the the Skorohod integral proved by Nualart and Pardoux [**N-P**].

**3.2. Stochastic integration with respect to fBm in the case $\mathbf{H < \frac{1}{2}}$.** The extension of the previous results to the case $H < \frac{1}{2}$ is not trivial and new difficulties appear. In order to illustrate these difficulties, let us first remark that the forward integral $\int_0^T B_t dB_t$ in the sense of Russo and Vallois (with the convergence in $L^2$) does not exists. In fact, a simple argument shows that, if $t_i = \frac{iT}{n}$, the expectation of the Riemann sums

$$
\sum_{i=1}^n B_{t_{i-1}}(B_{t_i} - B_{t_{i-1}})
$$

diverges:

$$
\begin{aligned}
\sum_{i=1}^n E\left(B_{t_{i-1}}(B_{t_i} - B_{t_{i-1}})\right) &= \frac{1}{2}\sum_{i=1}^n \left[t_i^{2H} - t_{i-1}^{2H} - (t_i - t_{i-1})^{2H}\right] \\
&= \frac{1}{2} T^{2H}\left(1 - n^{1-2H}\right) \to -\infty,
\end{aligned}
$$

as $n$ tends to infinity. Notice, however, that the expectation of symmetric Riemann sums is constant:

$$
\frac{1}{2}\sum_{i=1}^n E\left((B_{t_i} + B_{t_{i-1}})(B_{t_i} - B_{t_{i-1}})\right) = \frac{1}{2}\sum_{i=1}^n \left[t_i^{2H} - t_{i-1}^{2H}\right] = \frac{T^{2H}}{2}.
$$

We recall that for $H < \frac{1}{2}$ the operator $K_H^*$ given by (1.35) is an isometry between the Hilbert space $\mathcal{H}$ and $L^2(0,T)$. We have the estimate :

$$
\left| \frac{\partial K}{\partial t}(t,s) \right| \le c_H\left(\frac{1}{2} - H\right) (t-s)^{H-\frac{3}{2}}. \qquad (3.8)
$$

Consider the following seminorm on the set $\mathcal{E}$ of step functions on $[0,T]$:

$$\begin{aligned}\|\varphi\|_K^2 &= \int_0^T \varphi^2(s)K(T,s)^2 ds \\ &\quad + \int_0^T \left( \int_s^T |\varphi(t)-\varphi(s)| \, (t-s)^{H-\frac{3}{2}} dt \right)^2 ds.\end{aligned}$$

We denote by $\mathcal{H}_K$ the completion of $\mathcal{E}$ with respect to this seminorm. The space $\mathcal{H}_K$ is continuously embedded in $\mathcal{H}$.

The following result is the counterpart of Proposition 3.2 in the case $H<\frac{1}{2}$ and its has been proved in [**A-L-N**]:

PROPOSITION 3.9. *Let $u=\{u_t, t\in[0,T]\}$ be a stochastic process in the space $\mathbb{D}^{1,2}(\mathcal{H}_K)$. Suppose that the trace defined as the limit in probability*

$$\mathrm{Tr}Du := \lim_{\varepsilon\to 0} \frac{1}{2\varepsilon} \int_0^T \left\langle Du_s, \mathbf{1}_{[s-\varepsilon,s+\varepsilon]} \right\rangle_{\mathcal{H}} ds$$

*exists and*

$$E\left( \int_0^T u_s^2 \left( s^{2H-1} + (T-s)^{2H-1} \right) ds \right) < \infty,$$

$$E\left( \int_0^T \int_0^T (D_r u_s)^2 \left( s^{2H-1} + (T-s)^{2H-1} \right) ds dr \right) < \infty.$$

*Then the symmetric stochastic integral of $u$ with respect to fBm in the sense of Definition 3.1 exists and*

$$\int_0^T u_t dB_t = \delta(u) + \mathrm{Tr}Du.$$

Consider the particular case of the process $u_t = F(B_t)$, where $F$ is a continuously differentiable function satisfying the growth condition

$$\max\left\{ |F(x)|, |F'(x)|, |F''(x)| \right\} \le c e^{\lambda x^2},$$

where $c$ and $\lambda$ are positive constants such that $\lambda < \frac{1}{4T^{2H}}$. If $H>\frac{1}{4}$, the process $F(B_t)$ the process belongs to$\mathbb{D}^{1,2}(\mathcal{H}_K)$. Moreover, $\mathrm{Tr}Du$ exists and

$$\mathrm{Tr}Du = H \int_0^T F'(B_t) t^{2H-1} dt.$$

As a consequence we obtain

$$\int_0^T F(B_t) dB_t = \int_0^T F(B_t)\delta B_t + H \int_0^T F'(B_t) t^{2H-1} dt.$$

3.2.1. *Itô's formulas for the divergence integral in the case $H<\frac{1}{2}$.* An Itô's formula similar to (3.6) was proved in [**A-M-N2**] for general Gaussian processes of Volterra-type of the form $B_t = \int_0^t K(t,s) dW_s$, where $K(t,s)$ is a singular kernel. In particular, the process $B_t$ can be a fBm with Hurst parameter $\frac{1}{4} < H < \frac{1}{2}$. Moreover, in this paper, an Itô's formula for the indefinite divergence process $X_t = \int_0^t u_s \delta B_s$ similar to (3.7) was also proved.

On the other hand, in the case of the fractional Brownian motion with Hurst parameter $\frac{1}{4} < H < \frac{1}{2}$, an Itô's formula for the indefinite symmetric integral $X_t = \int_0^t u_s dB_s$ has been proved in [**A-L-N**] assuming again $\frac{1}{4} < H < \frac{1}{2}$.

Let us explain the reason for the restriction $\frac{1}{4} < H$. In order to define the divergence integral $\int_0^T F'(B_s)\delta B_s$, we need the process $F'(B_s)$ to belong to $L^2(\Omega; \mathcal{H})$. This is clearly true, provided $F$ satisfies the growth condition (3.4), because $F'(B_s)$ is Hölder continuous of order $H - \varepsilon > \frac{1}{2} - H$ if $\varepsilon < 2H - \frac{1}{2}$. If $H \leq \frac{1}{4}$, one can show (see [**C-N**]) that

$$P(B \in \mathcal{H}) = 0,$$

and the space $\mathbb{D}^{1,2}(\mathcal{H})$ is too small to contain processes of the form $F'(B_t)$.

Following the approach of [**C-N**] we are going to extend the domain of the divergence operator to processes whose trajectories are not necessarily in the space $\mathcal{H}$.

Using (1.35) and applying the integration by parts formula for the fractional calculus (1.9) we obtain for any $f, g \in \mathcal{H}$

$$\begin{aligned} \langle f, g \rangle_{\mathcal{H}} &= \langle K_H^* f, K_H^* g \rangle_{L^2(0,T)} \\ &= d_H^2 \left\langle s^{\frac{1}{2}-H} D_{T-}^{\frac{1}{2}-H} s^{H-\frac{1}{2}} f, s^{\frac{1}{2}-H} D_{T-}^{\frac{1}{2}-H} s^{H-\frac{1}{2}} g \right\rangle_{L^2(0,T)} \\ &= d_H^2 \left\langle f, s^{H-\frac{1}{2}} s^{\frac{1}{2}-H} D_{0+}^{\frac{1}{2}-H} s^{1-2H} D_{T-}^{\frac{1}{2}-H} s^{H-\frac{1}{2}} g \right\rangle_{L^2(0,T)}. \end{aligned}$$

This implies that the adjoint of the operator $K_H^*$ in $L^2(0,T)$ is

$$\left(K_H^{*,a} f\right)(s) = d_H s^{\frac{1}{2}-H} D_{0+}^{\frac{1}{2}-H} s^{1-2H} D_{T-}^{\frac{1}{2}-H} s^{H-\frac{1}{2}} f.$$

Set $\mathcal{H}_2 = (K_H^*)^{-1} \left(K_H^{*,a}\right)^{-1} (L^2(0,T))$. Denote by $\mathcal{S}_{\mathcal{H}}$ the space of smooth and cylindrical random variables of the form

$$F = f(B(\phi_1), \ldots, B(\phi_n)), \tag{3.9}$$

where $n \geq 1$, $f \in C_b^\infty(\mathbb{R}^n)$ ($f$ and all its partial derivatives are bounded), and $\phi_i \in \mathcal{H}_2$.

DEFINITION 3.10. *Let $u = \{u_t, t \in [0,T]\}$ be a measurable process such that*

$$E\left(\int_0^T u_t^2 dt\right) < \infty.$$

*We say that $u \in \mathrm{Dom}^*\delta$ if there exists a random variable $\delta(u) \in L^2(\Omega)$ such that for all $F \in \mathcal{S}_{\mathcal{H}}$ we have*

$$\int_{\mathbb{R}} E(u_t K_H^{*,a} K_H^* D_t F) dt = E(\delta(u) F).$$

This extended domain of the divergence operator satisfies the following elementary properties:

(1) $\mathrm{Dom}\delta \subset \mathrm{Dom}^*\delta$, and $\delta$ restricted to $\mathrm{Dom}\delta$ coincides with the divergence operator.
(2) If $u \in \mathrm{Dom}^*\delta$ then $E(u)$ belongs to $\mathcal{H}$.
(3) If $u$ is a deterministic process, then $u \in \mathrm{Dom}^*\delta$ if and only if $u \in \mathcal{H}$.

This extended domain of the divergence operator leads to the following version of Itô's formula for the divergence process, established by Cheridito and Nualart in [**C-N**].

THEOREM 3.11. *Suppose that $F$ is a function of class $C^2(\mathbb{R})$ satisfying the growth condition (3.4). Then for all $t \in [0,T]$, the process $\{F'(B_s)\mathbf{1}_{[0,t]}(s)\}$ belongs to* $\mathrm{Dom}^*\delta$ *and we have*

$$F(B_t) = F(0) + \int_0^t F'(B_s)\delta B_s + H\int_0^t F''(B_s)s^{2H-1}ds. \tag{3.10}$$

SKETCH OF THE PROOF. $F'(B_s)1_{[0,t]}(s) \in L^2(\Omega \times [0,T])$ and

$$F(B_t) - F(0) - H\int_0^t F''(B_s^H)s^{2H-1}ds \in L^2(\Omega)\,.$$

Hence, it suffices to show that for any $G \in \mathcal{S}_{\mathcal{H}}$ we have

$$\begin{aligned} & E\left(\langle F'(B_\cdot)1_{[0,t]}, D_\cdot G\rangle_{\mathcal{H}}\right) \\ = \; & E\left[G\left(F(B_t) - F(0) - H\int_0^t F''(B_s^H)s^{2H-1}ds\right)\right]. \end{aligned} \tag{3.11}$$

Equality (3.11) is proved by choosing smooth and cylindrical random variables of the form $G = H_n(B(\varphi))$, where $H_n$ denotes the $n$th Hermite polynomial, and applying an integration by parts formula. □

3.2.2. *Local time and Tanaka's formula for fBm.* Berman proved in [**B**] that that fractional Brownian motion $B = \{B_t, t \geq 0\}$ has a local time $l_t^a$ continuous in $(a,t) \in \mathbb{R} \times [0,\infty)$ which satisfies the occupation formula

$$\int_0^t g(B_s)ds = \int_{\mathbb{R}} g(a)l_t^a da. \tag{3.12}$$

for every continuous and bounded function $g$ on $\mathbb{R}$. Moreover, $l_t^a$ is increasing in the time variable. Set

$$L_t^a = 2H\int_0^t s^{2H-1}l^a(ds).$$

It follows from (3.12) that

$$2H\int_0^t g(B_s)s^{2H-1}ds = \int_{\mathbb{R}} g(a)L_t^a da.$$

This means that $a \to L_t^a$ is the density of the occupation measure

$$\mu(C) = 2H\int_0^T \mathbf{1}_C(B_s)s^{2H-1}ds,$$

where $C$ is a Borel subset of $\mathbb{R}$. Furthermore, the continuity property of $l_t^a$ implies that $L_t^a$ is continuous in $(a,t) \in \mathbb{R} \times [0,\infty)$.

As an extension of the Itô's formula (3.10), the following result has been proved in [**C-N**]:

THEOREM 3.12. *Let $0 < t < \infty$ and $a \in \mathbb{R}$. Then*

$$\mathbf{1}_{\{B_s > a\}}\mathbf{1}_{[0,t]}(s) \in \mathrm{Dom}^*\delta\,,$$

*and*

$$(B_t - a)^+ = (-a)^+ + \int_0^t \mathbf{1}_{\{B_s > a\}} \delta B_s + \frac{1}{2} L_t^a \,. \tag{3.13}$$

This result can be considered as a version of Tanaka's formula for the fBm. In [**C-N-T**] it is proved that for $H > \frac{1}{3}$, the process $\mathbf{1}_{\{B_s > a\}} \mathbf{1}_{[0,t]}(s)$ belongs to Dom$\delta$ and (3.13) holds.

The local time $\lambda_t^a$ has Hölder continuous paths of order $\delta < 1 - H$ in time, and of order $\gamma < \frac{1-H}{2H}$ in the space variable, provided $H \geq \frac{1}{3}$(see Table 2 in [**G-H**]). Moreover, $\lambda_t^a$ is absolutely continuous in $a$ if $H < \frac{1}{3}$ , it is continuously differentiable if $H < \frac{1}{5}$, and its smoothness in the space variable increases when $H$ decreases.

In a recent paper, Eddahbi, Lacayo, Solé, Tudor and Vives [**E-L-S-T-V**] have proved that $l_t^a \in \mathbb{D}^{\alpha,2}$ for all $\alpha < \frac{1-H}{2H}$. That means, the regularity of the local time $l_t^a$ in the sense of Malliavin calculus is the same order as its Hölder continuity in the space variable. This result follows from the Wiener chaos expansion (see [**C-N-T**]):

$$l_t^a = \sum_{n=0}^{\infty} \int_0^t s^{-nH} p(s^{2H}, a) H_n(a s^{-H}) I_n \left( K_H(s, \cdot)^{\otimes n} \right) ds,$$

where $p(t,a)$ denotes the Gaussian density with variance $t$. In fact, the series

$$\begin{aligned} &\sum_{n=0}^{\infty} (1+n)^{\alpha} E \left[ \left( \int_0^t s^{-nH} p(s^{2H}, a) H_n(a s^{-H}) I_n \left( K_H(s, \cdot)^{\otimes n} \right) ds \right)^2 \right] \\ = \;&\sum_{n=0}^{\infty} (1+n)^{\alpha} n! \int_0^t \int_0^t (sr)^{-nH} p(s^{2H}, a) p(r^{2H}, a) H_n(a s^{-H}) H_n(a r^{-H}) \\ &\times \langle K_H(s, \cdot), K_H(r, \cdot) \rangle_{\mathcal{H}} \, dr ds \end{aligned}$$

is equivalent to

$$\begin{aligned} &\sum_{n=1}^{\infty} n^{-\frac{1}{2}+\alpha} \int_0^t \int_0^t R_H(u,v)(uv)^{-nH-1} du dv \\ = \;&\sum_{n=0}^{\infty} n^{-\frac{1}{2}+\alpha} \int_0^1 R_H(1,z) z^{\ nH \ \ 1} dz. \end{aligned}$$

Then, the result follows from the estimate

$$\left| \int_0^1 R_H(1,z) z^{-nH-1} dz \right| \leq C n^{-\frac{1}{2H}}.$$

## 4. Stochastic differential equations driven by a fBm

Let $B = \{B_t, t \geq 0\}$ be an $m$-dimensional fractional Brownian motion of Hurst parameter $H \in \left(\frac{1}{2}, 1\right)$. This means that the components of $B$ are independent fBm with the same Hurst parameter $H$. Consider the equation on $\mathbb{R}^d$

$$X_t = X_0 + \int_0^t \sigma(s, X_s) \, dB_s + \int_0^t b(s, X_s) ds, \;\; t \in [0,T] \,, \tag{4.1}$$

where $X_0$ is a $d$-dimensional random variable. The integral with respect to $B$ is a path-wise Riemann-Stieltjes integral, and we know that this integral exists provided that the process $\sigma(s, X_s)$ has Hölder continuous trajectories of order larger that $1 - H$.

In [**Ly1**], Lyons considered deterministic integral equations of the form

$$x_t = x_0 + \int_0^t \sigma(x_s) dg_s,$$

$0 \le t \le T$, where the $g : [0, T] \to \mathbb{R}^d$ is a continuous functions with bounded $p$-variation for some $p \in [1, 2)$. This equation has a unique solution in the space of continuous functions of bounded $p$-variation if each component of $g$ has a Hölder continuous derivative of order $\alpha > p - 1$. Taking into account that fBm of Hurst parameter $H$ has locally bounded $p$-variation paths for $p > 1/H$, the result proved in [**Ly1**] can be applied to Equation (4.1) in the case $\sigma(s, x) = \sigma(x)$, and $b(s, x) = 0$, provided the coefficient $\sigma$ has a Hölder continuous derivative of order $\alpha > \frac{1}{H} - 1$.

Using the approach based on the notion of $p$-variation and the general limit theorem proved by Lyons in [**Ly2**] for differential equations driven by geometric rough paths, Coutin and Qian [**C-Q2**], [**C-Q1**] have established the existence of strong solutions and a Wong-Zakai type approximation limit for stochastic differential equations driven by a fractional Brownian motion with parameter $H > \frac{1}{4}$.

In [**Ru**] Ruzmaikina establishes an existence and uniqueness theorem for ordinary differential equations with Hölder continuous forcing. The global solution is constructed, first, in small time intervals where the contraction principle can be applied, provided the Hölder constant is small enough. The main estimates are deduced using Hölder norms.

In [**Z2**] the existence and uniqueness of solutions is proved for differential equations driven by a fractional Brownian motion with parameter $H > \frac{1}{2}$, in a small random interval, provided the diffusion coefficient is a contraction in the space $W_{2,\infty}^\beta$, where $\frac{1}{2} < \beta < H$. Here $W_{2,\infty}^\beta$ denotes the Besov-type space of bounded measurable functions $f : [0, T] \to \mathbb{R}$ such that

$$\int_0^T \int_0^T \frac{(f(t) - f(s))^2}{|t - s|^{2\beta+1}} dsdt < \infty.$$

In [**N-R**] Nualart and Răşcanu have established the existence and uniqueness of solution for Equation (4.1) using an a priori estimate based on the fractional integration by parts formula, following the approach of Zähle [**Z1**]. In this section we will survey the main ideas and results of [**N-R**].

**4.1. Generalized Stieltjes integrals.** Given a function $g : [0, T] \to \mathbb{R}$, set $g_{T-}(s) = g(s) - \lim_{\varepsilon \downarrow 0}(T - \varepsilon)$ provided this limit exists. Take $p, q \ge 1$ such that $\frac{1}{p} + \frac{1}{q} \le 1$ and $0 < \alpha < 1$. Suppose that $f$ and $g$ are functions on $[0, T]$ such that $g(T-)$ exists, $f \in I_{0+}^\alpha(L^p)$ and $g_{T-} \in I_{T-}^{1-\alpha}(L^q)$. Then the *generalized Stieltjes integral* of $f$ with respect to $g$ is defined by (see [**Z1**])

$$\int_0^T f dg = \int_0^T D_{0+}^\alpha f_{a+}(s) D_{T-}^{1-\alpha} g_{T-}(s) ds. \tag{4.2}$$

In [**Z1**] it is proved that this integral coincides with the Riemann-Stieltjes integral if $f$ and $g$ are Hölder continuous of orders $\alpha$ and $\beta$ with $\alpha + \beta > 1$.

Fix $0 < \alpha < \frac{1}{2}$. Denote by $W_0^{\alpha,\infty}(0, T)$ the space of measurable functions $f : [0, T] \to \mathbb{R}$ such that

$$\|f\|_{\alpha,\infty} := \sup_{t\in[0,T]} \left( |f(t)| + \int_0^t \frac{|f(t) - f(s)|}{(t-s)^{\alpha+1}} ds \right) < \infty.$$

We have, for all $0 < \varepsilon < \alpha$

$$C^{\alpha+\varepsilon}(0,T) \subset W_0^{\alpha,\infty}(0,T) \subset C^{\alpha-\varepsilon}(0,T).$$

Denote by $W_T^{1-\alpha,\infty}(0,T)$ the space of measurable functions $g : [0,T] \to \mathbb{R}$ such that

$$\|g\|_{1-\alpha,\infty,T} := \sup_{0<s<t<T} \left( \frac{|g(t) - g(s)|}{(t-s)^{1-\alpha}} + \int_s^t \frac{|g(y) - g(s)|}{(y-s)^{2-\alpha}} dy \right) < \infty.$$

We have, for all $0 < \varepsilon < \alpha$

$$C^{1-\alpha+\varepsilon}(0,T) \subset W_T^{1-\alpha,\infty}(0,T) \subset C^{1-\alpha}(0,T).$$

For $g \in W_T^{1-\alpha,\infty}(0,T)$define

$$\begin{aligned} \Lambda_\alpha(g) &:= \frac{1}{\Gamma(1-\alpha)} \sup_{0<s<t<T} \left| \left(D_{t-}^{1-\alpha} g_{t-}\right)(s) \right| \\ &\leq \frac{1}{\Gamma(1-\alpha)\Gamma(\alpha)} \|g\|_{1-\alpha,\infty,T}. \end{aligned}$$

Finally, denote by $W_0^{\alpha,1}(0,T)$ the space of measurable functions $f$ on $[0,T]$ such that

$$\|f\|_{\alpha,1} := \int_0^T \frac{|f(s)|}{s^\alpha} ds + \int_0^T \int_0^s \frac{|f(s) - f(y)|}{(s-y)^{\alpha+1}} dy\, ds < \infty.$$

If $f$ is a function in the space $W_0^{\alpha,1}(0,T)$, and $g$ belongs to $W_T^{1-\alpha,\infty}(0,T)$, then the generalized Stieltjes integral $\int_0^t f dg$ exists for all $t \in [0,T]$ and we have

$$\left| \int_0^t f dg \right| \leq \Lambda_\alpha(g) \|f\|_{\alpha,1}.$$

Indeed,

$$\begin{aligned} \left| \int_0^t f dg \right| &= \left| \int_0^t \left(D_{0+}^\alpha f\right)(s) \left(D_{t-}^{1-\alpha} g_t\right)(s)\, ds \right| \\ &\leq \sup_{0\leq s\leq t\leq T} \left| \left(D_{t-}^{1-\alpha} g_{t-}\right)(s) \right| \int_0^t \left| \left(D_{0+}^\alpha f\right)(s) \right| ds \\ &\leq \Lambda_\alpha(g) \|f\|_{\alpha,1}. \end{aligned}$$

**4.2. Main estimate.** Fix $0 < \alpha < \frac{1}{2}$. Given two functions $g \in W_T^{1-\alpha,\infty}(0,T)$ and $f \in W_0^{\alpha,1}(0,T)$ we set

$$h_t = \int_0^t f dg.$$

Then for all $s < t \leq T$ we have

$$\begin{aligned} |h_t| + \int_0^t \frac{|h_t - h_s|}{(t-s)^{\alpha+1}} ds &\leq \Lambda_\alpha(g)\, c_{\alpha,T}^{(1)} \int_0^t \left((t-r)^{-2\alpha} + r^{-\alpha}\right) \\ &\quad \times \left( |f_r| + \int_0^r \frac{|f_r - f_y|}{(r-y)^{\alpha+1}} dy \right) dr, \end{aligned} \tag{4.3}$$

where $c_{\alpha,T}^{(1)}$ is a constant depending on $\alpha$ and $T$.

As a consequence of this estimate, if $f \in W_T^{1-\alpha,\infty}(0,T)$ we have

$$\left|\int_s^t f dg\right| \leq \Lambda_\alpha(g)\, c_{\alpha,T}^{(2)}(t-s)^{1-\alpha}\, \|f\|_{\alpha,\infty},$$

and

$$\left\|\int_0^{\cdot} f dg\right\|_{\alpha,\infty} \leq \Lambda_\alpha(g)\, c_{\alpha,T}^{(3)}\, \|f\|_{\alpha,\infty}.$$

SKETCH OF THE PROOF OF (4.3). Using the definition and additivity property of the indefinite integral we obtain

$$\begin{aligned} |h_t - h_s| &= \left|\int_s^t f dg\right| = \left|\int_s^t D_{s+}^{\alpha}(f)(r)\left(D_{t-}^{1-\alpha} g_{t-}\right)(r)\, dr\right| \\ &\leq \Lambda_\alpha(g)\left(\int_s^t \frac{|f_r|}{(r-s)^\alpha} dr + \alpha \int_s^t \int_s^r \frac{|f_r - f_y|}{(r-y)^{\alpha+1}} dy dr\right). \end{aligned} \tag{4.4}$$

Taking $s = 0$ we obtain the desired estimate for $|h_t|$. Multiplying (4.4) by $(t-s)^{-\alpha-1}$ and integrating in $s$ yields

$$\begin{aligned} \int_0^t \frac{|h_t - h_s|}{(t-s)^{\alpha+1}} ds &\leq \Lambda_\alpha(g) \int_0^t (t-s)^{-\alpha-1} \\ &\quad \times \left(\int_s^t \frac{|f_r|}{(r-s)^\alpha} dr + \alpha \int_s^t \int_s^r \frac{|f_r - f_y|}{(r-y)^{\alpha+1}} dy dr\right) ds. \end{aligned} \tag{4.5}$$

By the substitution $s = r - (t-r)y$ we have

$$\int_0^r (t-s)^{-\alpha-1}(r-s)^{-\alpha} ds \leq (t-r)^{-2\alpha} \int_0^\infty (1+y)^{-\alpha-1} y^{-\alpha} dy \tag{4.6}$$

and, on the other hand,

$$\int_0^y (t-s)^{-\alpha-1} ds = \alpha^{-1}\left[(t-y)^{-\alpha} - t^{-\alpha}\right] \leq \alpha^{-1}(t-y)^{-\alpha}. \tag{4.7}$$

Substituting (4.6) and (4.7) into (4.5) yields

$$\begin{aligned} \int_0^t \frac{|h_t - h_s|}{(t-s)^{\alpha+1}} ds &\leq \Lambda_\alpha(g)\left[c_\alpha^{(1)} \int_0^t \frac{|f_r|}{(t-r)^{2\alpha}} dr \right. \\ &\quad \left. + \int_0^t \int_0^r \frac{|f(r) - f(y)|}{(r-y)^{\alpha+1}} (t-y)^{-\alpha} dy dr\right], \end{aligned}$$

where

$$c_\alpha^{(1)} = \int_0^\infty (1+y)^{-\alpha-1} y^{-\alpha} dy = \beta(2\alpha, 1-\alpha).$$

□

**4.3. Deterministic differential equations.** Let $0 < \alpha < \frac{1}{2}$ be fixed. Let $g^i \in W_T^{1-\alpha,\infty}(0,T;\mathbb{R}^m)$, $j = 1, \ldots, m$. Consider the deterministic differential equation on $\mathbb{R}^d$

$$x_t = x_0 + \int_0^t b(s, x_s) ds + \int_0^t \sigma(s, x_s)\, dg_s, \quad t \in [0,T], \tag{4.8}$$

where $x_0 \in \mathbb{R}^d$.

Let us introduce the following assumptions on the coefficients:

**H1:** $\sigma(t,x)$ is differentiable in $x$, and there exist some constants $0<\beta,\delta\le 1$ and for every $N\ge 0$ there exists $M_N>0$ such that the following properties hold:

$$
\begin{aligned}
&|\sigma(t,x)-\sigma(t,y)|\le M_0|x-y|, \quad \forall\, x\in\mathbb{R}^d,\ \forall\, t\in[0,T],\\
&\quad|\partial_{x_i}\sigma(t,x)-\partial_{x_i}\sigma(t,y)|\le M_N|x-y|^{\delta},\ \forall\,|x|,|y|\le N,\ \forall t\in[0,T],\\
&\qquad|\sigma(t,x)-\sigma(s,x)|+|\partial_{x_i}\sigma(t,x)-\partial_{x_i}\sigma(s,x)|\le M_0|t-s|^{\beta},\\
&\qquad\qquad\qquad\qquad\qquad\qquad\forall\, x\in\mathbb{R}^d,\ \forall\, t,s\in[0,T].
\end{aligned}
$$

for each $i=1,\dots,d$.

**H2:** The coefficient $b(t,x)$ satisfies for every $N\ge 0$

$$
\begin{aligned}
|b(t,x)-b(t,y)| &\le L_N|x-y|,\ \forall\,|x|,|y|\le N,\ \forall t\in[0,T],\\
|b(t,x)| &\le L_0|x|+b_0(t),\ \forall x\in\mathbb{R}^d,\ \forall t\in[0,T],
\end{aligned}
$$

where $b_0\in L^{\rho}\left(0,T;\mathbb{R}^d\right)$, with $\rho\ge 2$ and for some constant $L_N>0$.

THEOREM 4.1. *Suppose that the coefficients $\sigma$ and $b$ satisfy the assumptions H1 and H2 with $\rho=\frac{1}{\alpha}$, $0<\beta,\delta\le 1$ and $0<\alpha<\alpha_0=\min\left(\frac{1}{2},\beta,\frac{\delta}{\delta+1}\right)$. Then Equation (4.8) has a unique continuous solution such that $x^i\in W_0^{\alpha,\infty}(0,T)$ for all $i=1,\dots,d$.*

SKETCH OF THE PROOF. Suppose $d=m=1$. Fix $\lambda>1$ and define the seminorm in $W_0^{\alpha,\infty}(0,T)$ by

$$
\|f\|_{\alpha,\lambda}=\sup_{t\in[0,T]} e^{-\lambda t}\left(|f_t|+\int_0^t \frac{|f_t-f_s|}{(t-s)^{\alpha+1}}\,ds\right).
$$

Consider the operator $\mathcal{L}$ on defined by

$$
(\mathcal{L}f)_t=x_0+\int_0^t b(s,f_s)ds+\int_0^t \sigma\left(s,f_s\right)dg_s.
$$

There exists $\lambda_0$ such that for $\lambda\ge\lambda_0$ we have

$$
\|\mathcal{L}f\|_{\alpha,\lambda}\le|x_0|+1+\frac{1}{2}\|f\|_{\alpha,\lambda}.
$$

Hence, the operator $\mathcal{L}$ leaves invariant the ball $B_0$ of radius $2\left(|x_0|+1\right)$ in the norm $\|\cdot\|_{\alpha,\lambda_0}$ of the space $W_0^{\alpha,\infty}(0,T)$. Moreover, there exists a constant $C$ depending on $g$ such that for any $\lambda\ge 1$ and $u,v\in B_0$

$$
\|\mathcal{L}(u)-\mathcal{L}(v)\|_{\alpha,\lambda}\le\frac{C}{\lambda^{1-2\alpha}}\left(1+\Delta(u)+\Delta(v)\right)\|u-v\|_{\alpha,\lambda}, \tag{4.9}
$$

where

$$
\Delta(u)=\sup_{r\in[0,T]}\int_0^r \frac{|u_r-u_s|^{\delta}}{(r-s)^{\alpha+1}}\,ds.
$$

A basic ingredient in the proof of this inequality is the estimate

$$
\begin{aligned}
&|\sigma(r,f_r)-\sigma(s,f_s)-\sigma(r,h_r)+\sigma(s,h_s)|\\
&\le M_0|f_r-f_s-h_r+h_s|+M_0|f_r-h_r|(r-s)^{\beta}\\
&+M_N|f_r-h_r|\left(|f_r-f_s|^{\delta}+|h_r-h_s|^{\delta}\right),
\end{aligned}
$$

which is an immediate consequence of the properties of the function $\sigma$. The seminorm $\Delta$ is bounded on $\mathcal{L}(B_0)$, and, as a consequence, (4.9) implies that $\mathcal{L}$ is a contraction operator in $\mathcal{L}(B_0)$ with respect to a different norm $\|\cdot\|_{\alpha,\lambda_2}$ for a suitable value of $\lambda_2 > 1$. Finally, the the existence of a solution follows from a suitable fixed point argument (see Lemma 4.2). The uniqueness is proved again using the main estimate (4.3). □

The following lemma has been used in the proof of Theorem 4.1 (for its proof see [**N-R**]).

LEMMA 4.2. *Let $(X,\rho)$ be a complete metric space and $\rho_0, \rho_1, \rho_2$ some metrics on $X$ equivalent to $\rho$. If $\mathcal{L}: X \to X$ satisfies:*
*i) there exists $r_0 > 0$, $x_0 \in X$ such that if $B_0 = \{x \in X : \rho_0(x_0, x) \le r_0\}$ then*

$$\mathcal{L}(B_0) \subset B_0\,,$$

*ii) there exists $\varphi : (X,\rho) \to [0,+\infty]$ lower semicontinuous function and some positive constants $C_0, K_0$ such that denoting $N_\varphi(a) = \{x \in X : \varphi(x) \le a\}$*

$$\begin{array}{ll} a) & \mathcal{L}(B_0) \subset N_\varphi(C_0)\,, \\ b) & \rho_1(\mathcal{L}(x), \mathcal{L}(y)) \le K_0 \rho_1(x,y)\,,\ \forall x,y \in N_\varphi(C_0) \cap B_0, \end{array}$$

*iii) there exists $a \in (0,1)$ such that*

$$\rho_2(\mathcal{L}(x), \mathcal{L}(y)) \le a\, \rho_2(x,y)\,,\ \forall x,y \in \mathcal{L}(B_0)\,,$$

*then there exists $x^* \in \mathcal{L}(B_0) \subset X$ such that*

$$x^* = \mathcal{L}(x^*)\,.$$

4.3.1. *Estimates of the solution.* Suppose that the coefficient $\sigma$ satisfies the assumptions of the Theorem 4.1 and

$$|\sigma(t,x)| \le K_0 (1 + |x|^\gamma)\,, \tag{4.10}$$

where $0 \le \gamma \le 1$. Then, the solution $f$ of Equation (4.8) satisfies

$$\|f\|_{\alpha,\infty} \le C_1 \exp(C_2 \Lambda_\alpha(g)^\kappa)\,, \tag{4.11}$$

where

$$\kappa = \begin{cases} \frac{1}{1-2\alpha} & \text{if} \quad \gamma = 1 \\ > \frac{\gamma}{1-2\alpha} & \text{if} \quad \frac{1-2\alpha}{1-\alpha} \le \gamma < 1 \\ \frac{1}{1-\alpha} & \text{if} \quad 0 \le \gamma < \frac{1-2\alpha}{1-\alpha} \end{cases}$$

and the constants $C_1$ and $C_2$ depend on $T$, $\alpha$, and the constants that appear in conditions H1, H2 and (4.10).

The proof of (4.11) is based on the following version of Gronwall lemma:

LEMMA 4.3. *Fix $0 \le \alpha < 1$, $a, b \ge 0$. Let $x : [0,\infty) \to [0,\infty)$ be a continuous function such that for each $t$*

$$x_t \le a + bt^\alpha \int_0^t (t-s)^{-\alpha} s^{-\alpha} x_s ds. \tag{4.12}$$

*Then*

$$x_t \le a + a \sum_{n=1}^\infty b^n \frac{\Gamma(1-\alpha)^{n+1} t^{n(1-\alpha)}}{\Gamma[(n+1)(1-\alpha)]}.$$

$$\leq a d_\alpha \exp\left[c_\alpha t b^{1/(1-\alpha)}\right], \tag{4.13}$$

*where $c_a$ and $d_\alpha$ are positive constants depending only on $\alpha$ (as an example, one can set $c_\alpha = 2\left(\Gamma(1-\alpha)\right)^{1/(1-\alpha)}$ and $d_\alpha = 4e^2\frac{\Gamma(1-\alpha)}{1-\alpha}$ ).*

This implies that there exists a constants $c_\alpha, d_\alpha > 0$ such that

$$x_t \leq a d_\alpha \exp\left[c_\alpha t b^{1/(1-\alpha)}\right].$$

**4.4. Stochastic differential equations with respect to fBm.** Fix a parameter $\frac{1}{2} < H < 1$. Let $B = \{B_t, t \in [0,T]\}$ be a fractional Brownian motion with parameter $H$. Choose $\alpha$ such that $1 - H < \alpha < \frac{1}{2}$. By Fernique's theorem, for any $0 < \delta < 2$ we have

$$E\left(\exp\left(\Lambda_\alpha(B)^\delta\right)\right) < \infty.$$

As a consequence, if $u = \{u_t,\ t \in [0,T]\}$ is a stochastic process whose trajectories belong to the space $W_T^{\alpha,1}(0,T)$, almost surely, the path-wise generalized Stieltjes integral integral $\int_0^T u_s dB_s$ exists and we have the estimate

$$\left|\int_0^T u_s dB_s\right| \leq G \left\|u\right\|_{\alpha,1}.$$

Moreover, if the trajectories of the process $u$ belong to the space $W_0^{\alpha,\infty}(0,T)$, then the indefinite integral $U_t = \int_0^t u_s dB_s$ is Hölder continuous of order $1-\alpha$, and its trajectories also belong to the space $W_0^{\alpha,\infty}(0,T)$.

Consider the stochastic differential equation (4.1) on $\mathbb{R}^d$ where the process $B$ is an $m$-dimensional fBm with Hurst parameter $H \in \left(\frac{1}{2}, 1\right)$ and $X_0$ is a $d$-dimensional random variable. Suppose that the coefficients $\sigma^{i,j}, b^i : \Omega \times [0,T] \times \mathbb{R}^d \to \mathbb{R}$ are measurable functions satisfying conditions H1 and H2, where the constants $M_N$ and $L_N$ may depend on $\omega \in \Omega$, and $\beta > 1 - H$, $\delta > \frac{1}{H} - 1$. Fix $\alpha$ such that

$$1 - H < \alpha < \alpha_0 = \min\left(\frac{1}{2}, \beta, \frac{\delta}{\delta+1}\right)$$

and $\alpha < \frac{1}{\rho}$. Then the stochastic equation (4.3) has a unique continuous solution such that $X^i \in W_0^{\alpha,\infty}(0,T)$ for all $i = 1, \ldots, d$. Moreover the solution is Hölder continuous of order $1-\alpha$.

Assume that $X_0$ is bounded and the constants do not depend on $\omega$. Suppose that

$$\left|\sigma(t,x)\right| \leq K_0\left(1 + |x|^\gamma\right),$$

where $0 \leq \gamma \leq 1$. Then,

$$\left\|X\right\|_{\alpha,\infty} \leq C_1 \exp\left(C_2 \Lambda_\alpha(B)^\kappa\right).$$

Hence,for all $p \geq 1$

$$E\left(\left\|X\right\|_{\alpha,\infty}^p\right) \leq C_1^p E\left(\exp\left(p C_2 \Lambda_\alpha(B)^\kappa\right)\right) < \infty$$

provided $\kappa < 2$, that is,

$$\frac{\gamma}{4} + \frac{1}{2} \leq H$$

and

$$1 - H < \alpha < \frac{1}{2} - \frac{\gamma}{4}.$$

- If $\gamma = 1$ this means $\alpha < \frac{1}{4}$ and $H \leq \frac{3}{4}$
- If $\gamma < 2 - \frac{1}{H}$ we can take any $\alpha$ such that $1 - H < \alpha < \frac{1}{2}$.

## 5. Applications

In this section we will describe some applications of the stochastic calculus with respect to fBm.

**5.1. Vortex filaments based on fBm.** The observations of three–dimensional turbulent fluids indicate that the vorticity field of the fluid is concentrated along thin structures called vortex filaments. In his book Chorin [**Cho**] suggests probabilistic descriptions of vortex filaments by trajectories of self-avoiding walks on a lattice. Flandoli [**F**] introduced a model of vortex filaments based on a three–dimensional Brownian motion. A basic problem in these models is the computation of the kynetic energy of a given configuration.

Denote by $u(x)$ the velocity field of the fluid at point $x \in \mathbb{R}^3$, and let $\xi = \mathrm{curl} u$ be the associated vorticity field. The kynetic energy of the field will be

$$\mathbb{H} = \frac{1}{2}\int_{\mathbb{R}^3} |u(x)|^2 dx = \frac{1}{8\pi}\int_{\mathbb{R}^3}\int_{\mathbb{R}^3} \frac{\xi(x)\cdot\xi(y)}{|x-y|} dxdy. \tag{5.1}$$

We will assume that the vorticity field is concentrated along a thin tube centered in a curve $\gamma = \{\gamma_t, 0 \leq t \leq T\}$. Moreover, we will choose a random model and consider this curve as the trajectory of a stochastic process three-dimensional fractional Brownian motion $B = \{B_t, 0 \leq t \leq T\}$. This can be formally expressed as

$$\xi(x) = \Gamma \int_{\mathbb{R}^3} \left( \int_0^T \delta(x - y - B_s)\dot{B}_s ds \right) \rho(dy), \tag{5.2}$$

where $\Gamma$ is a parameter called the circuitation, and $\rho$ is a probability measure on $\mathbb{R}^3$ with compact support.

Substituting (5.2) into (5.1) we derive the following formal expression for the kynetic energy:

$$\mathbb{H} = \int_{\mathbb{R}^3}\int_{\mathbb{R}^3} \mathbb{H}_{xy}\rho(dx)\rho(dy), \tag{5.3}$$

where the so-called interaction energy $\mathbb{H}_{xy}$ is given by the double integral

$$\mathbb{H}_{xy} = \frac{\Gamma^2}{8\pi}\sum_{i=1}^3 \int_0^T\int_0^T \frac{1}{|x + B_t - y - B_s|} dB_s^i dB_t^i. \tag{5.4}$$

We are interested in the following problems: Is $\mathbb{H}$ a well defined random variable? Does it have moments of all orders and even exponential moments?

In order to give a rigorous meaning to the double integral (5.4) let us introduce the regularization of the function $|\cdot|^{-1}$:

$$\sigma_n = |\cdot|^{-1} * p_{1/n}, \tag{5.5}$$

where $p_{1/n}$ is the Gaussian kernel with variance $\frac{1}{n}$. Then, the smoothed interaction energy

$$\mathbb{H}_{xy}^n = \frac{\Gamma^2}{8\pi}\sum_{i=1}^3 \int_0^T \left( \int_0^T \sigma_n(x + B_t - y - B_s)\, dB_s^i \right) dB_t^i, \tag{5.6}$$

is well defined, where the integrals are path-wise Riemann-Stieltjes integrals. Set

$$\mathbb{H}^n = \int_{\mathbb{R}^3}\int_{\mathbb{R}^3} \mathbb{H}^n_{xy}\rho(dx)\rho(dy). \tag{5.7}$$

The following result has been proved in [**N-R-T**]:

THEOREM 5.1. *Suppose that the measure $\rho$ satisfies*

$$\int_{\mathbb{R}^3}\int_{\mathbb{R}^3} |x-y|^{1-\frac{1}{H}}\rho(dx)\rho(dy) < \infty. \tag{5.8}$$

*Let $\mathbb{H}^n_{xy}$ be the smoothed interaction energy defined by (5.6). Then $\mathbb{H}^n$ defined in (5.7) converges, for all $k \geq 1$, in $L^k(\Omega)$ to a random variable $\mathbb{H} \geq 0$ that we call the energy associated with the vorticity field (5.2).*

If $H = \frac{1}{2}$, fBm $B$ is a classical three-dimensional Brownian motion. In this case condition (5.8) would be $\int_{\mathbb{R}^3}\int_{\mathbb{R}^3} |x-y|^{-1}\rho(dx)\rho(dy) < \infty$, which is the assumption made by Flandoli [**F**] and Flandoli and Gubinelli [**F-G**]. In this last paper, using Fourier approach and Itô's stochastic calculus, the authors show that $Ee^{-\beta\mathbb{H}} < \infty$ for sufficiently small negative $\beta$.

The proof of Theorem 5.1 is based on the stochastic calculus of variations with respect to fBm and the application of Fourier transform.

SKETCH OF THE PROOF OF THEOREM 5.1. The proof will be done in two steps:

**Step 1** (Fourier transform) Using

$$\frac{1}{|z|} = \int_{\mathbb{R}^3} (2\pi)^3 \frac{e^{-i\langle \xi, z\rangle}}{|\xi|^2} d\xi$$

we get

$$\sigma_n(x) = \int_{\mathbb{R}^3} |\xi|^{-2} e^{i\langle \xi, x\rangle - |\xi|^2/2n}\, d\xi.$$

Substituting this expression in (5.6), we obtain the following formula for the smoothed interaction energy

$$\begin{aligned} \mathbb{H}^n_{xy} &= \frac{\Gamma^2}{8\pi}\sum_{j=1}^3 \int_0^T\int_0^T \left(\int_{\mathbb{R}^3} e^{i\langle \xi, x+B_t-y-B_s\rangle}\,\frac{e^{-|\xi|^2/2n}}{|\xi|^2}\right) dB_s^j dB_t^j \\ &= \frac{\Gamma^2}{8\pi}\int_{\mathbb{R}^3} |\xi|^{-2} e^{i\langle \xi, x-y\rangle - |\xi|^2/2n}\, \|Y_\xi\|^2_{\mathbb{C}}\, d\xi, \end{aligned} \tag{5.9}$$

where

$$Y_\xi = \int_0^T e^{i\langle \xi, B_t\rangle} dB_t$$

and $\|Y_\xi\|^2_{\mathbb{C}} = \sum_{i=1}^3 Y^i_\xi \overline{Y^i_\xi}$. Integrating with respect to $\rho$ yields

$$\mathbb{H}^n = \frac{\Gamma^2}{8\pi}\int_{\mathbb{R}^3} \|Y_\xi\|^2_{\mathbb{C}}\, |\xi|^{-2}\, |\widehat{\rho}(\xi)|^2\, e^{-|\xi|^2/2n} d\xi \geq 0. \tag{5.10}$$

From Fourier analysis and condition (5.8) we know that

$$\int_{\mathbb{R}^3}\int_{\mathbb{R}^3} |x-y|^{1-\frac{1}{H}}\rho(dx)\rho(dy) = C_H \int_{\mathbb{R}^3} |\widehat{\rho}(\xi)|^2\, |\xi|^{\frac{1}{H}-4} d\xi < \infty. \tag{5.11}$$

Then, taking into account (5.11) and (5.10), in order to show the convergence in $L^k(\Omega)$ of $\mathbb{H}^n$ to a random variable $\mathbb{H} \geq 0$ it suffices to check that

$$E\left(\|Y_\xi\|_{\mathbb{C}}^{2k}\right) \leq C_k \left(1 \wedge |\xi|^{k\left(\frac{1}{H}-2\right)}\right). \tag{5.12}$$

**Step 2** (Stochastic calculus) We will present the main arguments for the proof of the estimate (5.12) for $k = 1$. Relation (3.2) applied to the process $u_t = e^{i\langle \xi, B_t\rangle}$ allows us to decompose the path-wise integral $Y_\xi = \int_0^T e^{i\langle \xi, B_t\rangle} dB_t$ into the sum of a divergence plus a trace term:

$$Y_\xi = \int_0^T e^{i\langle \xi, B_t\rangle} \delta B_t + H \int_0^T i\xi e^{i\langle \xi, B_t\rangle} t^{2H-1} dt. \tag{5.13}$$

On the other hand, applying the three dimensional version of Itô's formula (3.6) we obtain

$$e^{i\langle \xi, B_T\rangle} = 1 + \sum_{j=1}^3 \int_0^T i\xi_j e^{i\langle \xi, B_t\rangle} \delta B_t^j - H \int_0^T t^{2H-1} |\xi|^2 e^{i\langle \xi, B_t\rangle} dt. \tag{5.14}$$

Multiplying both members of (5.14) by $i\xi|\xi|^{-2}$ and adding the result to (5.13) yields

$$Y_\xi = p_\xi \left( \int_0^T e^{i\langle \xi, B_t\rangle} \delta B_t \right) - \frac{i\xi}{|\xi|^2} \left( e^{i\langle \xi, B_T\rangle} - 1 \right) := Y_\xi^{(1)} + Y_\xi^{(2)},$$

where $p_\xi(v) = v - \frac{\xi}{|\xi|^2} \langle \xi, v\rangle$ is the orthogonal projection of $v$ on $\langle \xi\rangle^\perp$. It suffices to derive the estimate (5.12) for the term $Y_\xi^{(1)}$. Using the duality relationship (2.2) for each $j = 1, 2, 3$ we can write

$$E\left(Y_\xi^{(1),j} \overline{Y}_\xi^{(1),j}\right) = E\left( \left\langle e^{i\langle \xi, B_\cdot\rangle}, p_\xi^j D_\cdot \left( p_\xi^j \int_0^T e^{-i\langle \xi, B_t\rangle} \delta B_t \right) \right\rangle_{\mathcal{H}} \right). \tag{5.15}$$

The commutation relation $\langle D(\delta(u)), h\rangle_{\mathcal{H}} = \langle u, h\rangle_{\mathcal{H}} + \delta(\langle Du, h\rangle_{\mathcal{H}})$ implies

$$D_r^k \left( \int_0^T e^{-i\langle \xi, B_t\rangle} \delta B_t^j \right) = e^{-i\langle \xi, B_r^k\rangle} \delta_{k,j} + (-i\xi^k) \int_0^T \mathbf{1}_{[0,t]}(r) e^{-i\langle \xi, B_t\rangle} \delta B_t^j.$$

Applying the projection operators yields

$$\begin{aligned} p_\xi^j D_r \left( p_\xi^j \int_0^T e^{-i\langle \xi, B_t\rangle} \delta B_t \right) &= e^{-i\langle \xi, B_r\rangle} \left( I - \frac{\xi^* \xi}{|\xi|^2} \right)_{j,j} \\ &= e^{-i\langle \xi, B_r\rangle} \left( 1 - \frac{\left(\xi^j\right)^2}{|\xi|^2} \right) \end{aligned}$$

Notice that the term involving derivatives in the expectation (5.15) vanishes. This cancellation is similar to what happens in the computation of the variance of the divergence of an adapted process, in the case of the Brownian motion. Hence,

$$
\begin{aligned}
\sum_{j=1}^{3} E\left(Y_\xi^{(1),j}\overline{Y}_\xi^{(1),j}\right) &= 2\,E\left(\left\langle e^{-i\langle \xi, B_\cdot\rangle}, e^{-i\langle \xi, B_\cdot\rangle} \right\rangle_{\mathcal{H}}\right) \\
&= 2\alpha_H \int_0^T\int_0^T E\left(e^{i\langle \xi, B_s - B_r\rangle}\right) |s-r|^{2H-2}\, dsdr \\
&= 2\alpha_H \int_0^T\int_0^T e^{-\frac{|s-r|^{2H}}{2}|\xi|^2} |s-r|^{2H-2}\, dsdr,
\end{aligned}
$$

which behaves as $|\xi|^{\frac{1}{H}-2}$ as $|\xi|$ tends to infinity. This completes the proof of the desired estimate for $k=1$. □

In the general case $k \geq 2$ the proof makes use of the *local nondeterminism property* of fBm:

$$
\operatorname{Var}\left(\sum_i (B_{t_i} - B_{s_i})\right) \geq k_H \sum_i (t_i - s_i)^{2H}.
$$

5.1.1. *Decomposition of the interaction energy.* Assume $\frac{1}{2} < H < \frac{2}{3}$. For any $x \neq y$, set

$$
\widehat{\mathbb{H}_{xy}} = \sum_{i=1}^{3} \int_0^T \left(\int_0^t \frac{1}{|x + B_t - y - B_s|}\, dB_s^i\right) dB_t^i. \tag{5.16}
$$

Then $\widehat{\mathbb{H}_{xy}}$ exists as the limit in $L^2(\Omega)$ of the sequence $\widehat{\mathbb{H}_{xy}^n}$ defined using the approximation $\sigma_n(x)$ of $|x|^{-1}$ introduced in (5.5) and the following decomposition holds

$$
\begin{aligned}
\widehat{\mathbb{H}_{xy}} &= \sum_{i=1}^{3} \int_0^T\int_0^t \frac{1}{|x-y+B_t-B_r|}\delta B_r^i \delta B_t^i \\
&\quad -H^2 \int_0^T\int_0^t \delta_0(x-y+B_t-B_r)(t-r)^{2(2H-1)} drdt \\
&\quad +H(2H-1)\int_0^T \left(\int_0^t \frac{1}{|x-y+B_t-B_r|}(t-r)^{2H-2} dr\right) dt \\
&\quad +H\int_0^T \left(\frac{1}{|x-y+B_T-B_r|}(T-r)^{2H-2} + \frac{1}{|x-y+B_r|} r^{2H-1}\right) dr.
\end{aligned}
$$

Notice that in comparison with $\mathbb{H}_{xy}$, in the definition of $\widehat{\mathbb{H}_{xy}}$ we chose to deal with the half integral over the domain

$$
\{0 \leq s \leq t \leq T\},
$$

and to simplify the notation we have omitted the constant $\frac{\Gamma^2}{8\pi}$. Nevertheless, it holds that $\mathbb{H}_{xy} = \frac{\Gamma^2}{8\pi}\left(\widehat{\mathbb{H}_{xy}} + \widehat{\mathbb{H}_{yx}}\right)$, and we have proved using Fourier analysis that $\mathbb{H}_{xy}$ has moments of any order.

The following results have been proved in [**N-R-T**]:

(1) All the terms in the above decomposition of $\widehat{\mathbb{H}_{xy}}$ exists in $L^2(\Omega)$ for $x \neq y$.

(2) If $|x-y| \to 0$, then the terms behave as $|x-y|^{\frac{1}{H}-1}$, so they can be integrated with respect to $\rho(dx)\rho(dy)$.
(3) The bound $H < \frac{2}{3}$ is sharp: For $H = \frac{2}{3}$ the weighted self-intersection local time diverges.

**5.2. Application to financial mathematics.** Fractional Brownian motion has been applied to describe the behavior to prices of assets and volatilities in stock markets. The long-range dependence self-similarity properties make this process a suitable model to describe these quantities.

5.2.1. *Fractional Black Scholes model.* Assume the price of a stock is modelled as

$$S_t = S_0 e^{\mu t + \sigma B_t},$$

where $B_t$ is a fBm with Hurst parameter $H$ and $\mu$ and $\sigma > 0$ are constants. If $H \neq \frac{1}{2}$ this model admits arbitrage (see [**Ro**], Shiryaev [**Sh**], Cheridito [**Che2**]). In the case $H > \frac{1}{2}$, one can construct an arbitrage in the following way. Suppose $S_0 = \sigma = 1$ and $\mu$ coincides with the interest rate $r$, and define the strategy $(\alpha_t, \beta_t)$, where $\alpha_t$ is the number of bonds and $\beta_t$ is the number of assets, by

$$\begin{aligned} \alpha_t &= 1 - e^{2B_t}, \\ \beta_t &= 2(e^{B_t} - 1). \end{aligned}$$

The value of this strategy at time $t$ is

$$V_t = \alpha_t e^{rt} + \beta_t S_t = e^{rt}\left(e^{B_t} - 1\right)^2.$$

This strategy is self-financing because

$$\begin{aligned} dV_t &= re^{rt}\left(e^{B_t} - 1\right)^2 dt + 2e^{rt + B_t}\left(e^{B_t} - 1\right) dB_t \\ &= r\alpha_t e^{rt} + \beta_t dS_t, \end{aligned}$$

however, $V_0 = 0$ and $V_t > 0$ for all $t > 0$. So, this strategy is an arbitrage.

5.2.2. *Stochastic volatility models.* In Comte and Renault [**C-R**], and Hu [**H1**], the following model with stochastic volatility is considered. The price of an asset $S_t$ is given by

$$dS_t = \mu S_t dt + \sigma_t S_t dW_t,$$

where $\sigma_t = f(Y_t)$ and

$$dY_t = \alpha(m - Y_t)dt + \beta_t dB_t.$$

The process $W_t$ is an ordinary Brownian motion and $B_t$ is a fractional Brownian motion with Hurst parameter $H > \frac{1}{2}$, independent of $W$. Notice that $Y_t$ is a fractional Ornstein-Uhlenbeck process. Examples of functions $f$ are $f(x) = e^x$ and $f(x) = |x|$. Let us mention the following results on this model that are proved in [**H1**]:

1) The market is incomplete and martingale measures are not unique.
2) Set $\gamma_t = (r - \mu)/\sigma_t$ and

$$\frac{dQ}{dP} = \exp\left(\int_0^T \gamma_t dW_t - \frac{1}{2}\int_0^T |\gamma_t|^2 dt\right).$$

Then, $Q$ is the *minimal martingale measure* associated with $P$.

3) The risk minimizing-hedging price of an European call option is given by

$$v = e^{-rT} E_Q \left[ (S_T - K)^+ \right].$$

If $\mathcal{G}_t$ denotes the filtration generated by fBm, it holds that

$$\begin{aligned} v &= e^{-rT} E_Q \left[ E_Q \left( (S_T - K)^+ | \mathcal{G}_T \right) \right] \\ &= e^{-rT} E_Q \left[ C_{BS}(S_0, \sigma) \right], \end{aligned}$$

where $\sigma = \sqrt{\int_0^T \sigma_s^2 ds}$.

## References


[A-L-N] E. Alòs, J. A. León and D. Nualart, *Stratonovich stochastic calculus with respect to fractional Brownian motion with Hurst parameter less than 1/2,* Taiwanese J. Math. **5** (2001), pp. 609–632.

[A-M-N1] E. Alòs, O. Mazet and D. Nualart, *Stochastic calculus with respect to fractional Brownian motion with Hurst parameter lesser than $\frac{1}{2}$,* Stochastic Process. Appl. **86** (1999), pp. 121–139.

[A-M-N2] E. Alòs, O. Mazet and D. Nualart, *Stochastic calculus with respect to Gaussian processes,* Ann. Probab. **29** (2001), pp. 766–801.

[A-N] E. Alòs and D. Nualart, *Stochastic integration with respect to the fractional Brownian motion,* Stoch. Stoch. Rep. To appear.

[B] S. Berman, *Local nondeterminism and local times of Gaussian processes,* Indiana Univ. Math. J. **23** (1973), pp. 69–94.

[C-C-M] P. Carmona, L. Coutin and G. Montseny, *Stochastic integration with respect to fractional Brownian motion,* Ann. Inst. H. Poincaré Probab. Statist. **39** (2003), pp. 27–68.

[Che1] P. Cheridito, *Mixed fractional Brownian motion,* Bernoulli **7** (2001), pp. 913–934.

[Che2] P. Cheridito, *Regularizing Fractional Brownian Motion with a View Towards Stock Price Modelling,* PhD Dissertation, ETH, Zürich, (2001).

[C-N] P. Cheridito and D. Nualart, *Stochastic integral of divergence type with respect to the fractional Brownian motion with Hurst parameter $H < \frac{1}{2}$,* Preprint.

[Cho] A. Chorin, *Vorticity and Turbulence,* Springer–Verlag, (1994).

[C-R] F. Comte and E. Renault, *Long memory in continuous–time stochastic volatility models,* Math. Finance **8** (1998), pp. 291–323.

[C-N-T] L. Coutin, D. Nualart and C. A. Tudor, *Tanaka formula for the fractional Brownian motion,* Stochastic Process. Appl. **94** (2001), pp. 301–315.

[C-Q1] L. Coutin and Z. Qian, *Stochastic differential equations for fractional Brownian motions,* C. R. Acad. Sci. Paris Sér. I Math. **331** (2000), pp. 75–80.

[C-Q2] L. Coutin and Z. Qian, *Stochastic analysis, rough paths analysis and fractional Brownian motions,* Probab. Theory Related Fields **122** (2002), pp. 108–140.

[D-H] W. Dai and C. C. Heyde, *Itô's formula with respect to fractional Brownian motion and its application,* J. Appl. Math. Stochastic Anal. **9** (1996), pp. 439–448.

[D-U] L. Decreusefond and A. S. Üstünel, *Stochastic analysis of the fractional Brownian motion,* Potential Anal. **10** (1998), pp. 177–214.

[D-H-P] T. E. Duncan, Y. Hu and B. Pasik-Duncan, *Stochastic calculus for fractional Brownian motion I. Theory,* SIAM J. Control Optim. **38** (2000), pp. 582–612.

[E-L-S-T-V] M. Eddahbi, R. Lacayo, J. L. Solé, C. A. Tudor and J. Vives, *Regularity and asymptotic behaviour of the local time for the d-dimensional fractional Brownian motion with N-parameters,* Preprint.

[F] F. Flandoli, *On a probabilistic description of small scale structures in 3D fluids,* Ann. Inst. H. Poincaré Probab. Statist. **38** (2002), pp. 207–228.

[F-G] F. Flandoli and M. Gubinelli, *The Gibbs ensemble of a vortex filament,* Probab. Theory Related Fields **122** (2001), pp. 317–340.

[G-H] D. Geman and J. Horowitz, *Occupation densities,* Ann. Probab. **8** (1980), pp. 1–67.

[H1] Y. Hu, *Option pricing in a market where the volatility is driven by a fractional Brownian motion,* Preprint.

[H2] Y. Hu, *Integral transformations and anticipative calculus for fractional Brownian motion,* Preprint.

[Hur] H. E. Hurst, *Long-term storage capacity in reservoirs,* Trans. Amer. Soc. Civil Eng. **116** (1951), pp. 400–410.

[K] A. N. Kolmogorov, *Wienersche Spiralen und einige andere interessante Kurven im Hilbertschen Raum,* C. R. (Doklady) Acad. URSS (N.S.) **26** (1940), pp. 115–118.

[Li] S. J. Lin, *Stochastic analysis of fractional Brownian motions,* Stoch. Stoch. Rep. **55** (1995), pp. 121–140.

[L-S] R. Sh. Lipster and A. N. Shiryaev, *Theory of Martingales,* Kluwer Acad. Publ., Dordrecht, (1989).

[Ly1] T. Lyons, *Differential equations driven by rough signals (I): An extension of an inequality of L. C. Young,* Math. Res. Lett. **1** (1994), pp. 451–464.

[Ly2] T. Lyons, *Differential equations driven by rough signals,* Rev. Mat. Iberoamericana **14** (1998), pp. 215–310.

[M-N] B. B. Mandelbrot and J. W. Van Ness, *Fractional Brownian motions, fractional noises and applications,* SIAM Rev. **10** (1968), pp. 422–437.

[M-M-V] J. Memin, Y. Mishura and E. Valkeila, *Inequalities for the moments of Wiener integrals with respect to fractional Brownian motions,* Statist. Probab. Lett. **55** (2001), pp. 421–430.

[N-V-V] I. Norros, E. Valkeila and J. Virtamo, *An elementary approach to a Girsanov formula and other analytical results on fractional Brownian motion,* Bernoulli **5** (1999), pp. 571–587.

[N] D. Nualart, *The Malliavin calculus and related topics,* Probab. Appl. **21**, Springer–Verlag, New York, (1995).

[N-P] D. Nualart and E. Pardoux, *Stochastic calculus with anticipating integrands,* Probab. Theory Related Fields **78** (1988), pp. 535–581.

[N-R] D. Nualart and A. Răşcanu, *Differential equations driven by fractional Brownian motion,* Collect. Math. **53** (2002), pp. 55–81.

[N-R-T] D. Nualart, C. Rovira and S. Tindel, *Probabilistic models for vortex filaments based on fractional Brownian motion,* Ann. Probab. To appear.

[Pe-Tu] V. Pérez-Abreu and C. Tudor, *A transfer principle for multiple stochastic fractional integrals,* Bol. Soc. Mat. Mexicana **3**, no. 8 (2002), pp. 55–71.

[Pi-Ta1] V. Pipiras and M. S. Taqqu, *Integration questions related to fractional Brownian motion,* Probab. Theory Related Fields **118** (2000), pp. 121–291.

[Pi-Ta2] V. Pipiras and M. S. Taqqu, *Are classes of deterministic integrands for fractional Brownian motion on an interval complete?,* Bernoulli **7** (2001), pp. 873–897.

[Ro] L. C. G. Rogers, *Arbitrage with fractional Brownian motion,* Math. Finance **7** (1997), pp. 95–105.

[R-V] F. Russo and P. Vallois, *Forward, backward and symmetric stochastic integration,* Probab. Theory Related Fields **97** (1993), pp. 403–421.

[Ru] A. A. Ruzmaikina, *Stieltjes integrals of Hölder continuous functions with applications to fractional Brownian motion,* J. Statist. Phys. **100** (2000), pp. 1049–1069.

[S-K-M] S.G. Samko, A.A. Kilbas and O.I. Marichev, *Fractional Integrals and Derivatives. Theory and Applications,* Gordon and Breach, (1993).

[S-T] G. Samorodnitsky and M. S. Taqqu, *Stable non–Gaussian random processes,* Chapman and Hall, (1994).

[Sh] A. N. Shiryaev, *Essentials of Stochastic Finance: Facts, Models and Theory,* World Scientific, Singapore, (1999).

[Sk] A. V. Skorohod, *On a generalization of a stochastic integral,* Theory Probab. Appl. **20** (1975), pp. 219–233.

[St] E. M. Stein, *Singular Integrals and Differentiability Properties of Functions,* Princeton University Press, (1971).

[T] C. Tudor, *On the Wiener integral with respect to the fractional Brownian motion,* Bol. Soc. Mat. Mexicana **1**, no. 8 (2002), pp. 97–106.

[Y] L. C. Young, *An inequality of the Hölder type connected with Stieltjes integration,* Acta Math. **67** (1936), pp. 251–282.

[Z1] M. Zähle, *Integration with respect to fractal functions and stochastic calculus. I,* Probab. Theory Related Fields **111** (1998), pp. 333–374.

[Z2] M. Zähle, *On the link between fractional and stochastic calculus,* in Stochastic Dynamics. Conference on Random Dynamical Systems, Bremen, Germany, April 28–May 2, 1997. Dedicated to Ludwig Arnold on the occasion of his 60th birthday, pp. 305–325, Springer–Verlag, New York, (1999).

Facultat de Matemàtiques, Universitat de Barcelona, Gran Via, 585, 08007, Barcelona, Spain

*E-mail address*: nualart@mat.ub.es

Contemporary Mathematics
Volume **336**, 2003

# Entropy and Economic Equilibrium

Esa Nummelin

*In honor of Rabi Bhattacharya*

ABSTRACT. We will be concerned with mathematical principles which underlie stochastic economic equilibrium theory. To this end we shall formulate and prove *principles of large deviations* and ensuing *principles of minimum entropy* concerning equilibrium prices in random economic systems.

## 1. Preliminaries

**1.1. Introduction.** The equilibrium prices $p^*$ in an economic system are defined as price vectors at which the total excess demand $Z(p)$ by the participating economic agents vanishes:

$$Z(p^*) = 0.$$

We will deal with large economic systems; namely, we assume that the number $n$ of economic agents is "big". (In the mathematical theorems $n \to \infty$.)

In classical equilibrium theory the total excess demand $Z(p)$ is regarded as a deterministic function, see [**De**]. In the present study we assume that it is a random variable (for each fixed price $p$). It follows that the equilibrium prices $p^*$ form a random set.

Seminal early works concerning the existence and laws of large numbers for random equilibrium prices are due to Hildenbrand [**H**], Bhattacharya and Majumdar [**B-M1**] and Föllmer [**F**].

We will formulate and prove a *principle of large deviations (PLD) for the random equilibrium prices.* It expresses the (small) probabilities of observations of non-expected equilibrium prices in terms of an *entropy function.*

The PLD implies *principles of minimum entropy (PME's) for the random equilibrium prices*:

According to the *law of large numbers for the random equilibrium prices*, the random equilibrium prices are asymptotically equal to entropy minimizing, "expected" equilibrium prices.

---

2000 *Mathematics Subject Classification.* Primary: 90A14, 60F15.

*Key words and phrases.* Large deviations, random equilibrium prices, economic equilibrium, stochastic finance markets, minimum entropy.

According to the *conditional law of large numbers for the random equilibrium prices*, the entropy minimizing prices in a given subset of prices can be characterized as the aposteriori predictions for the random equilibrium prices (conditionally on observing a random equilibrium price in this subset).

We will also formulate and prove *conditional laws of large numbers concerning the probability laws governing random economies.* These results provide a formula for the transformation of an apriori (micro- or macroeconomic) probability law into an aposteriori probability law, conditionally on observing an apriori non-expected random equilibrium price.

The results will be illustrated in terms of a *random Cobb-Douglas exchange economy.* We also indicate *stochastic finance markets* as a potential application.

We draw on the *theory of large deviations.*

The subject matter of the theory of large deviations (LD's) is concerned with viewing large stochastic systems as analogs of the systems of *statistical mechanics.* In particular, the LD theory provides a mathematically solid ground for the definition and proper use of the concept of *entropy.* Entropy in LD theory arises as an abstraction from statistical mechanics; namely, it appears in the exponent in the entropy representation for the probabilities of non-expected "rare events" (non-equilibrium states) in the system under study, see [**D-Z**]: Section 1.1.

Since the pioneering classical work by H. Cramer in 1938 [**Cr**], the theory of LD's has emerged as a major subject in probability theory. Standard references to the theory and its applications are the monographs by Bucklew [**Bu**] and Dembo and Zeitouni [**D-Z**]. The former focuses on engineering problems and on statistics; the latter presents a wide coverage both of the general theory and of the applications. The more specialized monographs by Cover and Thomas [**C-T**] and Shwartz and Weiss [**S-W**] deal with applications to information theory and communication networks, respectively.

An important application concerns conditioning on the occurrence of a rare event. This is referred to as *Gibbs conditioning principle*, see ([**D-Z**]: Sections 1.1, 3.3, 7.3). As a mathematical theorem the Gibbs conditioning principle becomes a conditional law of large numbers, see [**Cs**].

Standard LD theory deals with "stochastic phenomena of cumulative type", of which the simplest examples are provided by sums of i.i.d. random variables. Many economic phenomena are cumulative in nature; the "accumulation" may occur e.g. in proportion to the number of economic agents, or it may occur temporally. (Consider e.g. the production by a set of producers within a given time interval or the random diffusive movements of the prices of financial assets due to the accumulation of consequtive "infinitesimal random shocks".) LD theory of stochastic processes has been used in the study of calibration problems occurring in the pricing of financial assets, see Avellaneda [**Av**]. Aoki [**Ao**] has applied LD theory of stochastic processes to stochastic dynamical macroeconomic models. Also economic risks (actuarial or financial) are cumulative in nature. Martin-Löf [**M1**] has utilized LD theory in the study of the classical problem of ruin. Stutzer [**S**] has analyzed financial risks in the framework of LD theory.

The theoretical framework for the present study is provided by the *theory of large deviations of zeros of random vector fields* ([**N1**],[**N2**],[**N3**]). In order to be able to present the theory in a simple context we will often deal with socalled *statistically homogeneous economies;* these are stochastic exchange economies, where the participating economic agents are statistically independent, and where the associated individual microeconomic probability distributions are identical. The underlying abstract mathematical theory in [**N1**], [**N2**], [**N3**], however, does not require statistical homogeneity of the economic agents but the results hold true with much wider generality.

The principal goal of these lectures is to present new theoretical concepts and results which ensue due to the entry of LD theory into the theory of random economic equilibrium. The presentation attempts to be self-contained in that no pre-knowledge of general LD theory will be assumed.

Parts of these lectures are sketchy, suggesting more careful further investigation. Thus, for example, in the future it is desirable to pursue further e.g. with *stochastic equilibrium models for finance markets*, cf. Example 1.2, or also, with the *survival model* discussed at the end of Section 4.

**1.2. Economic systems and their price equilibria.** We consider an *economic system* (shortly, *economy*), where certain *commodities* $j = 1, ..., l+1$ are traded by a set of *economic agents* $i = 1, ..., n$. The *prices* of the commodities are denoted as vectors

$$p = (p^j;\ j = 1, ..., l+1) \in R_+^{l+1}.$$

(Superscripts will refer to the commodities whereas subscripts refer to the agents.)

We denote by $\zeta_i^j(p)$ the *individual excess demand* by agent $i$ on the commodity $j$ at price $p$. Often the individual excess demand is written as the difference of the *individual demand* and *supply*:

$$\zeta_i^j(p) = \delta_i^j(p) - \sigma_i^j(p).$$

We assume that only "relative prices count", i.e., for any constant $a > 0$:

$$\delta_i^j(ap) \equiv \delta_i^j(p)$$

and

$$\sigma_i^j(ap) \equiv \sigma_i^j(p).$$

It follows that

$$\zeta_i^j(ap) \equiv \zeta_i^j(p),$$

too. This property is commonly referred to as *homogeneity of degree 0 (with respect to the prices)*.

Due to the homogeneity of degree 0, the price vectors can without loss of generality be normed in the following way: *We assume that the price vectors $p$ belong to the simplex*

$$S^l \doteq \{p \in R_+^{l+1} : \sum_{j=1}^{l+1} p^j = 1\}.$$

*Walras' law* expresses the *budget constraint*:

$$\sum_{j=1}^{l+1} p^j \delta_i^j(p) \equiv \sum_{j=1}^{l+1} p^j \sigma_i^j(p);$$

or, equivalently:

$$\sum_{j=1}^{l+1} p^j \zeta_i^j(p) \equiv 0.$$

*We assume that the individual excess demand functions* $\zeta_i^j : S^l \to R \cup \{+\infty\}$ *are continuous. The value* $+\infty$ *is allowed only at the boundary*

$$\partial S^l \doteq \{p \in S^l : p^j = 0 \text{ for some } j = 1, ..., l+1\}.$$

It follows that the excess demand on one commodity (say, the $l+1$'st) is determined by the excess demands on the others:

$$\zeta_i^{l+1}(p) = -(p^{l+1})^{-1} \sum_{j=1}^{l} p^j \zeta_i^j(p) \text{ when } p^{l+1} > 0,$$

$$\zeta_i^{l+1}(p) = \lim_{p_m \to p, p_m^{l+1} > 0} \zeta_i^{l+1}(p) \text{ when } p^{l+1} = 0.$$

We shall henceforth denote by

$$\zeta_i(p) \doteq (\zeta_i^1(p), ..., \zeta_i^l(p)) \in R^l$$

the vector comprising the individual excess demands by agent $i$ on the commodities $j = 1, ..., l$ at price $p$.

The *total excess demand* is defined as the sum of the individual excess demands:

$$Z^j(p) \doteq \sum_{i=1}^{n} \zeta_i^j(p).$$

By Walras' law

$$\sum_{j=1}^{l+1} p^j Z^j(p) \equiv 0.$$

*We assume that all commodities are desired:*

$$Z^j(p) > 0 \text{ whenever } p^j = 0. \tag{1.1}$$

We denote by

$$\begin{aligned} Z(p) &\doteq (Z^1(p), ..., Z^l(p)) \\ &= \sum_{i=1}^{n} \zeta_i(p) \end{aligned}$$

the vector comprising the total excess demands on the commodities $j = 1, ..., l$ at price $p$.

Any price $p^* \in S^l$ which satisfies

$$Z(p^*) = 0$$

is called an *equilibrium price*. Note that, due to the postulated desirability (1.1), the equilibrium prices belong to the interior of the simplex $S^l$:

$$p^* \in \mathring{S}^l \doteq \{p \in S^l : p^j > 0 \text{ for all } j = 1, .., l+1\}. \tag{1.2}$$

We denote by

$$\pi^* \doteq \{p^* \in \mathring{S}^l : Z(p^*) = 0\}$$

the *set of equilibrium prices*. It is known that under the stated hypotheses there exists at least one equilibrium price (Corollary 1.5 of Theorem 1.3).

EXAMPLE 1.1. In a *Cobb-Douglas (CD) economy* the individual supply by the agent $i$ on the commodity $j$ is given by his *initial endowment* $e_i^j$ on this commodity:

$$\sigma_i^j(p) \equiv e_i^j.$$

The demand by $i$ is obtained via maximization of the *Cobb-Douglas utility function*:

$$u_i(x^1, ..., x^{l+1}) \doteq (x^1)^{a_i^1} \cdots (x^{l+1})^{a_i^{l+1}}.$$

Here it is assumed that

$$a_i \doteq (a_i^1, ..., a_i^{l+1}) \in S^l.$$

This leads to the individual demand on commodity $j$:

$$\delta_i^j(p) = (p^j)^{-1} a_i^j w_i(p),$$

where

$$w_i(p) \doteq \sum_{k=1}^{l+1} p^k e_i^k$$

denotes the *initial wealth* possessed by $i$. Thus the individual excess demand on commodity $j$ becomes:

$$\zeta^j(p) = (p^j)^{-1} a_i^j w_i(p) - e_i^j. \tag{1.3}$$

The total excess demand in a CD economy on the commodity $j$ thus becomes:

$$Z^j(p) = (p^j)^{-1} \sum_{k=1}^{l+1} \Theta^{kj} p^k - E^j, \tag{1.4}$$

where

$$\Theta^{kj} \doteq \sum_{i=1}^{n} a_i^j e_i^k$$

and

$$E^k \doteq \sum_{i=1}^{n} e_i^k = \sum_{j=1}^{l+1} \Theta^{kj}.$$

It is an easy exercise to show that the equilibrium prices in a CD economy are given by

$$(p^*)^j = (E^j)^{-1} (W^*)^j, \ j = 1, ..., l+1, \tag{1.5}$$

where $W^*$ is a left eigenvector of the stochastic matrix

$$\begin{aligned} A^{kj} &\doteq (E^k)^{-1} \Theta^{kj} \\ &= (\sum_{i=1}^{n} e_i^k)^{-1} \sum_{i=1}^{n} a_i^j e_i^k. \end{aligned}$$

Subject to the normalization

$$\sum_{j=1}^{l+1} (E^j)^{-1} (W^*)^j = 1.$$

EXAMPLE 1.2. In a *(one-period) financial market* there are $l+1$ *assets* $j = 1, ..., l+1$ traded by a set $i = 1, ..., n$ of *financial agents*. We assume that there are $S$ different *states* $s = 1, ..., S$ for the market at the end of the period. The *subjective probabilities* by agent $i$ for the different states are denoted by $q_i(s)$, $s = 1, ..., S$. Let $\psi^j(s)$ denote the (monetary) value of the $j$'th asset at state $s$ at the end of the period. Let $e_i^1, ..., e_i^{l+1}$ be the *initial endowments* by $i$ on the assets.

Let us assume that the agent's *utility* (for money) has the simple parametric form

$$u_i(x) = x - \frac{a_i x^2}{2}.$$

(The parameter $a_i = -\frac{u_i''(0)}{u_i'(0)}$ describes the agent's *risk aversion*.) The agent's individual demands $\delta_i^j(p)$ on the assets $j = 1, ..., l+1$ are obtained by maximizing the *expected utility*

$$U_i(x^1, ..., x^{l+1}) \doteq \sum_{s=1}^{S} u_i(\sum_{j=1}^{l+1} x^j \psi^j(s)) q_i(s),$$

subject to the budget constraint

$$\sum_{j=1}^{l+1} p^j x^j = \sum_{j=1}^{l+1} p^j e_i^j.$$

Let

$$\mu_{\psi;i}^k \doteq \sum_{s=1}^{S} \psi^k(s) q_i(s) \text{ and}$$

$$\Sigma_{\psi;i}^{jk} \doteq \sum_{s=1}^{S} \psi^j(s) \psi^k(s) q_i(s)$$

denote the *agent's subjective expectations* of the values of the assets and their correlations. The demand can in fact be solved as a function of the parameters $a_i, \mu_{\psi;i}, \Sigma_{\psi;i}, e_i$ and of the price $p$:

$$\delta_i(p) = \delta(a_i, \mu_{\psi;i}, \Sigma_{\psi;i}, e_i; p).$$

(We omit the details.)

We shall now formulate a theorem concerning general economies satisfying the stated hypotheses, which gives sufficient conditions for the existence of (at least one) equilibrium price in a given subset of prices.

To this end, suppose that $D \subset R^l$ is compact and convex and has non-empty topological interior: $\mathring{D} \neq \emptyset$. Let $\partial D \doteq$ the topological boundary of $D$. We denote $\overline{R}^l \doteq (R \cup \{\pm\infty\})^{\times l}$.

We say that a *function* $f : D \to \overline{R}^l$ *satisfies the boundary condition on* $D$, if

$$f(x) \cdot u < 0 \text{ for all } x \in \partial D, \text{ all outward unit normals } u \text{ for } D \text{ at } x.$$

(This condition means that the vector field $f(p)$ points "inwards" at the boundary of the domain $D$.)

THEOREM 1.3. *Suppose that* $f : D \to \overline{R}^l$ *is continuous and satisfies the boundary condition on* $D$. *Then*

$$\exists x^* \in D : \ f(x^*) = 0.$$

PROOF. Set

$$\begin{aligned}\tilde{f}(x) &= f(x) \quad \text{if} \quad |f(x)| \le 1,\\ &= \frac{f(x)}{|f(x)|} \quad \text{if} \quad |f(x)| \ge 1.\end{aligned}$$

For $\varepsilon > 0$ define the bounded continuous map $g_\varepsilon : D \to R^l$ by

$$g_\varepsilon(x) \doteq x + \varepsilon \tilde{f}(x).$$

For small $\varepsilon$:

$$g_\varepsilon : D \to D.$$

Now, according to *Brouwer's fixed point theorem*, there exists $x^* \in D$ such that $g_\varepsilon(x^*) = x^*$, viz. $f(x^*) = 0$ as required.

For $p \in S^l$, let

$$p^{(l)} \doteq (p^1, ..., p^l) \in S^{(l)},$$

where $S^{(l)} \doteq \{x \in R^l_+ : \sum_{j=1}^{l} x^j \le 1\}$. We can regard the total excess demand function $Z : S^l \to \overline{R}^l$ as well as a function $Z : S^{(l)} \to \overline{R}^l$ by setting

$$Z(p^{(l)}) \doteq Z(p^{(l)}, 1 - \sum_{j=1}^{l} p^j).$$

Suppose now that $B \subset S^l$ is compact and convex and has non-empty interior $\mathring{B} \neq \emptyset$. Let

$$B^{(l)} = \{p^{(l)} \in S^{(l)} : p \in B^l\}$$

denote the *projection of* $B$ *to* $S^{(l)}$. We say that $Z : S^l \to \overline{R}^l$ *satisfies the boundary condition on* $B$, if $Z : S^{(l)} \to \overline{R}^l$ satisfies the boundary condition on $B^{(l)}$. □

COROLLARY 1.4. *Suppose that $Z : S^l \to \overline{R}^l$ satisfies the boundary condition on $B$. Then*

$$\pi^* \cap B \neq \emptyset$$

*(i.e., $\exists p^* \in B$).*

Note that the desirability hypothesis (1.1) implies that $Z$ satisfies the boundary condition on $S^l$. Therefore there exists at least one equilibrium price $p^*$ in the simplex $S^l$ (whence in its interior $\mathring{S}^l$, see (1.2)):

COROLLARY 1.5.

$$\pi^* \cap \mathring{S}^l \neq \emptyset.$$

**1.3. Random economies.** An economy is *random*, if the parameters determining the economic behaviour of the individual agents are random variables. It follows in particular, that in a random economy the total excess demand $Z(p)$ is a random variable (for each fixed price $p$).

In formal terms, we assume that all the encountered random variables are defined on a *probability space*

$$(\Omega, \mathcal{F}, P).$$

The elements $\omega$ of $\Omega$ will be called *realizations*. For each fixed realization $\omega \in \Omega$, the map $Z(\omega; \cdot) : S^l \to \overline{R}^l$ will be called a *realization of the excess demand function*.

**Basic Hypothesis.** We assume that for each price $p \in S^l$, the map $Z(\cdot;p) : \Omega \to \overline{R}^l$ is a random variable (viz. $\mathcal{F}$-measurable), and that each realization $Z(\omega;\cdot) : S^l \to \overline{R}^l$ of the total excess demand function is continuous and satisfies the desirability condition:

$$Z^j(\omega;p) > 0 \text{ when } p^j = 0.$$

The values $\pm\infty$ are allowed for $Z^j(\omega;p)$ only at the boundary $\partial S$.

The *random equilibrium prices (r.e.p.'s)* are defined as price vectors $p^* \in S^l$ at which the random total excess demand function vanishes:

$$Z(p^*) = 0.$$

In exact terms this means the following: For a fixed realization $\omega \in \Omega$, any price $p^*(\omega) \in S^l$, which satisfies

$$Z(\omega;p^*(\omega)) = 0$$

will be called an *equilibrium price for the realization* $\omega$.

We denote by

$$\pi^*(\omega) \doteq \{p^*(\omega) \in S^l : \ Z(\omega;p^*(\omega)) = 0\}$$

the *set of equilibrium prices for the realization* $\omega$. Note that, due to the Corollary 1.5 of Theorem 1.3,

$$\emptyset \neq \pi^*(\omega) \subset \mathring{S}^l \text{ for all } \omega.$$

Let $B \subset S^l$ be a compact subset of prices. By an *observation of the random equilibrium price in* $B$ we mean the occurrence of a realization $\omega$ such that

$$p^*(\omega) \in B,$$

i.e., the occurrence of the event

$$\{\pi^* \cap B \neq \emptyset\} \doteq \{\omega \in \Omega : \ \exists p^*(\omega) \in B\}.$$

These observations are measurable:

LEMMA 1.6.

$$\{\pi^* \cap B \neq \emptyset\} \ \in \mathcal{F}.$$

PROOF. Let $C$ be a countable dense subset of $B$. In view of the Basic Hypothesis we have:

$$\begin{aligned} \{\pi^* \cap B \neq \emptyset\} &= \{\min_{p\in B} |Z(p)| = 0\} \\ &= \{\inf_{p\in C} |Z(p)| = 0\} \in \mathcal{F}. \end{aligned}$$

The *expected individual excess demand function* $\mu(p)$ is defined as the limit (provided that it exists):

$$\mu(p) \doteq \lim_{n\to\infty} n^{-1} Z_n(p).$$

(We shall henceforth use the subscript $n$ to indicate the number of agents.) The zeros $p_e^*$ of the expected excess demand constitute the *set of expected equilibrium prices*:

$$\pi_e^* \doteq \{p_e^* \in S^l : \ \mu(p_e^*) = 0\}.$$

□

*Example 1.1* (continuation):

In a *random Cobb-Douglas economy* the defining parameters $a_i^j$ and $e_i^j$ are random variables.

The random equilibrium prices are obtained as

$$(p^*)^j = (E^j)^{-1}(W^*)^j,$$

where $W^*$ is a (random) eigenvector of the (random) stochastic matrix

$$A^{kj} \doteq (E^k)^{-1}\Theta^{kj}$$
$$= (E^k)^{-1}\sum_{i=1}^{n} a_i^j e_i^k,$$

cf. (1.5).

*Example 1.2* (continuation):

In a *random financial market* the defining parameters $a_i, \mu_{\psi;i}, \Sigma_{\psi;i}$ and the initial endowments $e_i$ are random variables. (There is "double stochasticity" in that the agent's subjective expectations $\mu_\psi^j$ and $\Sigma_\psi^{jk}$ are regarded as random variables.)

We shall now formulate a theorem which gives a set of sufficient conditions for the (eventual) existence of random equilibrium prices in a given subset of prices:

As before (cf. Corollary 1 of Theorem 1) we assume that $B \subset S^l$ is a compact and convex set having non-empty interior $\mathring{B} \neq \emptyset$. (The abbreviation "w.p.1" means "with probability one" whereas "eventually" means "for all sufficiently big $n$".)

THEOREM 1.7 (cf.[**K**],[**N1**]). *Suppose that (i)* $\forall p \in \partial B : \exists \mu(p) \doteq \lim_{n\to\infty} n^{-1}Z_n(p)$, *w.p.1; (ii)* $\sup_{p\in\partial B, i\geq 1} |\zeta_i'(p)| < \infty$, *w.p.1; and (iii)* $\mu$ *satisfies the boundary condition on* $B$. *Then*

$$\pi_n^* \cap B \neq \emptyset, \text{ eventually, w.p.1.}$$

*(In other words, there exists a random variable* $N$ *which is finite w.p.1 and such that there exists a r.e.p.* $p_n^*$ *in* $B$ *for all* $n \geq N$.*)*

PROOF. The compactness of $B$ and the hypotheses (i) and (ii) imply that the convergence in (i) is uniform on $\partial B$ (w.p.1). This and (iii) in turn imply that the mean total excess demand function $n^{-1}Z_n(p)$ satisfies the boundary condition on B (eventually and w.p.1). The final assertion follows now from this fact, the postulated continuity of $Z_n(p)$ and Corollary 1.4 of Theorem 1.3. □

**1.4. On statistically homogenous exchange economies.** We shall often focus on the special case of a *(statistically) homogeneous exchange economy.* By this we mean the following:

We assume that with each agent $i$ there is associated the agent's *economic characteristics* $\theta_i$; namely, a random vector of $m$ parameters

$$\theta_i = (\theta_i^1, ..., \theta_i^m) \in R^m.$$

*We assume that the economic characteristics* $\theta_1, \theta_2, ...$ *form a sequence of i.i.d. random variables.*

It is assumed that the individual excess demand $\zeta_i(p)$ is obtained as a deterministic function, referred to as the *structure function*, of the economic characteristics $\theta_i$ and of the price $p$:

$$\zeta_i(p) = \zeta(\theta_i; p).$$

Since independence is preserved under deterministic transformations, it follows that the individual excess demands are i.i.d., too (for each fixed price $p$).

*We assume that $\zeta(\theta; p)$ is continuous as a function of the price $p \in S^l$, for each fixed parameter value $\theta_i = \theta$.*

We denote the common probability distribution function (p.d.f.) of the economic characteristics by $f(\theta)$, i.e.,

$$P(\theta_i \in A) = \int_A f(\theta)d\theta \quad \text{for } i = 1, 2, ..., \ A \subset R^m,$$

and call it the *(apriori) microeconomic p.d.f.*.

Due to the LLN, the *expected individual excess demand* is equal to the expectation of the individual excess demand:

$$\begin{aligned} \mu(p) &\doteq \lim_{n\to\infty} n^{-1} Z_n(p) \\ &= E\zeta_i(p) \\ &= \int \zeta(\theta; p) f(\theta) d\theta. \end{aligned}$$

We denote by

$$\Theta \doteq cl\{\theta \in R^m; \ f(\theta) > 0\}$$

the *support* of the microeconomic p.d.f. $f(\theta)$. (The symbol "cl" means "topological closure".) The probability space $\Omega$ will be the standard Cartesian product

$$\Omega \ = \ \Theta^{\times\infty} \doteq \{\omega : \omega = (\theta_1, \theta_2, ...) \text{ with } \theta_1, \theta_2, ... \in \Theta\}.$$

It will be equipped with the standard product Borel-$\sigma$-algebra $\mathcal{F} = \mathcal{B}(\Omega)$. The *random total excess demand function* $Z_n(p) = Z_n(\omega; p)$ becomes thus

$$Z_n(\omega; p) \doteq \sum_{i=1}^{n} \zeta(\theta_i; p).$$

It follows from the postulated continuity of the structure function $\zeta(\theta; p)$ that each realization $Z_n(\omega; \cdot)$ of the total excess demand function is continuous.

*Example 1.1* (continuation):

In a *homogeneous random CD economy* the natural choice for the economic characteristics of an agent $i$ is the pair comprising the parameter $a_i \in S^{l+1}$ and the initial endowment $e_i \in R^{l+1}$:

$$\theta_i = (a_i, e_i) \doteq (a_i^1, ..., a_i^{l+1}; e_i^1, ..., e_i^{l+1}) \in S^l \times R^{l+1} \subset R^{2l+1}.$$

(Thus the dimension $m = 2l + 1$.) The structure function is

$$\zeta^j(\theta; p) = \zeta^j(a, e; p) = (p^j)^{-1} a^j \sum_{k=1}^{l+1} p^k e^k - e^j, \ j = 1, ..., l, \ \text{see (1.3)}.$$

It follows that the expected individual excess demand on the commodity $j$ is given by

$$\mu^j(p) = (p^j)^{-1} \sum_{k=1}^{l+1} \mu_{a;e}^{kj} p^k - \mu_e^j,$$

where

$$\mu_{a;e}^{kj} \doteq E(a_i^j e_i^k) = \int \int a^j e^k f(a,e) dade,$$

$$\mu_e^k \doteq Ee_i^k = \sum_{j=1}^{l+1} \mu_{a;e}^{kj},$$

and $f(\theta) = f(a,e)$ denotes the *microeconomic distribution.*

It is easy to see that the expected equilibrium prices are obtained as

$$(p_e^*)^j = (\mu_e^j)^{-1} (w_e^*)^j,$$

where $w_e^*$ is a left eigenvector of the stochastic matrix

$$a_e^{kj} \doteq (\mu_e^k)^{-1} \mu_{a;e}^{kj}, \ k,j = 1, ..., l+1,$$

subject to the normalization

$$\sum_{j=1}^{l+1} (\mu_e^j)^{-1} (w_e^*)^j = 1,$$

cf. (1.5).

In particular, it follows that, if the matrix $(a_e^{kj})$ is *irreducible* (see e.g. [**Se**]), then there will be only one unique expected equilibrium price $p_e^*$.

*Example 1.2* (continuation):

A *homogeneous random financial market* is comprised of $n$ statistically independent and identical financial agents. The natural choice for the economic characteristics will be the $m$-dimensional ($m = l^2 + 2l + 2$) vector comprising the risk parameter $a_i$, the agents subjective expectations $\mu_{\psi;i}^j$ and $\Sigma_{\psi;i}^{jk}$ $(j,k = 1, ..., l)$, and the initial endowments $e_i^j$ $(j = 1, ..., k+1)$:

$$\theta_i \doteq (a_i, \mu_{\psi;i}, \Sigma_{\psi;i}, e_i), \ i = 1, ..., n.$$

The structure function is

$$\zeta(\theta;p) = \delta(a, \mu_\psi, \Sigma_\psi, e; p) - e.$$

## 2. The Principle of Large Deviations for the Random Equilibrium Prices

**2.1. A review of large deviation theory.** We draw on the *theory of large deviations (LD's).* LD theory is concerned with estimating the probabilities of occurrences of (apriori) rare events in large random systems.

*An illustrative example:*

Suppose that $X_n$, $n = 1, 2, ...$, are $Bin(\mu, n)$-distributed random variables, for some $0 < \mu < 1$.

By the *law of large numbers for i.i.d. random variables* the *apriori means* are asymptotically equal to the parameter $\mu$:

$$n^{-1} X_n \approx \mu.$$

Suppose now that an observation $\hat{X}_n$ of the r.v. $X_n$ leads to an *aposteriori mean*, which differs from the apriori mean:

$$\hat{\mu} \doteq n^{-1}\hat{X}_n \neq \mu.$$

When $\hat{\mu} - \mu$ is of the "small order" $n^{-\frac{1}{2}}$, then by the *central limit theorem* for i.i.d. random variables the apriori probability of this observation is of the order

$$P(n^{-1}X_n \approx \hat{\mu}) \approx e^{-n\frac{(\hat{\mu}-\mu)^2}{2\sigma^2}},$$

where

$$\sigma^2 \doteq \mu(1-\mu)$$

is the variance of a $Bin(\mu, 1)$ random variable.

However, if the apriori model is "bad", then the observed aposteriori mean $\hat{\mu}$ may represent a *large deviation* (within the apriori model), i.e., fall outside the narrow range of the validity of the CLT. By approximating the binomial probabilities with the aid of the standard *Stirling's formula* we obtain the *principle of large deviations for binomial random variables*; namely, an estimate which holds true for *arbitrary values* of $\hat{\mu} \in (0, 1)$:

(2.1) $$P(n^{-1}X_n \approx \hat{\mu}) \approx e^{-nK(\hat{\mu},\mu)},$$

where

$$K(\hat{\mu}, \mu) \doteq \hat{\mu}\log\frac{\hat{\mu}}{\mu} + (1-\hat{\mu})\log\frac{1-\hat{\mu}}{1-\mu}$$

denotes the *Kullback-Leibler entropy* of a $Bin(\hat{\mu}, 1)$ random variable w.r.t. a $Bin(\mu, 1)$ random variable, see [**D-Z**]. (Note that the LD and CLT estimates are consistent, because

$$K(\hat{\mu}, \mu) \approx \frac{(\hat{\mu}-\mu)^2}{2\sigma^2}$$

for small distances $|\hat{\mu} - \mu|$.)

The LD estimate (2.1) can be rephrased in terms of information contents:

To this end, recall that the *information content* $\mathcal{I}(A)$ of an arbitrary event $A$ is defined as the logarithm of the inverse of its probability $P(A)$:

$$\mathcal{I}(A) \doteq \log\frac{1}{P(A)} = -\log P(A).$$

Thus "common" events (having apriori probability $\approx 1$) have small information contents whereas "surprising" (apriori rare) events have big information contents.

The LD estimate for the binomial r.v.'s can thus be written also as:

$$\mathcal{I}(n^{-1}X \approx \hat{\mu}) \approx nK(\hat{\mu}, \mu).$$

So the *information content of the aposteriori observation of a non-expected mean is proportional to the size parameter* $n$, *with Kullback-Leibler entropy as the asymptotic coefficient of proportionality.* Thus the Kullback-Leibler entropy tells "how surprising" the observation $n^{-1}\hat{X}_n \approx \hat{\mu}$ apriori was. On the other hand, the actual realization of an apriori rare event indicates the "badness" of the apriori model. The Kullback-Leibler entropy provides in this sense a "measure of goodness" for the apriori model.

*Informal description of the PLD for random equilibrium prices:*

The random equilibrium prices in a large random economic system obey (under appropriate regularity conditions) classical statistical limit laws:

The *law of large numbers* (due to Bhattacharya and Majumdar [**B-M1**]) states that, apriori, the random equilibrium prices $p_n^*$ are "near to" expected equilibrium prices:

$$\lim_{n\to\infty} p_n^* = p_e^*.$$

The *central limit theorem for the r.e.p.'s* ([**B-M1**]) characterizes the "small deviations" of the r.e.p.'s from their expected values as asymptotically normal:

$$n^{\frac{1}{2}}(p_n^* - p_e^*) \to \mathcal{N} \text{ in distribution},$$

where $\mathcal{N}$ is a multinormal random variable.

Thus, if a price $p$ is at a distance of order $n^{-\frac{1}{2}}$ from its expected value $p_e^*$, then by the CLT the apriori probability of observing a r.e.p. "near" $p$ is of the exponential order

$$P(p_n^* \approx p) \approx e^{-n(p-p_e^*)^T(2\Sigma)^{-1}(p-p_e^*)},$$

where $\Sigma$ denotes the covariance matrix of the multinormal limit r.v. $\mathcal{N}$.

Thus, in analogy with the Binomial example, the use of the CLT requires the apriori statistical model to be "good" in the sense that the observed ("true") random equilibrium price ought to fall within the small range $n^{-\frac{1}{2}}$ from its apriori expected value.

The *principle large deviations (PLD) for the random equilibrium prices* will yield an exponential estimate for the (apriori small) probabilities of observations of large deviations in the r.e.p.'s in an exponential form which is *valid also outside the region of validity of the CLT:*

$$P(\exists \text{a r.e.p. } p_n^* \approx p) \approx e^{-ni(p)}.$$

The rate $i(p)$ will be referred to as the *entropy.*

In terms of *information contents* the PLD becomes

$$\mathcal{I}(\exists \text{a r.e.p. } p_n^* \approx p) \approx ni(p).$$

Thus *for a large economy, the information content of an observation of the equilibrium price near a given non-expected equilibrium price $p$ is proportional to the number $n$ of economic agents with the entropy $i(p)$ as the asymptotic coefficient of proportionality.* The entropy provides in this sense a "measure of goodness" for the apriori random economic model.

**2.2. Formulation of the principle of large deviations for the random equilibrium prices.** We shall now formulate the PLD for the r.e.p.'s in an exact mathematical form.

To this end, let $p \in S^l$ be an arbitrary price, and let

$$U(p,\delta) \doteq \{q \in S^l \mid |q-p| < \delta\}$$

denote its *$\delta$-neighborhood*, $\delta > 0$. By an *observation of the r.e.p. in the neighborhood* $U(p,\delta)$ we mean the occurrence of the event

$$\pi_n^* \cap U(p,\delta) \neq \emptyset,$$

i.e.,

$$\exists p_n^* : \ |p_n^* - p| < \delta.$$

*These observations are measurable.* Namely, recall from Lemma I.3.1 that

$$\{\pi_n^* \cap B \neq \emptyset\} \in \mathcal{F}$$

whenever $B \subset S^l$ is compact. Therefore

$$\{\pi_n^* \cap U(p,\delta) \neq \emptyset\} = \bigcup_{k=1}^{\infty} \{\pi_n^* \cap \overline{U}(p, \delta - k^{-1}) \neq \emptyset\} \in \mathcal{F}.$$

DEFINITION 2.1. We say that the *principle of large deviations (PLD) holds true at the price* $p \in S^l$, if there is a constant $0 \leq i(p) < \infty$ such that

$$\lim_{\delta \to 0} \limsup_{n \to \infty} |n^{-1}\mathcal{I}(\pi_n^* \cap U(p,\delta) \neq \emptyset) - i(p)| = 0.$$

The price-dependent variable $i(p)$ will be called the *entropy function.*

The PLD can be stated equivalently as:

$$e^{-n(i(p)+\varepsilon(\delta))} < P(\pi_n^* \cap U(p,\delta) \neq \emptyset) < e^{-n(i(p)-\varepsilon(\delta))} \text{ eventually,}$$

where $\lim_{\delta \to 0} \varepsilon(\delta) = 0$.

**2.3. A new theorem of large deviations for homogeneous random economies.** We formulate a set of hypotheses under which the PLD holds true in a homogeneous random economy.

The entropy function will be identified in terms of the *cumulating generating function (c.g.f.) of the individual excess demand* defined by

$$\begin{aligned} c(\alpha; p) &\doteq \log E e^{\alpha \cdot \zeta_1(p)} \\ &= \log \int e^{\alpha \cdot \zeta(\theta; p)} f(\theta) d\theta, \ \alpha \in R^l. \end{aligned}$$

The first hypothesis for the LD theorem is:

(i) $\exists \alpha = \alpha(p) \in R^l : \ \frac{\partial c}{\partial \alpha}(\alpha(p); p) = 0.$

REMARK: The c.g.f. of a random variable is known to be convex. Therefore

$$c(\alpha(p); p) = \min_{\alpha \in R^l} c(\alpha; p) = 0.$$

According to the second hypothesis the second derivatives $\zeta_i''(p)$ of the individual excess demands are uniformly bounded on a neighborhood of $p$:

(ii) $\exists A_2(p) < \infty, \ \varepsilon_2(p) > 0 : \ |\zeta_i''(q)| \leq A_2(p)$ w.p.1, for all $q \in U(p, \varepsilon_2(p))$.

According to the third hypothesis the derivatives $n^{-1}Z_n'(p)$ of the mean total excess demands are uniformly non-singular at $p$:

(iii) $\exists A_{-1}(p) < \infty : \ |(n^{-1}Z_n'(p))^{-1}| \leq A_{-1}(p)$ w.p.1., for all $n \geq 1$.

(This condition can in fact be replaced by the simpler condition

(iii') $\mu'(p)$ is non-singular.

But the proof becomes more complicated, see [**N3**].)

THEOREM 2.2. *Suppose that the hypotheses (i-iii) hold true. Then there is a constant $M(p) < \infty$ such that*

$$e^{-n(-c(\alpha(p);p)+M(p)\delta)} < P(\pi_n^* \cap U(p,\delta) \neq \emptyset) < e^{-n(-c(\alpha(p);p)-M(p)\delta)} \text{ eventually.}$$

*Thus the PLD holds true with entropy*

$$\begin{aligned} i(p) &= -c(\alpha(p);p) \\ &= -\min_{\alpha \in R^l} c(\alpha;p). \end{aligned}$$

*Example 1.1* (continuation):
For a *homogeneous random CD economy:*

$$c(\alpha;p) = -\log \int\int e^{\Sigma_{j=1}^l \alpha^j((p^j)^{-1}a^j w(e;p)-e^j)} f(a,e)dade.$$

Thus

$$\begin{aligned} i(p) &= -c(\alpha(p);p) \\ &= -\log \int e^{\Sigma_{j=1}^l \alpha^j(p)((p^j)^{-1}a^j w(e;p)-e^j)} f(a,e)dade, \end{aligned}$$

where $\alpha(p)$ is the solution of the equation system

$$\frac{\partial c}{\partial \alpha^k}(\alpha(p);p) = 0, \; k = 1,...,l,$$

viz. of

$$\int (a^k w(e;p) - p^k e^k) e^{\Sigma_{j=1}^l \alpha^j(p)((p^j)^{-1}a^j w(e;p)-e^j)} f(a,e)dade = 0.$$

PROOF. (Theorem 2.2)
Let

$$f(\theta|p) \doteq e^{\alpha(p)\cdot\zeta(\theta;p)+i(p)} f(\theta) = e^{\alpha(p)\cdot\zeta(\theta;p)-c(\alpha(p);p)} f(\theta). \tag{2.2}$$

Note that $f(\theta|p)$ is a p.d.f.. It will be called the *canonical microeconomical p.d.f.* (assoociated with the price $p$).

Note that hypothesis (i) implies that *$p$ is an expected equilibrium price under the canonical p.d.f.*. Namely,

$$E(\zeta_1(p)|p) \doteq \int \zeta(\theta;p)f(\theta|p)d\theta = \frac{\partial c}{\partial \alpha}(\alpha(p);p) = 0.$$

We denote by $P(\,\cdot\,|p)$ the product probability measure on $\Omega = \Theta^{\times\infty}$ generated by the canonical p.d.f. $f(\theta|p)$.

For the proof of the upper bound LD inequality we need two lemmas which are of standard type in LD theory:

LEMMA 2.3. *For each $\varepsilon > 0$, there exists a constant $\eta = \eta(\varepsilon;p) > 0$ such that*

$$P(|Z_n(p)| \geq n\varepsilon|p) < e^{-n\eta(\varepsilon;p)} \text{ for all } n \geq 1.$$

PROOF. Let $t > 0$ be arbitrary. By Chebyshev's inequality we have for the $j$'th component of the total excess demand:

$$\begin{aligned} P(Z_n^j(p) \geq n\varepsilon|p) &\leq e^{-tn\varepsilon} E\left(e^{tZ_n^j(p)}|p\right) \\ &= e^{nc(\alpha(p)+te_j;p)-nc(\alpha(p);p)-n\varepsilon t}, \end{aligned}$$

where $e_j$ denotes the $j$'th unit vector in $R^l$. Due to (i),

$$\begin{aligned} \sup_{n\geq 1} n^{-1} \log P(Z_n^j(p) \geq n\varepsilon|p) &\leq c(\alpha(p)+te_j;p) - c(\alpha(p);p) - \varepsilon t \\ &= \varepsilon(t)t - \varepsilon t, \end{aligned}$$

where $\varepsilon(t) \to 0$ as $t \to 0$. By choosing $t$ small enough we thus see that

$$\sup_{n\geq 1} n^{-1} \log P(Z_n^j(p) \geq n\varepsilon|p) < 0.$$

By symmetry, we have also

$$\sup_{n\geq 1} n^{-1} \log P(Z_n^j(p) \leq -n\varepsilon|p) < 0,$$

which completes the proof of Lemma 2.3. □

LEMMA 2.4. *For all* $\varepsilon > 0$, *we have:*

$$e^{-n(i(p)+2|\alpha(p)|\varepsilon)} < P(|Z_n(p)| < n\varepsilon) < e^{-n(i(p)-2|\alpha(p)|\varepsilon)} \text{ eventually.}$$

PROOF. It suffices to prove that

$$\limsup_{n\to\infty} |n^{-1} \log P(|Z_n(p)| < n\varepsilon) - c(\alpha(p);p)| \leq |\alpha(p)|\varepsilon. \tag{2.3}$$

Due to Lemma 2.3,

$$\frac{1}{2} < 1 - e^{-n\eta(\varepsilon;p)} < P(|Z_n(p)| < n\varepsilon|p) \leq 1 \text{ eventually,}$$

and hence, in view of the definition of the probability measure $P(\cdot|p)$:

$$\frac{1}{2} < e^{-nc(\alpha(p);p)} E(e^{\alpha(p)\cdot Z_n(p)}; |Z_n(p)| < n\varepsilon) \leq 1 \text{ eventually.}$$

Now clearly,

$$|\log E(e^{\alpha(p)\cdot Z_n(p)}; |Z_n(p)| < n\varepsilon) - \log P(|Z_n(p)| < n\varepsilon)| \leq |\alpha(p)|n\varepsilon,$$

whence

$$-\log 2 - |\alpha(p)|n\varepsilon < \log P(|Z_n(p)| < n\varepsilon) - nc(\alpha(p);p) \leq |\alpha(p)|n\varepsilon \text{ eventually,}$$

from which the claim (2.3) follows by letting $n \to \infty$. □

Now we are able to prove the upper bound LD inequality.

First, note that Hypothesis (ii) implies the following hypothesis:

(ii)' $\exists A_1(p) < \infty,\ \varepsilon_1(p) > 0 : |\zeta_1'(q)| \leq A_1(p)$ w.p.1, for $|q-p| < \varepsilon_1(p)$.

Now, due to this and the *mean value theorem*, we can conclude that the event

$$\pi_n^* \cap U(p,\delta) \neq \emptyset$$

implies the event

$$|Z_n(p)| \leq A_1(p)n\delta \text{ w.p.1, for all } n \geq 1,\ 0 < \delta < \varepsilon_1(p). \tag{2.4}$$

Thus, in view of Lemma 2.4,

$$P(\pi_n^* \cap U(p,\delta) \neq \emptyset) \leq P(|Z_n(p)| < A_1(p)n\delta) < e^{-n(i(p)-M_0(p)\delta)} \text{ eventually,}$$

where the constant $M_0(p) = 2A_1(p)|\alpha(p)|$.

For the lower bound we need the following lemma which is a straightforward corollary of Theorem XIV in [**La**].

LEMMA 2.5. *Suppose that $f : R_+^l \to R^l$ has bounded second derivative in an $\varepsilon$-neighborhood of the price $p$:*

$$|f''(q)| \leq M < \infty \text{ for } |q-p| < \varepsilon.$$

*Moreover, suppose that the derivative $f'(p) \in R^{l\times l}$ is non-singular, and*

$$|f'(p)^{-1}| < \min\{\frac{\varepsilon}{2|f(p)|}, \frac{1}{4M\varepsilon}\}.$$

*Then*

$$f(q) = 0 \text{ for some } |q-p| < \varepsilon.$$

PROOF. Let

$$g(h) \doteq f'(p)^{-1}(f(p+h) - f(p)), \; |h| < \varepsilon.$$

Then $g(0) = 0$, $g'(0) = I$ (= the identity), and

$$|g''(h)| \leq M|f'(p)^{-1}|.$$

It follows that

$$|g'(h_1) - g'(h_2)| < 2\varepsilon M|f'(p)^{-1}| < \frac{1}{2}.$$

Let

$$z \doteq -f'(p)^{-1}f(p).$$

Then $|z| = |f'(p)^{-1}||f(p)| < \frac{\varepsilon}{2}$ and hence by setting $s = \frac{1}{2}$ in [[**La**]: Lemma XIV.1.3] we can conclude that there exists a unique $|h| < \varepsilon$ satisfying $g(h) = z$, viz. $f(p+h) = 0$. □

Now we are able to prove the lower bound LD inequality. To this end, let

$$f(p) = n^{-1}Z_n(p)$$

in Lemma 2.5. Due to the hypotheses (ii) and (iii), we have

$$M = A_2(p)$$

and

$$|f'(p)^{-1}| \leq A_{-1}(p)$$

in Lemma 2.5.

Note that, by monotonicity, it suffices to prove the assertion for small $\delta > 0$ only. Thus we may assume that

$$\delta < \min\{\varepsilon_2(p), \frac{1}{4A_2(p)A_{-1}(p)}\},$$

where $\varepsilon_2(p)$ is as in (iii). Now, in view of Lemma 2.5 it follows that, if

$$|n^{-1}Z_n(p)| < \frac{\delta}{2A_{-1}(p)},$$

then

$$n^{-1}Z_n(q) = 0 \ \text{ for some } |q-p| < \delta,$$

viz.

$$\pi_n^* \cap U(p,\delta) \neq \emptyset.$$

Finally, by Lemma 2.4

$$\begin{aligned} P(\pi_n^* \cap U(p,\delta) \neq \emptyset) &\geq P(|n^{-1}Z_n(p)| < \frac{\delta}{2A_{-1}(p)}) \\ &> e^{-n(i(p)+M_1(p)\delta)} \text{ eventually,} \end{aligned}$$

where the constant

$$M_1(p) = \frac{|\alpha(p)|}{A_{-1}(p)}.$$

This completes the proof of the theorem. □

**2.4. A theorem of large deviations for general random economies.** We formulate a set of hypotheses under which the PLD holds true for a sequence of general random economies having random total excess demands $Z_n(p)$, respectively.

The entropy function will be identified in terms of the *cumulating generating functions of the total excess demands:*

$$C_n(\alpha;p) \doteq \log Ee^{\alpha \cdot Z_n(p)}, \ \alpha \in R^l.$$

*We assume that there exists the limit*

$$c(\alpha;p) \doteq \lim_{n\to\infty} n^{-1}C_n(\alpha;p).$$

Note that, as a limit of convex functions $c(\,\cdot\,;p) : R^l \to R$, too, is convex.

*We assume that the limit $c(\alpha;p)$ and the excess demand functions $\zeta_i(p)$ and $Z_n(p)$ satisfy exactly the same hypotheses (i)-(iii) as were made for Theorem 1.7 in the preceding Section 1.*

The statement of the LD theorem in the general case is exactly the same as for homogeneous economies:

THEOREM 2.6. *Suppose that the hypotheses (i-iii) hold true. Then there is a constant $M(p) < \infty$ such that*

$$e^{-n(-c(\alpha(p);p)+M(p)\delta)} < P(\pi_n^* \cap U(p,\delta) \neq \emptyset) < e^{-n(-c(\alpha(p);p)-M(p)\delta)} \ \textit{eventually.}$$

*Thus the PLD holds true with entropy*

$$\begin{aligned} i(p) &= -c(\alpha(p);p) \\ &= -\min_{\alpha\in R^l} c(\alpha;p). \end{aligned}$$

The proof will follow that of Theorem 2.2 with few exceptions:

We now define the following *sequence of canonical probability measures*:

(2.5) $$P_n(d\omega|p) = e^{\alpha(p)\cdot Z_n(\omega;p)-C_n(\alpha(p);p)}P(d\omega), \ n = 1,2,\dots .$$

The statement of Lemma 2.3 is now:

LEMMA 2.7. *For each $\varepsilon > 0$, there exists a constant $\eta = \eta(\varepsilon;p) > 0$ such that*

$$P_n(|Z_n(p)| \geq n\varepsilon|p) < e^{-n\eta(\varepsilon;p)} \ \textit{eventually.}$$

PROOF. Let $t > 0$ be arbitrary. By Chebyshev's inequality we have for the $j$'th component of the total excess demand:

$$\begin{aligned} P_n(Z_n^j(p) \geq n\varepsilon|p) &\leq e^{-tn\varepsilon}E\left(e^{tZ_n^j(p)}|p\right) \\ &= e^{C_n(\alpha(p)+te_j;p)-C_n(\alpha(p);p)-n\varepsilon t}, \end{aligned}$$

where $e_j$ denotes the $j$'th unit vector in $R^l$. Due to (i) and (ii),

$$\begin{aligned} \limsup_{n\to\infty} n^{-1}\log P_n(Z_n^j(p) \geq n\varepsilon|p) &\leq c(\alpha(p)+te_j;p) - c(\alpha(p);p) - \varepsilon t \\ &= \varepsilon(t)t - \varepsilon t, \end{aligned}$$

where $\varepsilon(t) \to 0$ as $t \to 0$. By choosing $t$ small enough we thus see that

$$\limsup_{n\to\infty} n^{-1}\log P_n(Z_n^j(p) \geq n\varepsilon|p) < 0.$$

By symmetry, we have also

$$\limsup_{n\to\infty} n^{-1}\log P_n(Z_n^j(p) \leq -n\varepsilon|p) < 0,$$

which completes the proof of Lemma 2.7. □

The rest of the proof of Theorem 2.6 follows exactly the same steps as the proof of Theorem 2.2, and it can therefore be omitted.

**2.5. Regularity properties of the entropy function.** For later purposes we formulate a set of sufficient conditions for the *continuity* of the entropy function

$$i(p) = -c(\alpha(p);p)$$

at a given fixed price $p \in S^l$.

PROPOSITION 2.8. *Suppose that the following three conditions hold true. (The first condition is as before.)*
*(i)* $\exists\alpha = \alpha(p) \in R^l : \frac{\partial c}{\partial\alpha}(\alpha(p);p) = 0$;
*(iv) the matrix* $\frac{\partial^2 c}{\partial\alpha^2}(\alpha(p);p) \in R^{l\times l}$ *is strictly positive definite; and*
*(v)* $\frac{\partial c}{\partial\alpha}(\,\cdot\,;\,\cdot\,) : R^l \times S^l \to R^l$ *is a* $C^1$*-map on an open neighborhood of* $(\alpha(p);p)$.
*Then the entropy function* $i(\cdot)$ *is a* $C^1$*-map on an open neighborhood of the price* $p$.

PROOF. The proof is short: Namely, by the standard *implicit function theorem* $\alpha(\cdot)$ is a $C^1$-map on an open neighborhood of $\alpha(p)$, and, consequently, (as a joint function) the entropy function $i(\cdot)$ is a $C^1$-map on an open neighborhood of $p$. □

Let again $p \in S^l$ be fixed and suppose that
(vi) $\exists\frac{\partial c}{\partial\alpha}(0;p)$.

REMARK 2.9. Condition (vi) implies that

$$\mu(p) \doteq \lim_{n\to\infty} n^{-1}Z_n(p) = \frac{\partial c}{\partial\alpha}(0;p),$$

see [**D-Z**]. Thus, in particular (due to the LLN), if the economy is homogeneous, then $\mu(p)$ is equal to the expected individual excess demand.

According to the second proposition, *prices having zero entropy are exactly the same as the expected equilibrium prices:*

PROPOSITION 2.10. *Assume the hypotheses (i), (iv) and (vi). Then*

$$i(p) = 0 \iff \mu(p) = 0.$$

PROOF. Assume $i(p) = 0$, i.e.,

$$c(\alpha(p); p) = \min_{\alpha \in R^l} c(\alpha; p) = 0.$$

Since $c(0; p) \equiv 0$, it follows from the uniqueness of $\alpha(p)$ due to (iv) that $\alpha(p) = 0$. Therefore

$$\mu(p) \doteq \frac{\partial c}{\partial \alpha}(0; p) = 0.$$

Suppose conversely that

$$\mu(p) \doteq \frac{\partial c}{\partial \alpha}(0; p) = 0.$$

Again, due to uniqueness, $\alpha(p) = 0$ so that

$$i(p) = -c(\alpha(p); p) = -c(0; p) = 0,$$

indeed. □

## 3. Principles of Minimum Entropy for the Random Equilibrium Prices

In this part we formulate three *principles of minimum entropy (PME's)* concerning *partial observations of the random equilibrium prices.* All of them are corollaries of the principle of large deviations.

**3.1. Partial observation of the r.e.p.'s.** By a *partial observation of the random equilibrium price* we mean its observation in some subset $B \subset S^l$ of the prices, i.e., the occurence of the event

$$\pi^* \cap B \neq \emptyset,$$

viz. of the event

$$\exists\, p^* \in B.$$

Throughout this Part III we assume the following basic hypotheses:
(i) the observation set $B \subset S^l$ is compact and convex and $\mathring{B} \neq \emptyset$;
(ii) the PLD holds true at each price $p \in B$; and
(iii) the entropy function $i(p)$ is continuous on $B$.

For sufficient conditions for (ii) and (iii) see Theorems 1.3 and 1.7 and Corollaries 1.4 and 1.5.

We denote

$$i(B) \doteq \min_{p \in B} i(p)$$

and

$$\pi_B^* \doteq \{p \in B : \ i(p) = i(B)\}.$$

We say that the observation set $B$ is *regular* if there exists a unique entropy minimizing price $p_B^*$ in $B$, i.e.,

$$i(p) > i(p_B^*) \text{ for all } p \in B \text{ such that } p \neq p_B^*.$$

or, equivalently, if

$$\pi_B^* = \{p_B^*\}.$$

(Cf. the concept of *dominating point* in LD theory, see e.g. [**Ne**], [**Me**].)

EXAMPLE 3.1. A partial observation could e.g. mean that the equilibrium prices of some $l'$ commodities, say of $1, ..., l'$, are observed, i.e.,

$$(p^*)^j = q^j \text{ for some known prices } q^1, \ldots, q^{l'}$$

whereas the equilibrium prices $(p^*)^{l'+1}, \ldots, (p^*)^{l+1}$ of the commodities $l'+1, \ldots, l+1$ remain unknown. The observation set $B$ is in this case

$$B = \{p \in S^l : p^1 = q^1, ..., p^{l'} = q^{l'}\}.$$

Let $q_1 \doteq (q^1, \ldots, q^{l'})$ denote the (known) price vector of the commodities $1, ..., l'$, and let $p_2 \doteq (p^{l'+1}, \ldots, p^{l+1})$ denote an arbitrary price vector for the unobserved commodities $l' + 1, ..., l + 1$. Let $p$ denote the combined price vector

$$p \doteq (q_1, p_2) = (q^1, \ldots, q^{l'}, p^{l'+1}, \ldots, p^{l+1}) \in S^l.$$

Under appropriate conditions the entropy minimizing prices

$$(p_B^*)_2 \doteq ((p_B^*)^{l'+1}, \ldots, (p_B^*)^l)$$

for the commodities $l' + 1, ..., l + 1$ and the corresponding parameter

$$\alpha_B^* \doteq \alpha(q_1, (p_B^*)_2)$$

are obtained as the solutions of the equation pair

$$\frac{\partial c}{\partial \alpha}(\alpha_B^*; q_1, (p_B^*)_2)) = 0,$$

$$\frac{\partial c}{\partial p_2}(\alpha_B^*; q_1, (p_B^*)_2)) = 0.$$

EXAMPLE 3.2. Suppose that in a *multi-period asset market* the prices of the assets $j = 1, ..., l + 1$ at the end of the first period $t = 1$ are observed:

$$(p^*)^{j,1} = q^j, \text{ for some known prices } q^1, \ldots, q^{l+1}.$$

By treating the assets at different periods $t = 1, 2, ..., T$ as separate, doubly-indexed commodities $(j, t)$ this becomes a special case of the previous example.

EXAMPLE 3.3. If we observe the price of a given *commodity bundle* $b = (b^1, ..., b^{l+1}) \in R^l$, i.e.,

$$b \cdot p^* \doteq \sum_{j=1}^{l+1} b^j (p^*)^j = q$$

for some known price $q \in R$, then the associated observation set is

$$B = \{p \in S^l : b \cdot p^* = q\}.$$

(In the context of an asset market one would rather speak of a *(weighted) index* rather than of a commodity bundle.)

The entropy minimization over this observation set $B$ can be solved with the aid of the Lagrangian function

$$\Lambda(p; \beta) = i(p) - \beta(b \cdot p - q).$$

Since

$$i'(p) = -\frac{\partial c}{\partial p}(\alpha(p); p),$$

the entropy minimizing price $p_B^*$, the corresponding conjugate parameter

$\alpha_B^* \doteq \alpha(p_B^*)$ and the Lagrangian coefficient $\beta_B^*$ are obtained as the solutions of the following equation system:

$$\begin{aligned} \frac{\partial c}{\partial \alpha}(\alpha_B^*; p_B^*) &= 0, \\ \frac{\partial c}{\partial p}(\alpha_B^*; p_B^*) &= -\beta_B^* b, \\ b \cdot p_B^* &= q. \end{aligned}$$

**3.2. A theorem of large deviations for partial observations.** The first PME is concerned with probabilities of *LD's of partial observations of the random equilibrium prices.*

THEOREM 3.4. *Suppose that the hypotheses (i)-(iii) hold true. Then*

$$\lim_{n\to\infty} n^{-1}\mathcal{I}(\pi_n^* \cap B \neq \emptyset) = i(B).$$

*An equivalent statement is:*

$$P(\exists\ p_n^* \in B) = e^{-n(i(B)+o(n))},$$

*where*

$$\lim_{n\to\infty} o(n) = 0.$$

PROOF. Let $\varepsilon > 0$ and $p \in B$ be arbitrary. Then, due to the PLD, there exists $\delta = \delta(\varepsilon; p) > 0$ such that

$$e^{-n(i(p)+\varepsilon)} < P(\pi_n^* \cap U(p,\delta) \neq \emptyset) < e^{-n(i(p)-\varepsilon)} \text{ eventually.} \tag{3.1}$$

Cover $B$ by a finite collection of these neighborhoods $U(p_k, \delta_k)$ (where $\delta_k \doteq \delta(\varepsilon; p_k)$) to see that, eventually:

$$\begin{aligned} P(\pi_n^* \cap B \neq \emptyset) &\leq P(\cup_k\{\pi_n^* \cap U(p_k,\delta_k) \neq \emptyset\}) \\ &< \sum_k e^{-n(i(p_k)-\varepsilon)} \\ &< \max_k e^{-n(i(p_k)-2\varepsilon)} \\ &\leq e^{-n(i(B)-2\varepsilon)}. \end{aligned}$$

On the other hand, due to the hypotheses (i) and (ii), there exists $p \in \mathring{B}$ such that

$$i(p) < i(B) + \varepsilon.$$

From (3.1) we see, if $U(p,\delta) \subset B$, then eventually:

$$\begin{aligned} P(\pi_n^* \cap B \neq \emptyset) &\geq P(\pi_n^* \cap U(p,\delta) \neq \emptyset) \\ &> e^{-n(i(p)+\varepsilon(\delta))} \\ &> e^{-n(i(B)+2\varepsilon(\delta))}, \end{aligned}$$

where $\varepsilon(\delta) \to 0$ as $\delta \to 0$. □

**3.3. The LLN for the r.e.p.'s.** The second PME is a *law of large numbers for the random equilibrium prices.* According to it the random equilibrium prices in a subset $B$ converge towards the zero entropy prices in this subset. We denote

$$\pi^*_{B;e} \doteq \{p \in B : \ i(p) = 0\}.$$

Under appropriate regularity conditions this set is equal to the set of expected equilibrium prices in $B$:

$$\pi^*_{B;e} = \{p \in B : \ \mu(p) = 0\},$$

see Proposition 2.10.

THEOREM 3.5. *Suppose that the hypotheses (i)-(iii), and in addition, the following hypothesis hold true:*
*(iv)* $\pi^*_n \cap B \neq \emptyset$ *eventually, w.p.1.*
*Then* $\pi^*_{B;e} \neq \emptyset$, *and for any* $\varepsilon > 0$ *we have*

$$\pi^*_n \cap B \subset U(\pi^*_{B;e}, \varepsilon) \text{ eventually, w.p.1.}$$

Thus in particular, if there exists a unique expected equilibrium price $p^*_e \in B$, then all the random equilibrium prices $p^*_n$ in $B$ converge towards this expected price:

$$\lim_{n\to\infty} p^*_n = p^*_e \text{ w.p.1.}$$

For sufficient conditions for (iv) see Theorem 1.7.

PROOF. If $i(p) \equiv 0$ on $B$, then the result is trivial. So let $p \in B$ with $i(p) > 0$ be arbitrary. Due to the continuity of $i(p)$ and the PLD, there exists $\delta = \delta(p)$ such that

$$U(p, \delta) \cap \pi^*_B = \emptyset$$

and

$$P(\pi^*_n \cap U(p, \delta) \neq \emptyset) < e^{-n\frac{i(p)}{2}} \text{ eventually.}$$

By using the standard *Borel-Cantelli lemma* we see that

$$\pi^*_n \cap U(p, \delta) = \emptyset \text{ eventually, w.p.1.} \tag{3.2}$$

If $\pi^*_{B;e}$ were empty, i.e., $i(p) > 0$ for all $p \in B$, then we could cover $B$ by a finite number of the neighborhoods $U(p, \delta)$ implying that

$$\pi^*_n \cap B = \emptyset \text{ eventually, w.p.1,}$$

which would contradict the hypothesis (iv). Thus $\pi^*_{B;e} \neq \emptyset$, indeed.

In view of (3.2) we can cover the compact set

$$B \cap U(\pi^*_{B;e}p, \varepsilon)^c \tag{3.3}$$

by a finite union of neighborhoods $U(p_i, \delta_i)$, in each of which (eventually and w.p.1) there are no r.e.p.'s. Consequently, neither in the set (3.3) there are any r.e.p.'s:

$$\pi^*_n \cap B \cap U(\pi^*_B, \varepsilon)^c = \emptyset \text{ eventually, w.p.1.}$$

Thus

$$\pi^*_n \cap B \subset U(\pi^*_B, \varepsilon) \text{ eventually, w.p.1,}$$

as asserted. □

REMARK 3.6. Under appropriate regularity conditions there exists a constant $\eta > 0$ such that

$$\pi_n^* \subset \{p \in S^l : d(p, \partial S) \geq \eta\} \text{ eventually, w.p.1,}$$

cf. the Basic Hypothesis in Subsection 1.3 and Theorem 1.7. Let

$$\pi_e^* \doteq \{p \in S^l : i(p) = 0\}.$$

Recall again that under appropriate regularity conditions this set is the same as the set

$$\{p \in S^l : \mu(p) = 0\},$$

see Proposition 2.10. Thus by applying Theorem 3.5 to the subset

$$B = \{p \in S^l : d(p, \partial S) \geq \eta\}$$

we see that under appropriate conditions we have the following "global version" of the LLN:

$$\pi_n^* \subset U(\pi_e^*; \varepsilon) \text{ eventually, w.p.1.}$$

**3.4. The conditional law of large numbers for the equilibrium prices.** The third PME is a *conditional law of large numbers for the random equilibrium prices*. It states that, *conditionally on observing a r.e.p.* $p_n^*$ *in* $B$, *this r.e.p. is with probability* $\approx 1$ *in a small neighborhood of the set* $\pi_B^*$ *of entropy minimizing prices in* $B$*:*

THEOREM 3.7. *Suppose that the hypotheses (i)-(iii) hold true. Then, for all* $\varepsilon > 0$*:*

$$\lim_{n\to\infty} P(\pi_n^* \cap B \subset U(\pi_B^*, \varepsilon) | \pi_n^* \cap B \neq \emptyset) = 1.$$

*Thus, in particular, if* $B$ *is regular, i.e., there exists a unique entropy minimizing price* $p_B^* \in B$*, then*

$$\lim_{n\to\infty} P(\forall p_n^* \in \pi_n^* : p_n^* \in B \Rightarrow |p_n^* - p_B^*| < \varepsilon) | \exists p_n^* \in B) = 1.$$

PROOF. Let $\varepsilon > 0$ and

$$B_\varepsilon \doteq \{p \in B : d(p, \pi_B^*) \geq \varepsilon\}.$$

Then, due to the continuity of the entropy function,

$$i(B_\varepsilon) > i(B),$$

whereas due to the PLD, for all $p \in B_\varepsilon$ there exists $\delta = \delta(p, \varepsilon)$ such that

$$P(\pi_n^* \cap U(p, \delta) \neq \emptyset) < e^{-n(i(p) - \frac{i(B_\varepsilon) - i(B)}{2})} \text{ eventually.}$$

Using the covering argument again we now see that, eventually,

$$P(\pi_n^* \cap B_\varepsilon \neq \emptyset) < e^{-n(i(B_\varepsilon) - \frac{i(B_\varepsilon) - i(B)}{2})} = e^{-n\frac{i(B) + i(B_\varepsilon)}{2}},$$

cf. the proof of Theorem 1. On the other hand, by Theorem 1,

$$P(\pi_n^* \cap B \neq \emptyset) = e^{-n(i(B) + o(n))},$$

whence

$$P(\pi_n^* \cap B_\varepsilon \neq \emptyset | \pi_n^* \cap B \neq \emptyset) < e^{-n\frac{i(B_\varepsilon) - i(B)}{2}} \text{ eventually}$$

and therefore

$$\lim_{n\to\infty} P(\pi_n^* \cap B \subset U(\pi_B^*, \varepsilon) | \pi_n^* \cap B \neq \emptyset) = 1 \tag{3.4}$$

(with a geometric rate). This completes the proof of Theorem 3.7. □

REMARK 3.8. Note that, due to the geometric rate in (3.4) and the Borel-Cantelli lemma, we can conclude that *the unconditional LLN* (Theorem 3.5) *is in fact a direct corollary of the conditional LLN.*

## 4. Canonical Probability Laws as a posteriori Probability Laws

**4.1. Calibration of microeconomic models.** Suppose that a random equilibrium price $p^*$ is observed to belong to the non-expected prices. (Recall that, if the apriori model is "bad", then contradictory observations having small apriori pro babilities may well occur.)

The following *calibration problem* arises:

*What is the aposteriori statistical structure of the economic model after this observation of the realized equilibrium price $p^*$?*

*Throughout this Subsection 4.1 we are concerned with statistically homogeneous random economies.*

As a mathematical theorem the calibration problem can be formulated as a *conditional law of large numbers (CLLN) concerning the microeconomic random characteristics* of the individual agents.

Let us recall the definition of the *canonical microenomic p.d.f.*

$$f(\theta|p) \doteq e^{\alpha(p)\cdot\zeta(\theta;p)+i(p)}f(\theta) \quad = \quad e^{\alpha(p)\cdot\zeta(\theta;p)-c(\alpha(p);p)}f(\theta),$$

see (subsection 2.3). Here $\alpha(p)$ is as in Hypothesis (i) for Theorem 2.2:

$$\text{(i)} \qquad \frac{\partial c}{\partial \alpha}(\alpha(p);p) = 0.$$

In fact, we will show that the canonical p.d.f. which is associated with the price $p$ represents the aposteriori microeconomic p.d.f., conditionally on observing a r.e.p. "near to" the price $p$. Symbolically,

$$f(\theta|p) = f(\theta|\exists p^* \approx p). \tag{4.1}$$

(This justifies the use of the symbol $f(\theta|p$ for the canonical p.d.f..)

*Example 1.1* (continuation):

For a *homogeneous CD economy*

$$\begin{aligned} f(\theta|p) &= f(a,e|p) \\ &= e^{\Sigma_{j=1}^{l}\alpha^j(p)((p^j)^{-1}a^j w(e;p)-e^j)+i(p)}f(a,e). \end{aligned}$$

REMARK: The terminology to be used comes from statistical mechanics. Namely, in a formal sense the subject matter of *statistical mechanics* is also concerned with the "calibration of an apriori model subject to an aposteriori observation".

This is due to the fact that, according to *Liouville's theorem*, in the "hypothetical apriori model" the positions and velocities of the particles are "uniformly distributed" over the phase space. (In exact mathematical terms this can be phrased as the *invariance of Lebesgue's measure under the Hamiltonian dynamics*, see e.g. [**M2**].)

Now, e.g., the measurement of the temperature $T$ corresponds to an "aposteriori observation"; this observation will restrict the thermodynamical system to a *compact energy shell* and the system obeys an "aposteriori distribution" which is concentrated on this energy shell. Rather than as an aposteriori distribution, this distribution is referred to in statistical mechanics as the *microcanonical distribution.* At the *thermodynamical limit*, viz. for an "infinitely large system", the microcanonical distribution becomes the *canonical distribution.* The canonical distribution belongs to an *exponential family* generated by the apriori uniform distribution. For example, the canonical distribution of the *ideal gas* is the multivariate uncorrelated Gaussian distribution, see [**M2**] or [**L-N**].

Let $g : \Theta \to R$ be an arbitrary bounded measurable function. We shall call $g$ a *sampling function* and

$$\hat{\mu}_{g;n} \doteq n^{-1} \sum_{i=1}^{n} g(\theta_i), \ n = 1, 2, \dots ,$$

the corresponding *sample means.*

By the classical *law of large numbers* for i.i.d. random variables, the *sample means are apriori asymptotically equal to the the apriori expectations*:

$$\lim_{n \to \infty} \hat{\mu}_{g;n} = Eg(\theta_1) \doteq \int g(\theta) f(\theta) d\theta.$$

The *conditional law of large numbers* to be formulated in Theorem 4.1 states that the *sample means are aposteriori asymptotically equal to the expectation of the samples* $g(\theta_i)$ *w.r.t. the canonical p.d.f.* $f(\theta|p)$:

$$\lim_{n \to \infty} \hat{\mu}_{g;n} = E(g(\theta_1)|p) \doteq \int g(\theta) f(\theta|p) d\theta, \text{ conditionally on } p^* \approx p.$$

In other words, the aposteriori sample means converge towards a limit, which one would obtain, if the economic characteristics $\theta_i$ were $f(\theta|p)$-distributed random variables.

*For Theorem 4.1 we make the same hypotheses (i)-(iii) as were made for the LD theorem for homogeneous economies, see Theorem 2.2.*

THEOREM 4.1. *Under the stated hypotheses, for all* $\gamma > 0$ *there exists* $\delta = \delta(\gamma, p) > 0$ *such that for all* $0 < \delta < \delta(\gamma, p)$*:*

$$\lim_{n \to \infty} P(|\hat{\mu}_{g;n} - E(g(\theta_1)|p)| < \gamma | \pi_n^* \cap U(p, \delta) \neq \emptyset) = 1.$$

PROOF. The *joint cumulant generating function* of the pair

$$(\zeta(\theta_i; p), g(\theta_i)) \in R^{l+1}$$

is defined by

$$\begin{aligned} c_g(\alpha, \beta; p) &\doteq \log E e^{\alpha \cdot \zeta(\theta_i; p) + \beta g(\theta_i)} \\ &= \log \int e^{\alpha \cdot \zeta(\theta; p) + \beta g(\theta)} f(\theta) d\theta, \ \alpha \in R^l, \beta \in R. \end{aligned}$$

Note that, since

$$c_g(\alpha, 0; p) \equiv c(\alpha; p),$$

it follows from the hypothesis (i) that

$$\frac{\partial c_g}{\partial \alpha}(\alpha(p), 0; p) = 0. \tag{4.2}$$

By differentiating under the integral sign we see that

$$x(p) \doteq \frac{\partial c_g}{\partial \beta}(\alpha(p), 0; p) = E(g(\theta_1)|p). \tag{4.3}$$

Let $X_n$ denote the r.v.

$$X_n(\omega) \doteq \sum_{i=1}^{n} g(\theta_i)$$

so that

$$n^{-1} X_n = \hat{\mu}_{g;n}. \tag{4.4}$$

Due to the independence of the pairs $(\zeta(\theta_i; p), g(\theta_i))$, we have

$$\log E e^{\alpha \cdot Z_n(p) + \beta X_n} \equiv n c_g(\alpha, \beta; p).$$

The proof of the theorem is again based on a *centering argument* (cf. the proof of Theorem 2.2):

LEMMA 4.2. *For each $\gamma > 0$, there exists a constant $\eta = \eta(\gamma; p) > 0$ such that*

$$P(|n^{-1} X_n(p) - x(p)| \geq \gamma | p) < e^{-n\eta(\gamma;p)} \text{ for all } n \geq 1.$$

PROOF. By Chebyshev's inequality:

$$\begin{aligned} P(X_n \geq n(x(p) + \gamma) | p) &\leq e^{-tn(x(p)+\gamma)} E(e^{tX_n} | p) \\ &= e^{nc_g(\alpha(p),t;p) - nc_g(\alpha(p),0;p) - n(x(p)+\gamma)t}. \end{aligned}$$

Due to (4.2),

$$\begin{aligned} \log P(X_n \geq n(x(p) + \gamma) | p) &\leq nc_g(\alpha(p), te; p) - nc_g(\alpha(p), 0; p) - n(x(p) + \gamma)t \\ &= -n\gamma t + n\varepsilon(t)t, \text{ where } \varepsilon(t) \to 0 \text{ as } t \to 0, \\ &< -\frac{n\gamma t_0}{2} \text{ for some sufficiently small } t_0 > 0. \end{aligned}$$

By symmetry, we have also

$$\log P(X_n \leq -n(x(p) + \gamma) | p) < -\frac{n\gamma t_1}{2} \text{ for some } t_1 > 0,$$

which completes the proof of Lemma 4.2. □

LEMMA 4.3.

$$P(|n^{-1} Z_n(p)| < \frac{\eta(\gamma; p)}{2|\alpha(p)|}, |n^{-1} X_n - x(p)| \geq \gamma) < e^{-n(i(p) + \frac{\eta(\gamma;p)}{2})} \textit{ eventually.}$$

PROOF.

$$P(|n^{-1} Z_n(p)| < \frac{\eta(\gamma; p)}{2|\alpha(p)|}, |n^{-1} X_n - x(p)| \geq \gamma)$$

$$\begin{aligned} &= e^{nc(\alpha(p);p)} E(e^{\alpha(p) \cdot Z_n(p)}; |n^{-1} Z_n(p)| < \frac{\eta(\gamma; p)}{2|\alpha(p)|}, |n^{-1} X_n - x(p)| \geq \gamma | p) \\ &\leq e^{-ni(p) + n|\alpha(p)| \frac{\eta(\gamma;p)}{2|\alpha(p)|}} P(|n^{-1} X_n - x(p)| \geq \gamma | p) \\ &< e^{-ni(p) + n\frac{\eta(\gamma;p)}{2} - n\eta(\gamma;p)} \text{ eventually, by Lemma 4.2,} \\ &= e^{-n(i(p) + \frac{\eta(\gamma;p)}{2})}. \end{aligned}$$

□

LEMMA 4.4. *There exists $\delta(\gamma;p)>0$ such that for all $0<\delta<\delta(\gamma;p)$ we have*

$$P(\pi_n^*\cap U(p,\delta)\neq\emptyset,|n^{-1}X_n-x(p)|\geq\gamma)<e^{-n(i(p)+\frac{\eta(\gamma;p)}{2})}\ \textit{eventually.}$$

PROOF. Recall that

$$\{\pi_n^*\cap U(p,\delta)\neq\emptyset\}\subset\{|n^{-1}Z_n|<\varepsilon\}\text{ for }0<\delta<\frac{\varepsilon}{A_1(p)},$$

see (Section 2). Thus in view of Lemma 4.3 we have the asserted inequality for all $0<\delta<\delta(\gamma;p)\doteq\frac{\eta(\gamma;p)}{2|\alpha(p)|A_1(p)}$ eventually. □

*Proof of Theorem 4.1* (completion):
Due to the PLD we have for an arbitrary $\delta>0$:

$$P(\pi_n^*\cap U(p,\delta)\neq\emptyset)>e^{-n(i(p)+\frac{\eta(\gamma;p)}{4})}\text{ eventually.}$$

Therefore, due to Lemma 4.3, we have for all $\gamma>0$, all $0<\delta<\delta(\gamma,p)$:

$$P(|n^{-1}X_n-x(p)|\geq\gamma|\pi_n^*\cap U(p,\delta)\neq\emptyset)<e^{-n\frac{\eta(\gamma;p)}{4}}\text{ eventually,}$$

or,

$$P(|n^{-1}X_n-x(p)|<\gamma|\pi_n^*\cap U(p,\delta)\neq\emptyset)>1-e^{-n\frac{\eta(\gamma;p)}{4}}\text{ eventually.}$$

In view of (1.3) and (1.4) this becomes

$$P(|\hat{\mu}_{g;n}-E(g(\theta_1)|p)|<\gamma|\pi_n^*\cap U(p,\delta)\neq\emptyset)>1-e^{-n\frac{\eta(\gamma;p)}{4}}\text{ eventually,}$$

which completes the proof of the theorem. □

We can write the statement of Theorem 4.1 symbolically as:

$$\exists p_n^*\approx p\Rightarrow n^{-1}\sum_{i=1}^{n}g(\theta_i)\approx E(g(\theta_1)|p),\text{ w.p.}\approx 1.$$

More generally, as is easy to prove, under the stated hypotheses, for any $k=1,2,\ldots$, and any bounded measurable $k$-variate function

$$g:\Theta^{\times k}\to R$$

we have:

$$\exists p_n^*\approx p\Rightarrow n^{-1}\sum_{i=1}^{n}g(\theta_i,\theta_{i+1},\ldots,\theta_{i+k+1})\approx E(g(\theta_1,\ldots,\theta_k)|p),\text{ w.p.}\approx 1.\tag{4.5}$$

indicating that, aposteriori, conditionally on the observation $\exists p_n^*\approx p$, the economic characteristics are (in the above sense) i.i.d. $f(\theta|p)$-distributed r.v.'s.

Let us assume that the hypotheses (i-iii) for the LD theorem for partial observations (Theorem 3.4) hold true. Moreover, suppose that the hypotheses (i-iii) for Theorem 2.2 hold true at the entropy minimizing price $p_B^*$.

THEOREM 4.5. *Under these assumptions, for all $\gamma>0$:*

$$\lim_{n\to\infty}P(|\hat{\mu}_{g;n}-E(g(\theta_1)|p_B^*)|<\gamma|\pi_n^*\cap B)\neq\emptyset)=1.$$

PROOF. Let us apply the result of Lemma 4.4 at $p_B^*$ to see that there exists a constant $\delta > 0$ such that

$$(4.6)\quad P(\pi_n^* \cap U(p_B^*, \delta) \neq \emptyset,\ |n^{-1}X_n - x(p_B^*)| \geq \gamma) < e^{-n(i(B)+\frac{\eta(\gamma;p_B^*)}{2})} \text{ eventually.}$$

On the other hand, due to the LD theorem for partial observations (Theorem 3.4), we can estimate as follows:

Let

$$B_\delta \doteq B \cap U(p_B^*, \delta)^c.$$

Then $\varepsilon(\delta) \doteq i(B_\delta) - i(B) > 0$, and eventually we have:

$$(4.7)\qquad P(\pi_n^* \cap B_\delta \neq \emptyset) < e^{-n(i(B_\delta)-\frac{\varepsilon(\delta)}{2}))} = e^{-n(i(B)+\frac{\varepsilon(\delta)}{2})}.$$

Therefore, by summing the two inequalities (4.6) and (4.7) we see that

$$(4.8)\quad P(\pi_n^* \cap B \neq \emptyset,\ |n^{-1}X_n - x(p_B^*)| \geq \gamma) < e^{-n(i(B)+\min\{\frac{\eta(\gamma;p_B^*)}{2},\frac{\varepsilon(\delta)}{2}\})} \text{ eventually.}$$

On the other hand, by the LD theorem again, we have

$$P(\pi_n^* \cap B) \neq \emptyset) > e^{-n(i(B)+\min\{\frac{\eta(\gamma;p_B^*)}{4},\frac{\varepsilon(\delta)}{4}\})} \text{ eventually,}$$

and so the claim follows from this and (4.8) after division. □

In symbolic terms Theorem 4.5 can be written as

$$(4.9)\qquad f(\theta|p_B^*) = f(\theta|\exists p^* \in B).$$

I.e., *the canonical p.d.f. associated with the entropy minimizing price in the observation set represents the aposteriori microeconomic p.d.f., conditionally on a partial observation of the r.e.p..*

Again there is a multivariate generalization to Theorem 4.5 written symbolically as:

$$\exists p_n^* \in B \Rightarrow \theta_1, \theta_2, \ldots \text{ are i.i.d. } f(\theta|p_B^*) \text{ distributed r.v.'s.}$$

**4.2. Calibration of macroeconomic probability laws.** Let now $Z_n(p)$, $n = 1, 2, \ldots$, denote the total excess demands in a sequence of general random economies, and let $X_n$ denote an arbitrary sequence of real valued random variabes associated with this sequence. Let $P_n$ denote the probability measure associated with the $n$'th economy. We shall call it the *apriori probability law* (of the $n$'th economy).

We define the sequence of *canonical probability measures* as in (subsection 2.3):

$$P_n(d\omega|p) \quad = \quad e^{\alpha(p)\cdot Z_n(\omega;p)-C_n(\alpha(p);p)}P_n(d\omega), \; n = 1, 2, \ldots.$$

Under appropriate regularity conditions the sequence $X_n$ satisfies the *(weak) law of large numbers:*

$$\lim_{n\to\infty} P_n(n^{-1}|X_n - E_n(X_n)| < \gamma) = 1.$$

Here $E_n(X_n)$ denotes the (apriori) expectation of the random variable $X_n$.

The following Theorems 4.6 and 4.7 are *conditional laws of large numbers* for the sequence $X_n$. According to Theorem 3 the canonical probability measure $P_n(d\omega|p)$ can be interpreted as the *aposteriori macroeconomic probability law* of the $n'$th economy (for big $n$), in that the mean values $n^{-1}X_n$ converge towards the expectations of $X_n$ w.r.t. the canonical probability law $P_n(d\omega|p)$, conditionally on observing a r.e.p. in a small neighborhood of the price $p$. Theorem 4.5 extends this result to concern partial observations of r.e.p.'s.

We omit the proofs of these theorems. (The arguments generalize those used in the proofs of Theorems 4.1 and 4.5 in an analogous way as the arguments used in the proof of Theorem 2.6 generalized the arguments of the proof of Theorem 2.2.)

Let

$$C_{X;n}(\alpha, \beta; p) \doteq \log E_n e^{\alpha \cdot Z_n(p) + \beta X_n}, \quad \alpha \in R^l, \ \beta \in R^l$$

denote the *cumulating generating function of the pair* $Z_n(p)$ and $X_n$.

*We assume that there exists the limit*

$$c_X(\alpha, \beta; p) \doteq \lim_{n \to \infty} n^{-1} C_{X;n}(\alpha, \beta; p).$$

As before, let

$$C_n(\alpha; p) \doteq \log E_n e^{\alpha \cdot Z_n(p)} = C_{X;n}(\alpha, 0; p),$$

and

$$c(\alpha; p) \doteq \lim_{n \to \infty} n^{-1} C_n(\alpha; p) = c_X(\alpha, 0; p).$$

*We assume that the limit* $c(\alpha; p)$ *and the total excess demand functions* $Z_n(p)$ *satisfy exactly the same hypotheses (II:i-iii) as were made for Theorem 2.6.*

In addition, we assume that

(iv) $\exists x(p) \doteq \frac{\partial c_X}{\partial \beta}(\alpha(p), 0; p)$,

(v) $\exists x_n(p) \doteq n^{-1} \frac{\partial C_{X;n}}{\partial \beta}(\alpha(p), 0; p)$,

and that these mean derivatives converge:

(vi) $\lim_{n \to \infty} x_n(p) = x(p)$.

Note that $x_n(p) = n^{-1} E_n(X_n(p)|p)$.

THEOREM 4.6. *Under the stated hypotheses, for any* $\gamma > 0$ *there exists* $\delta = \delta(\gamma, p) > 0$ *such that for all* $0 < \delta < \delta(\gamma, p)$:

$$\lim_{n \to \infty} P_n(n^{-1}|X_n - E_n(X_n|p)| < \gamma | \pi_n^* \cap U(p, \delta) \neq \emptyset) = 1.$$

We can write the statement of Theorem 4.6 symbolically as

$$P_n(d\omega|p) = P_n(d\omega|\exists p^* \approx p).$$

Similarly, we obtain the following CLLN, conditionally on a partial observation of the r.e.p., cf. Theorem 4.5.

THEOREM 4.7. *Assume the same hypotheses as for Theorem 4.5. Then for all* $\gamma > 0$:

$$\lim_{n \to \infty} P(n^{-1}|X_n - E_n(X_n|p_B^*)| < \gamma | \pi_n^* \cap B \neq \emptyset) = 1.$$

In symbolic terms:

$$P_n(d\omega|p_B^*) = P_n(d\omega|\exists p_n^* \in B).$$

**4.3. The principle of large deviations for composite equilibria.** The calibration problem is a special case of a more general scheme (see [**N2**]):

Let $X_n(p)$ denote a sequence of price-depending $R^d$–valued random variables (for some $d$), like e.g. the total demand, supply, production or share of these by the whole economy or by some of its sectors, and let

$$\mu_X(p) \doteq \lim_{n \to \infty} n^{-1} X_n(p)$$

denote the *asymptotic mean* of the *auxiliary variables* $X_n(p)$ at price $p \in S^l$ (provided that the limit exists).

For any r.e.p. $p_n^* \in S^l$, let

$$x_n^* \doteq n^{-1} X_n(p_n^*)$$

denote the *mean value of the auxiliary variable* $X_n(p)$ *at the equilibrium price* $p = p_n^*$. The pair $(p_n^*, x_n^*)$ will be called a *composite equilibrium.* We denote the *random set of composite equilibria* by

$$\begin{aligned} \pi_n^*(X) &\doteq \{(p_n^*, x_n^*) : \ p_n^* \in \pi_n^*\} \\ &= \{(p, n^{-1}X_n(p)); \ Z_n(p) = 0\} \subset S^l \times R^d. \end{aligned}$$

In informal terms the *principle of large deviations for the composite equilibria means an asymptotic exponential rate for the probabilities of observing them:*

$$P(\exists p_n^* \approx p, \ x_n^* \approx x) \approx e^{-n i_X(p,x)}.$$

In order to state this in an exact mathematical, let $(p, x) \in S^l \times R^d$ be an arbitrary price-variable pair, and let

$$U(p, x; \delta) \doteq \{(q, y) \in S^l \times R^d | \ |q - p| < \delta, \ |y - x| < \delta\}$$

denote its (partially cubic) *$\delta$-neighborhood,* $\delta > 0$. By an *observation of the composite random equilibrium in this neighborhood* we mean the occurrence of the event

$$\pi_n^*(X) \cap U(p, x; \delta) \neq \emptyset,$$

i.e.,

$$\exists p_n^* : \ |p_n^* - p| < \delta, \ |x_n^* - x| < \delta.$$

DEFINITION 4.8. We say that the *principle of large deviations (PLD) holds true for the composite random equilibrium at* $(p, x) \in S^l \times R^d$ if there is a number $0 \leq i_X(p, x) < \infty$ such that

$$\lim_{\delta \to 0} \limsup_{n \to \infty} |n^{-1} \mathcal{I}(\pi_n^*(X) \cap U(p, x; \delta) \neq \emptyset) - i_X(p, x)| = 0.$$

The *entropy function for the composite equilibrium* $i_X(p, x)$ can be expressed in terms of the limit

$$c_X(\alpha, \beta; p) \doteq \lim_{n \to \infty} n^{-1} C_{X;n}(\alpha, \beta; p)$$

*joint cumulating generating functions*

$$C_{X;n}(\alpha, \beta; p) \doteq \log E e^{\alpha \cdot Z_n(p) + \beta \cdot X_n(p)}, \ \alpha \in R^l, \ \beta \in R^d.$$

Namely,

$$i_X(p, x) \doteq - \min_{(\alpha, \beta) \in R^{l+d}} (c_X(\alpha, \beta; p) - \beta \cdot x) = \beta(p, x)x - c_X(\alpha(p, x), \beta(p, x); p),$$

where $\alpha(p, x)$ and $\beta(p, x)$ are solutions of the equations

$$\frac{\partial c_X}{\partial \alpha}(\alpha(p, x), \beta(p, x); p) = 0,$$

and

$$\frac{\partial c_X}{\partial \beta}(\alpha(p, x), \beta(p, x); p) = x.$$

Under appropriate regularity conditions the PLD holds true as a LD theorem [[**N2**]:Corollary 4.2]; and, in fact, *the whole theory of random equilibrium prices can be generalized to concern random composite equilibria.*

We illustrate the concept of random composite equilibrium in terms of the following *survival model*, recently studied by Bhattacharya and Majumdar [**B-M2**]. Our approach here will be only sketchy in nature, and we suggest a more precise analysis of this example as a subject of future research for the LD method.

*Example 1.1* (continuation):

Consider a homogeneous random CD economy as described before. We assume that at each price $p$ there is *survival level* $\underline{w}(p)$ such that an agent $i$ having initial endowment $e_i = (e_i^1, ..., e_i^{l+1})$ can survive only if his initial wealth exceeds this level: $w_i(e;p) \geq \underline{w}(p)$. Thus, letting $\xi_i(p)$ denote the *indicator of non-survival*, i.e.,

$$\xi_i(p) = 1, \text{ if } \sum_{j=1}^{l+1} p^j e_i^j < \underline{w}(p),$$

$$\xi_i(p) = 0, \text{ if } \sum_{j=1}^{l+1} p^j e_i^j \geq \underline{w}(p),$$

we can write the total number of non-surviving agents as the sum

$$X_n(p) = \sum_{i=1}^{n} \xi_i(p).$$

Let

$$\nu(p) \doteq P(\xi_i(p) = 1)$$

denote the probability of non-survival. By the LLN for i.i.d. r.v.'s we have (apriori):

$$\lim_{n\to\infty} n^{-1} X_n(p) = \nu(p).$$

Let $p_n^*$ denote the r.e.p.. Due to the LLN for r.e.p.'s we have

$$\lim_{n\to\infty} p_n^* = p_e^*,$$

and therefore, under appropriate regularity conditions, the *proportion* $x_n^* \doteq n^{-1} X_n(p_n^*)$ *of non-surviving agents at the (random) equilibrium* converges also:

$$\lim_{n\to\infty} x_n^* = \lim_{n\to\infty} n^{-1} X_n(p_n^*) = \nu(p_e^*).$$

In analogy with the calibration formula (4.1), under appropriate conditions we have the following calibration formula

$$f(a, e|p_n^* \approx p, x_n^* \approx x) = f(a, e|p, x),$$

where

$$f(a, e|p_n^* \approx p, x_n^* \approx x)$$

denotes the *aposteriori microeconomic p.d.f., conditionally on observing a r.e.p.* $p_n^* \approx p$ *and proportion* $x_n^* \approx x$ *of non-surviving agents,* and

$$f(a, e|p, x) \doteq e^{\alpha(p,x)\cdot\zeta(a,e;p)+\beta(p,x)\xi(a,e;p)+i(p,x)} f(a, e)$$

denotes the *canonical p.d.f. associated with the price* $p$ *and proportion* $x$.

In analogy with the formula (4.9) we see that, if only the proportion of non-surving agents is observed, i.e., we have only a partial observation of the composite equilibrium, then, under appropriate conditions, we have a calibration formula which is of the form

$$f(a, e|x_n^* \approx x) = f(a, e|p(x), x),$$

where $p(x)$ denotes the price which minimizes the entropy $i_X(p, x)$ over the associated observation set

$$C \doteq \{(p, x) : p \in S^l\};$$

namely,

$$i_X(p(x), x) = \min_p i_X(p, x).$$

## References


[Ao] M. Aoki, *New Approaches to Macroeconomic Modeling. Evolutionary Stochastic Dynamics, Multiple Equilibria, and Externalities as Field Effects*, Cambridge University Press, Cambridge (1996).

[Av] M. Avellaneda, *The minimum-entropy algorithm and related methods for calibrating asset-pricing models*, In Proceedings of the International Congress of Mathematicians, vol. III, Berlin (1998), pp. 545-563.

[B-M1] R.N. Bhattacharya and M. Majumdar, *Random exchange economies*, J. Economic Theory **6** (1973), pp. 37-67.

[B-M2] R.N. Bhattacharya and M. Majumdar, *On characterizing the probability of survival in a large competitive economy*, Rev. of Economic Design **6** (2001), pp. 133-153.

[Bi] P. Billingsley, *Probability and Measure*, second edition, John Wiley & Sons, New York, (1986).

[Bu] J.A. Bucklew, *Large Deviations Techniques in Decision, Simulation and Estimation*, John Wiley & Sons, New York, (1990).

[C-T] T.M. Cover and J.A. Thomas, *Elements of Information Theory*, John Wiley & Sons, New York, (1991).

[Cr] H. Cramér, *Sur une nouveau théorème–limite de la theorie des probabilités*, In Actualités Scientifiques et Industrielle, no. 736 (1938), pp. 5-23.

[Cs] I. Csiszar, *Sanov property, generalized I-projection and a conditional limit theorem*, Ann. of Probab. **12** (1984), pp. 768-793.

[De] G. Debreu, *Theory of Value*, John Wiley & Sons, New York, (1959).

[D-Z] A. Dembo and O. Zeitouni, *Large Deviations and Applications*, Jones & Bartlett, Boston, (1993).

[F] H. Föllmer, *Random economies with many interacting agents*, J. Math. Economics **1** (1974), pp. 52-62.

[H] W. Hildenbrand, *Random preferences and equilibrium analysis*, J. Economic Theory **3** (1971), pp 414-429.

[K] M. Kuusela, *On Price Equilibria of Random Economies*, Ph. D. Thesis, University of Helsinki, (2003).

[La] S. Lang, *Real and Functional Analysis*, Springer-Verlag, New York, (1993).

[L-N] T. Lehtonen and E. Nummelin, *Level I theory of large deviations in the ideal gas*, Int. J. of Theor. Physics **29** (1990), pp. 621-635.

[M1] A. Martin-Löf, *Entropy, a useful concept in risk theory*, Scand. Actuarial J. (1986), pp. 223-235.

[M2] A. Martin-Löf, *Statistical Mechanics and the Foundations of Thermodynamics*, Springer-Verlag, Berlin, (1979).

[Me] A. Meda, *Conditional Laws and Dominating Points*, Ph. D. Thesis, University of Wisconsin, (1998).

[Ne] P. Ney, *Convexity and large deviations*, Ann. of Probab. **12** (1984), pp. 903-906.

[N1] E. Nummelin, *On the existence and convergence of price equilibria for random economies,* Ann. of Appl. Probab. **10** (2000), pp. 268-282.

[N2] E. Nummelin, *Large deviations of random vector fields with applications to economics,* Adv. in Appl. Math. **24** (2000), pp. 222-259.

[N3] E. Nummelin, Technical report, (2002).

[Se] E. Seneta, *Non-negative Matrices and Markov Chains*, second edition, Springer-Verlag, New York, (1981).

[S-W] A. Shwartz and A. Weiss, *Large Deviations for Performance Analysis: Queues, Communications and Computing*, Chapman and Hall, New York, (1995).

[S] M. Stutzer, *An information theoretic index of risk in financial markets*, In Bayesian Analysis in Statistics and Econometrics, ed. D. Perry et al, John Wiley & Sons, New York, (1996), pp. 191-200.

DEPARTMENT OF MATHEMATICS, P. O. BOX 4 (YLIOPISTONKATU 5), FIN–00014, UNIVERSITY OF HELSINKI, FINLAND

*E-mail address*: `esa.nummelin@helsinki.fi`

Contemporary Mathematics
Volume **336**, 2003

# Credit Risk - A Survey

Thorsten Schmidt and Winfried Stute

ABSTRACT. This paper presents a review of the developments in the area of credit risk. Starting in 1974, Merton developed a pricing method for a bond facing default risk, which was mainly settled in the framework of Black and Scholes [**B-S**]. Certain attempts have been made to relax the assumptions, giving rise to a class of models called structural models. A second class, called hazard rate models, was first addressed in Pye [**Py**] and more recently reached attention with the works of, e.g., Lando [**La**]. There are extensions in different directions, e.g., models which incorporate ratings, models for a portfolio of bonds or market models. The so called commercial models are readily implemented models which are widely accepted in practice. Finally we describe certain credit derivatives.

## 1. Structural Models

The first class of models tries to measure the credit risk of a corporate bond by relating the firm value of the issuing company to its liabilities. If the firm value at maturity $T$ is below a certain level, the company is not able to pay back the full amount of money, so that a default event occurs.

**1.1. Merton (1974).** In his landmark paper Merton [**Me**] applied the framework of Black and Scholes [**B-S**] to the pricing of a corporate bond. A corporate bond promises the repayment $F$ at maturity $T$. Since the issuing company might not be able to pay the full amount of money back, the payoff is subject to default risk.

Let $V_t$ denote the firm's value at time $t$. If, at time $T$, the firm's value $V_T$ is below $F$, the company is not able to make the promised repayment so that a default event occurs. In Merton's model it is assumed that there are no bankruptcy costs and that the bond holder receives the remaining $V_T$, thus facing a financial loss.

2000 *Mathematics Subject Classification.* Primary: 60-02, 91B28, 91B26, Secondary: 60H30, 60J75.

*Key words and phrases.* Credit Risk, Structural Models, Hazard Rate Models, Commercial Models, Market Models, Dependent Defaults.

If we consider the payoff of the corporate bond in this model, we see that it is equal to $F$ in the case of no default ($V_T \geq F$) and $V_T$ otherwise, i.e.,

$$1_{\{V_T > F\}} F + 1_{\{V_T \leq F\}} V_T = F - (F - V_T)^+.$$

If we split the single liability into smaller bonds with face value 1, then we can replicate the payoff of this bond by a portfolio of a riskless bond $B(t,T)$ with face value 1 (long) and $1/F$ puts with strike $F$ (short).

Consequently the price of the corporate bond at time $t$, which we denote by $\bar{B}(t,T)$, equals the price of the replicating portfolio:

$$\begin{aligned} \bar{B}(t,T) &= B(t,T) - 1/F \cdot P(F, V_t, t, T, \sigma_V) \\ &= e^{-r(T-t)} - \frac{1}{F}\Big(Fe^{-r(T-t)}\Phi(-d_2) - V_t\Phi(-d_1)\Big) \\ &= e^{-r(T-t)}\Phi(d_2) + \frac{V_t}{F}\Phi(-d_1), \end{aligned} \tag{1.1}$$

where $\Phi(\cdot)$ is the cumulative distribution function of a standard normal random variable.

Furthermore, $P(F, V_t, t, T, \sigma_V)$ denotes the price of a European put on the underlying $V$ with strike $F$, evaluated at time $t$, when maturity is $T$ and the volatility of the underlying is $\sigma_V$. This price is calculated using the Black and Scholes option pricing formula. The constants $d_1$ and $d_2$ are

$$\begin{aligned} d_1 &= \frac{\ln \frac{V_t}{Fe^{-r(T-t)}} + \frac{1}{2}\sigma^2(T-t)}{\sigma\sqrt{T-t}} \\ d_2 &= d_1 - \sigma\sqrt{T-t}. \end{aligned}$$

If the current firm value $V_t$ is far above $F$ the put is worth almost nothing and the price of the corporate bond equals the price of the riskless bond. If, otherwise, $V_t$ approaches $F$ the put becomes more valuable and the price of the corporate bond reduces significantly. This is the premium the buyer receives as a compensation for the credit risk included in the contract. Price reduction implies a higher yield for the bond. The excess yield over the risk-free rate is directly connected to the creditworthiness of the bond and is called the *credit spread*. In this model the credit spread at time $t$ equals

$$\begin{aligned} s(t,T) &= -\frac{1}{T-t}\ln\left[\bar{B}(t,T)e^{r(T-t)}\right] \\ &= -\frac{1}{T-t}\ln\left(\Phi(d_2) + \frac{V_t}{F \cdot e^{-r(T-t)}}\Phi(-d_1)\right), \end{aligned}$$

see Figure 1.

The question of *hedging* the corporate bond is easily solved in this context, as hedging formulas for the put are readily available. To replicate the bond the hedger has to trade the risk-free bond and the firm's share simultaneously[1]. This reveals

[1]The hedge consists primarily of hedging $\frac{1}{F}$ put and is a straightforward consequence of the Black-Scholes Delta-Hedging.

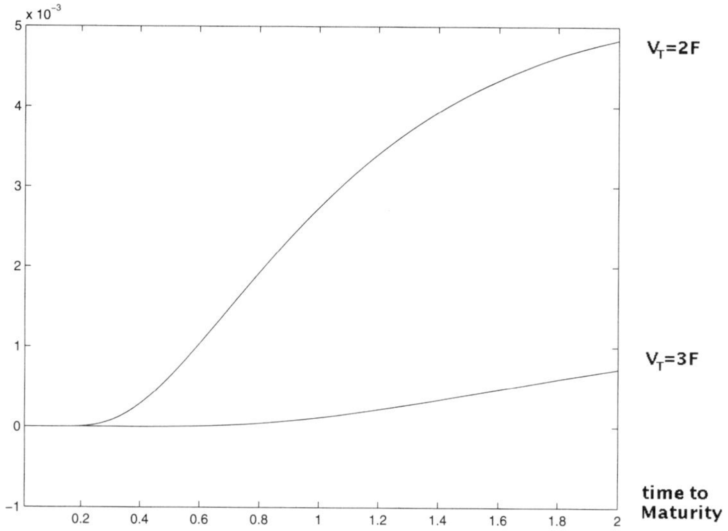


FIGURE 1. This plot shows the credit spread versus time to maturity in the range from zero to two years. The upper line is the price of a bond issued by a company whose firm value equals twice the liabilities while for the second the liabilities are three times as high. Note that if maturity is below 0.3 years the credit spreads approach zero.

the fact that in Merton's model the corporate bond is a derivative on the risk-free bond and the firm's share.

We face the following problems within this model:

- The credit spreads for short maturity are close to zero if the firm value is far above $F$. This is in contrast to observations in the credit markets, where these *short maturity spreads* are not negligible because even close to maturity the bond holder is uncertain whether the full amount of money will be paid back or not; cf. [**W-G**] and [**J-M-R**].
  The reason for this are the assumptions of the model, in particular continuity and log-normality of the firm value process. On the other hand, the intrinsic modeling of the default event may also be questionable. In reality there can be many reasons for a default which are not covered by this model.
- The model is not designed for different bonds with different maturities. Also it can happen that not all bonds default at the same time (*seniority*).
- In practice not all liabilities of a firm have to be paid back at the same time. One distinguishes between short-term and long-term liabilities. To determine the critical level where the company might default Vasiček [**V1**] introduced the *default point* as a mixture of the level of outstandings. This concept is discussed in Section 7.1.
- The interest rates are assumed to be constant. This assumption is relaxed, for example, by [**K-R-S**], as discussed in Section 1.4.

- As there are only few parameters which determine the price of the bond, this model cannot be calibrated to all traded bonds on the market, which reveals arbitrage possibilities.

Geske and Johnson [**Ge-J**] extended the Merton model to coupon-bearing bonds while Shimko, Tejima and van Deventer [**S-T-D**] considered stochastic interest rates using the interest rate model proposed in [**V**]. The second extension is essentially equivalent to pricing a European put option with Vasiček interest rates, where closed-form solutions are available. Of course, any other interest rate model can be used in this framework, like Cox, Ingersoll and Ross [**C-I-R**] or Heath, Jarrow and Morton [**H-J-M**].

**1.2. Longstaff and Schwartz (1995).** As already mentioned defaults in the Merton model are restricted to happen only at maturity, if at all. In practice defaults may happen at any time. Also, when a company offers more than one bond with different maturities or seniorities, inconsistencies in the Merton model show up which can be solved by the following approach.

Black and Cox [**B-C**] first used *first passage time models* in the context of credit risk. This means that a default happens at the first time, when the firm value falls below a pre-specified level. They used a time dependent boundary, $F(t) = ke^{-\gamma(T-t)}$, which resulted in a random default time $\tau$. Unfortunately, this framework proves to be unsatisfactory.

Longstaff and Schwartz [**L-S**] extended the Merton, respectively Black and Cox, framework with respect to the following issues:

- Default may happen at the first time, denoted by $\tau$, when the firm value $V_t$ drops below a certain level $F$.
- Interest rates are stochastic and assumed to follow the Vasiček model.

As a consequence, the firm value at default equals $F$. In the Merton model the value of the defaulted bond was assumed to be $V_T/F$ which equals 1 in this context. The recovery value of the bond is therefore assumed to be a pre-specified constant $(1-w)$. This is the fraction of the principal the bond holder receives at maturity. Since further defaults are excluded in this model, the bond value at default equals $\bar{B}(\tau,T) = (1-w)B(\tau,T)$, where $B(t,T)$ is the value of a risk-free bond maturing at $T$. This assumption is often referred to as *recovery of treasury value*.

In the following, we present the model of Longstaff and Schwartz [**L-S**] in greater detail. The firm value is assumed to follow the stochastic differential equation

$$\frac{dV(t)}{V(t)} = \mu(t)\,dt + \sigma\,dW_V(t),$$

and the spot rate is modeled according to the model of Vasišek [**V**]:

$$dr(t) = \nu(\theta - r(t))\,dt + \eta\,dW_r(t). \tag{1.2}$$

Moreover,

$$\mathbb{E}(W_V(s)\,W_r(t)) = \rho\cdot(s\wedge t) \qquad \text{for all } t \text{ and } s.$$

The last equation reveals a possible correlation between the two Brownian Motions $W_V$ and $W_r$.

The Vasiček model exhibits a mean-reversion behavior at level $\theta$ and easily allows for an explicit representation of $r_t$. It is a classical model used in interest rate theory and often taken as a starting point for more sophisticated models. A drawback of this model is the fact that it may exhibit negative interest rates with positive probability. See, for example, [**B-M**] and the discussions therein.

For the price of the defaultable bond they obtain

$$\begin{aligned} \bar{B}_{\mathrm{LS}}(t,T) &= B(t,T)\cdot \mathbb{E}^{Q^T}\left[1_{\{\tau>T\}} + (1-w)1_{\{\tau\leq T\}}\middle|\mathcal{F}_t\right] \\ &= B(t,T)\cdot\left[w\,Q^T(\tau>T|\mathcal{F}_t) + (1-w)\right]. \qquad (1.3) \end{aligned}$$

Note that $Q^T(\tau > T|\mathcal{F}_t)$ is the conditional probability (under the $T$-forward measure[2]) that the default does not happen before $T$.

To the best of our knowledge, a closed-form solution for this probability is not available[3]. Nevertheless there are certain quasi-explicit results provided by [**L-S**]. See also [**Le**] for an implementation of the model.

In the empirical investigation of Wei and Guo [**W-G**], the Longstaff and Schwartz model reveals a performance worse than the Merton model. According to these authors this is mainly due to the exogenous character of the recovery rate.

**1.3. Jump Models - Zhou (1997).** Another approach to solve the problem of *short maturity spreads* is to extend the firm value process to allow for jumps. Mason and Bhattacharya [**M-B**] extended the Black and Cox [**B-C**] model to a pure jump process for the firm value. The size of the jumps has a binomial distribution. In this model there is some considerable probability for the default to happen even just before maturity.

Alternatively, Zhou [**Z**] extended the Merton model by assuming the firm value to follow a jump-diffusion process. The immediate consequence is that defaults are not predictable. The model is formulated directly under an equivalent martingale measure $Q$, and the firm value is assumed to follow

$$dV_t/V_{t-} = (r_t - \lambda\nu)dt + \sigma dW_V(t) + (\Pi_t - 1)dN_t. \qquad (1.4)$$

$N_t$ is a Poisson process with constant intensity $\lambda$. The jumps are $\Pi_t = U_{N_t}$, where $U_1, U_2, \dots$ are i.i.d. and assumed to be independent of $N$, $r_t$ and $W_V$. Denote $\nu := \mathbb{E}(U_i) - 1$. Note that the integral of $(\Pi_t - 1)\,dN_t$ is shorthand for

$$Y_s := \int_0^s (\Pi_t - 1)\,dN_t = \sum_{i=1}^{N_s}(U_i - 1),$$

so that $Y_t$ is a marked point process. It can be proved[4] that $Y_t - \lambda\nu t$ is a martingale so that consequently the discounted firm value is a martingale under the measure $Q$.

---

[2]The $T$-forward measure is the risk neutral measure which has the risk-free bond with maturity $T$ as numeraire. For details see [**Bj**].

[3]See discussions in [**B-R**] and [**Go**].

[4]See, for example, [**Br**].

The interest rate is assumed to be stochastic and follow the Vasiček model; see (1.2). The recovery rate is determined by a deterministic function $w$, so that the bond holder receives

$$\big(1 - w(V_\tau/F)\big)$$

at default. The function $w$ represents the loss of the bond's value due to the reorganization of the firm. For $w = 1$ we have the zero recovery case.

Zhou considers two models. The first, more general model, assumes that default happens at the first time when the firm value falls below a certain threshold. See the previous chapter for more examples of this class of models. Since in this case no closed-form solutions are available, the author proposes an implementation via Monte-Carlo techniques.

In the second, more restrictive model, the author obtains closed form solutions. For this a constant interest rate and log-normality of the $U_i$'s is assumed and default happens only at maturity $T$, when $V_T < F$. Furthermore $w$ is assumed to be linear, i.e., $w(x) = 1 - \tilde{w}\, x$. For $\tilde{w} = 1$ we obtain the recovery structure of the Merton model.

Equation (1.4) takes the form of a Doleans-Dade exponential and can be explicitly solved under these assumptions, cf. [**Pr**], p. 77:

$$V_t = V_0 \exp\Big[\sigma_V W_V(t) + (r - \frac{1}{2}\sigma_V^2 - \lambda\nu)t\Big] \prod_{i=1}^{N_t} U_i.$$

Denote by $\sigma_U^2$ the variance of $\ln U_1$. We then have the following

PROPOSITION 1.1 (Zhou). *Denote $\tilde{\nu} := 1 + \nu$. Then the price of a defaultable bond in the above model equals*

$$\begin{aligned}\bar{B}_{ZH}(0,T) = &\\ &\frac{\tilde{w}}{F} V_0 e^{-\lambda T \tilde{\nu}} \sum_{j=0}^{\infty} \frac{(\lambda\nu T)^j}{j!} \Phi\Big(\frac{\ln\frac{F}{V_0} - (r + \frac{1}{2}\sigma_V^2 - \lambda\nu)T - j(\ln\tilde{\nu} + \frac{1}{2}\sigma_U^2)}{\sqrt{\sigma_V^2 T + j\sigma_U^2}}\Big)\\ &- e^{-(r+\lambda)T} \sum_{J=0}^{\infty} \frac{(\lambda T)^j}{j!} \Phi\Big(-\frac{\ln\frac{F}{V_0} - (r - \frac{1}{2}\sigma_V^2 - \lambda\nu)T - j(\ln\tilde{\nu} - \frac{1}{2}\sigma_U^2)}{\sqrt{\sigma_V^2 T + j\sigma_U^2}}\Big).\end{aligned}$$

PROOF. The payoff of the bond equals

$$\begin{aligned}\bar{B}_{ZH}(t,T) &= 1_{\{\tau > T\}} + 1_{\{\tau \le T\}}\big(1 - w(V_T/F)\big)\\ &= 1_{\{\tau > T\}} + 1_{\{\tau \le T\}} \tilde{w} \frac{V_T}{F} = 1 + 1_{\{\tau \le T\}}\big(\tilde{w}\frac{V_T}{F} - 1\big).\end{aligned}$$

To compute the present value of the bond we consider the expectation of the discounted payoff

$$\begin{aligned}
\bar{B}_{ZH}(t,T) &= \mathbb{E}^Q\Big[e^{-r(T-t)}\cdot\Big(1+1_{\{\tau\le T\}}\big(\tilde{w}\frac{V_T}{F}-1\big)\Big)\Big|\mathcal{F}_t\Big] \\
&= e^{-r(T-t)}\Big[1+\mathbb{E}^Q\Big(1_{\{V_T<F\}}\big(\tilde{w}\frac{V_T}{F}-1\big)\Big)\Big|\mathcal{F}_t\Big] \\
&= e^{-r(T-t)}\Big[1+\frac{\tilde{w}}{F}\mathbb{E}^Q\Big(1_{\{V_T<F\}}V_T\Big|\mathcal{F}_t\Big)-\mathbb{E}^Q\Big(1_{\{V_T<F\}}\Big|\mathcal{F}_t\Big)\Big].
\end{aligned}$$

Note that conditionally on $\{N_T = j\}$ we obtain a log-normal distribution for $V_T$:

$$\begin{aligned}
\mathbb{P}(V_T<F|N_T=j) &= \mathbb{P}\Big(V_0e^{(r-\frac{1}{2}\sigma_V^2-\lambda\nu)T}\exp[\sigma_V W_V(T)]\prod_{i=1}^{N_T}U_i<F\Big|N_T=j\Big) \\
&= \mathbb{P}\Big(\ln V_0+(r-\frac{1}{2}\sigma_V^2-\lambda\nu)T+\sigma_V W_V(T)+\sum_{i=1}^{j}\ln U_i<\ln F\Big) \\
&=: \mathbb{P}(\xi_j<\ln F),
\end{aligned}$$

where $\sigma_V W(T) + \sum_{i=1}^{j}\ln U_i$ as a sum of independent normally distributed random variables is again normally distributed. Recall $\sigma_U^2$, the variance of $\ln U_1$. As $\mathbb{E}(\ln U) = \ln(1+\nu) - \frac{1}{2}\sigma_U^2$, we get

$$\begin{aligned}
\xi_j &\sim \mathcal{N}\Big(\ln V_0+(r-\frac{1}{2}\sigma_V^2-\lambda\nu)T+j(\ln\nu-\frac{1}{2}\sigma_U^2),\sigma_V^2T+j\sigma_U^2\Big) \\
&=: \mathcal{N}(\tilde{\mu}(j),\tilde{\sigma}^2(j)).
\end{aligned}$$

It is an easy exercise to verify that for $\xi \sim \mathcal{N}(\mu, \sigma_V^2)$

$$\mathbb{E}(e^{\xi}1_{\{e^{\xi}<F\}}) = e^{\mu+\frac{1}{2}\sigma_V^2}\Phi\Big(\frac{\ln F-\mu}{\sigma_V}-\sigma_V\Big).$$

Conclude that

$$\begin{aligned}
\mathbb{E}^Q\Big[1_{\{V_T<F\}}V_T\Big] &= \sum_{j=0}^{\infty}Q(N_T=j)\mathbb{E}^Q(1_{\{V_T<F\}}V_T|N_T=j) \\
&= \sum_{j=0}^{\infty}e^{-\lambda T}\frac{(\lambda T)^j}{j!}\exp(\frac{1}{2}\tilde{\sigma}^2(j)+\tilde{\mu}(j))\Phi\Big(\frac{\ln F-\tilde{\mu}(j)}{\tilde{\sigma}(j)}-\tilde{\sigma}(j)\Big) \\
&= e^{-\lambda T}V_0e^{(r-\lambda\nu)T}\sum_{j=0}^{\infty}\frac{(\lambda\nu T)^j}{j!} \\
&\quad\cdot\Phi\Big(\frac{\ln\frac{F}{V_0}-(r+\frac{1}{2}\sigma_V^2-\lambda\nu)T-j(\ln(1+\nu)+\frac{1}{2}\sigma_U^2)}{\sqrt{\sigma_V^2T+j\sigma_U^2}}\Big).
\end{aligned}$$

We therefore obtain

$$\begin{aligned}
\bar{B}_{ZH}(0,T) = \\
& e^{-rT} + \frac{\tilde{w}}{F} V_0 e^{-\lambda T(1+\nu)} \\
& \cdot \sum_{j=0}^{\infty} \frac{(\lambda \nu T)^j}{j!} \Phi\Big( \frac{\ln \frac{F}{V_0} - (r + \frac{1}{2}\sigma_V^2 - \lambda\nu)T - j(\ln(1+\nu) + \frac{1}{2}\sigma_U^2)}{\sqrt{\sigma_V^2 T + j\sigma_U^2}} \Big) \\
& - e^{-(r+\lambda)T} \sum_{j=0}^{\infty} \frac{(\lambda T)^j}{j!} \Phi\Big( \frac{\ln \frac{F}{V_0} - (r - \frac{1}{2}\sigma_V^2 - \lambda\nu)T - j(\ln(1+\nu) - \frac{1}{2}\sigma_U^2)}{\sqrt{\sigma_V^2 T + j\sigma_U^2}} \Big).
\end{aligned}$$

Noting that

$$e^{-rT} = e^{-(r+\lambda)T} \sum (\lambda T)^j / (j!),$$

the proof is complete.

□

In the case where no jumps are present, i.e., $\lambda = 0$, the sum reduces to the summand with $j = 0$ so that the bond price formula of Merton (1.1) is obtained as a special case.

This model features some properties which are also found in empirical investigations on credit risk:

- The term structure of the credit spreads can be "upward-sloping", flat, humped or "downward-sloping".
- The "short maturity spreads" can be significantly higher than in the Merton model.
- As the firm value at default is random, especially not equal to $F$ as in the Longstaff and Schwartz [**L-S**] model, the recovery is more realistic.
- The recovery rate is correlated with the firm value also just before default.

**1.4. Further Structural Models.** Kim et al [**K-R-S**] extended the first passage time models to also incorporate stochastic interest rates following the model of Cox, Ingersoll and Ross [**C-I-R**]. In their model there is an additional possibility for a default to happen at maturity. The payoff they considered equals $\min(F, V)$. Possibly the company is not able to meet its liabilities at maturity but did not face a default up to this time.

Nielsen, Saà-Requejo and Santa-Clara [**N-S-S**] extended these models to incorporate a stochastic default boundary. For the interest rate they used the model of Hull and White [**H-W**] but were only able to obtain explicit formulas in the special case of the Vasiček model, cf. formula (1.2).

In the work of Ammann [**Am**] *vulnerable claims* are considered. These are possibly stochastic payoffs which face a counterparty risk. Counterparty risk plays a role if the buyer of a claim considers the default probability of the seller as significant. He therefore will ask for a risk premium which compensates for the possible loss in case of a default. The default is assumed to happen if $V_T < F$, similar to Merton's model. In that case the buyer of the claim $X$ receives the fraction $\frac{V_T}{F} \cdot X$.

Explicit prices are derived for the Heath, Jarrow and Morton [**H-J-M**] forward rate structure and Merton-like firm dynamics.

This section on structural models heavily relies on the assumption that the firm's value is observable or even tradeable. From a practical point of view this seems not justifiable as the firm's value is not tradeable and even difficult to observe. This difficulty is discussed by Buffett [**Bu**] and also solved in the KMV-model; see Section 7.1.

## 2. Hazard Rate Models

In comparison to structural models, intensity based models or hazard rate models use a totally different approach for modeling the default. In the structural approach default occurs when the firm value falls below a certain boundary. The hazard rate approach takes the default time as an exogenous random variable and tries to model or fit its probability to default. The main tool for this is a Poisson process with possibly random intensity $\lambda_t$, and jumps denoting the default events. As in the first passage time models recovery is not intrinsic to this model and is often assumed to be a somehow determined constant.

The reason for this new approach lies in the very different causes for default. Precise determination as done in structural models seems to be very difficult. Furthermore, in structural models the calibration to market prices often causes difficulties, while intensity based models allow for a better fit to available market data.

In some approaches basic ideas of these model classes are combined, for example by Madan and Unal [**M-U**] and Ammann [**Am**] where the default intensity explicitly depends on the firm value. These models are called *hybrid models* and will be discussed in Section 5. As the firm value approaches a certain boundary, intensity increases sharply and default becomes very likely. So basic features of the structural models are mimicked.

A more involved hybrid model is presented by Duffie and Lando [**D-L**] where a firm value model with incomplete accounting data is considered.

Basically we may distinguish three types of hazard rate models. In the first approach the default process is assumed to be independent of most economic factors, sometimes it is even modeled independently from the underlying.

The *rating based approach* incorporates the firm's rating as this constitutes readily available information on the company's creditworthiness. In principle one tries to model the company's way through different rating classes up to a possible fall to the lowest rating class which determines the default.

A third and very recent class is in the line of the famous market models of Jamshidian [**Ja**] and Brace, Gatarek and Musiela [**B-G-M**], see Chapter 6.

**2.1. Mathematical Preliminaries.** In this section we consider the modeling of the default process in greater detail. The approach is mainly based on Lando

[**La**] and also discussed in many articles and books like [**Je**] and [**B-R**]. We first present a brief introduction to *Cox processes.*

As already mentioned different stopping times denoting the default events need to be modeled. The Poisson process is taken as a starting point. Constant intensity seems too restrictive so one uses Cox processes, which can be considered as Poisson processes with random intensities[5]. A special case which suits well for our purposes is the following:

Consider a stochastic process $\lambda_t$ which is adapted to some filtration $\mathcal{G}_t$. For a Poisson process $N_t$ with intensity 1 independent of $\sigma(\lambda_s : 0 \leq s \leq T^*)$ set

$$\tilde{N}_t := N\Big(\int_0^t \lambda_u \, du\Big), \qquad t \leq T^*.$$

$\tilde{N}_t$ is a Cox process. Observe that for positive $\lambda_t$ the process $\int_0^t \lambda_u \, du$ is strictly increasing and so $\tilde{N}$ can be viewed as a Poisson process under a random change of time. This reveals a very powerful concept for the problems considered in credit risk.

If just one default time $\tau$ is considered, this will be equal to the first jump $\tau_1$ of $\tilde{N}_t$. If more default events are considered, for example, transition to other rating classes, further jumps $\tau_i$ are taken into account. The bigger $\lambda$ is, the sooner the next jump may be expected to occur. We obtain, for any $t < T^*$,

$$\begin{aligned} \mathbb{P}(\tau > t) &= \mathbb{E}\big[\mathbb{P}\left(\tau > t | (\lambda_s)_{0 \leq s \leq t}\right)\big] \\ &= \mathbb{E}\Big[\exp\big(-\int_0^t \lambda_u \, du\big)\Big]. \end{aligned}$$

Conclude that conditionally on $\sigma(\lambda_s : 0 \leq s \leq T^*)$ the jumps are exponentially distributed with parameter $\int_0^t \lambda_u \, du$ .

It may be recalled that a fundamental assumption to obtain this is the independence of $\lambda$ and $N$.

**2.2. Jarrow and Turnbull (1995-2000).** In the work of Jarrow and Turnbull [**J-T**] a binomial model is considered. In extension of the classical Cox, Ross and Rubinstein [**C-R-R**] approach the authors also modeled the non-default and the default state. So for every time period four possible states may be attained: {up,down} × {non-default,default}. They discovered an analogy to the foreign-exchange markets. As the intensity of the model is assumed to be constant we do not discuss it in greater detail.

In Jarrow and Turnbull [**J-T1**] a Vasiček model for the spot rate is used and the hazard rate is explicitly modeled. Correlation of the hazard rate and spot rates are allowed. Denote by $Z_t$ and $W_t$ Brownian motions under the risk neutral measure $Q$, with constant correlation $\rho$. $Z_t$ can be some economic factor, like an index or the logarithm of the firm value.

[5]For a full treatment of Cox processes see [**Br**] and [**Gr**].

Assume the following dynamics

$$\begin{aligned} dr_t &= \kappa(\theta - r_t)\, dt + \sigma dW_t, \\ \lambda_t &= a_0(t) + a_1(t) r_t + a_2(t) Z_t. \end{aligned}$$

Note that $\lambda$ may take on negative values with positive probability.

Recovery must be modeled exogenously and the authors use the already mentioned *recovery of treasury value*[6]. This means if default happens prior to maturity of the bond, the bond holder receives a fraction $(1 - w)$ of the principal at maturity. For the value of the bond we calculate the expectation of the discounted payoff under the risk-neutral measure $Q$. For ease of notation we consider $t = 0$. By equation (1.3),

$$\bar{B}(0,T) = (1-w)B(0,T) + w\mathbb{E}^Q\Big[\exp\Big(-\int_0^T r_s\, ds\Big)1_{\{\tau > T\}}\Big].$$

In the model of Jarrow and Turnbull we obtain

$$\begin{aligned} \bar{B}(0,T) &= (1-w)B(0,T) + w\mathbb{E}^Q\Big[\exp(-\int_0^T r_u\, du) Q(\tau \le T | \lambda_s : 0 \le s \le T)\Big] \\ &= (1-w)B(0,T) + w\mathbb{E}^Q\Big[\exp[-\int_0^T (r_u + \lambda_u)\, du]\Big] \\ &= (1-w)B(0,T) + w\exp(-\mu_T + \frac{1}{2}v_T). \end{aligned}$$

In the last equation $\mu_T$ and $v_T$ denote expectation and variance of $\int_0^T (r_u + \lambda_u)\, du$. Under the stated assumptions this integral is normally distributed and $\mu$ and $v$ can be easily calculated.

The flexibility of the model leads to a good fit to market data, which is not obtained by most structural models. Also the model incorporates economic factors $(Z_t)$.

**2.3. Duffie and Singleton (1999).** The paper by Duffie and Singleton [**D-S**] combines two very successful model classes in interest rate modeling to access Credit Risk: exponential affine models and the Heath, Jarrow and Morton [**H-J-M**] methodology.

For the exponential affine model the authors model a vector of hidden factors which underlie the term structure of interest rates. This vector is assumed to follow a multidimensional Cox-Ingersoll-Ross model:

$$d\mathbf{y}(t) = \mathbf{K}(\boldsymbol{\Theta} - \mathbf{y}(t))dt + \boldsymbol{\Sigma}\ \mathrm{diag}(\mathbf{y}(t))^{1/2} d\mathbf{W}(t).$$

Consequently the components of $\mathbf{y}$ are nonnegative random numbers. Spot and hazard rate are assumed to be linear in $\mathbf{y}(t)$:

$$\begin{aligned} r(t) &= \delta_0 + \delta' \mathbf{y}(t), \\ \lambda(t)(1 - \theta(t)) &= \gamma_0 + \gamma' \mathbf{y}(t). \end{aligned}$$

[6] See the Longstaff and Schwartz model, Section 1.2.

A main feature of the exponential affine models is that the solution of the above SDE can be explicitly expressed in an exponential affine form. Hence we obtain deterministic functions $a(), b()$ such that

$$\mathbb{E}\left[\exp\left(i\boldsymbol{\xi}'\int_0^t \mathbf{y}(u)\,du\right)\right] = \exp[a(t,\boldsymbol{\xi}) + b(t,\boldsymbol{\xi}'\mathbf{y}(0))].$$

Thus the price of the defaultable bond can be calculated in closed form as the value of the characteristic function at a proper point.

The second approach uses the well known Heath-Jarrow-Morton model of forward rates. Denote by $\bar{f}(t,T)$ the forward rates determined by the term structure of the defaultable bond prior to default[7] and by $\mathbf{W}(t,T)$ a $d$-dimensional standard Brownian motion. Assume the dynamics of the forward rate to be

$$\bar{f}(t,T) = \bar{f}(0,T) + \int_0^t \mu(u,T)\,du + \int_0^t \boldsymbol{\sigma}(u,T)\,d\mathbf{W}(u).$$

Similar to Heath, Jarrow and Morton [**H-J-M**] the authors specify the dynamics under the objective measure and consider an equivalent measure $Q$. For arbitrage-freeness it is sufficient - see the work of Harrison and Pliska [**H-P**] - that all discounted price processes are martingales. Naturally this heavily relies on the recovery assumption.

Duffie and Singleton [**D-S**] introduced the *recovery of market value* which means that immediately at default the bond loses a fraction of its value. This setup is particularly well suited for working with SDEs. The loss rate $w_t$ is assumed to be an adapted process. Hence

$$\bar{B}(\tau,T) = (1-w_t)\bar{B}(\tau-,T).$$

Under these assumptions the authors derived the following *drift condition* for $\mu$ and $\sigma$:

$$\mu(t,T) = \sigma(t,T)\Big(\int_t^T \sigma(u,T)\,du\Big)'.$$

On the other hand, using the above mentioned *recovery of treasury value* (cf. 1.2) and denoting the riskless forward rate by $f(t,T)$, the authors obtained

$$\mu(t,T) = \sigma(t,T)\Big(\int_t^T \sigma(u,T)\,du\Big)' + \theta(t)\lambda(t)\frac{v(t,T)}{p(t,T)}(\bar{f}(t,T) - f(t,T)).$$

## 3. Credit Ratings Based Methods

Simple hazard rate models are often criticized because they do not incorporate available economic fundamental information like firm value or credit ratings. This section reveals some models which incorporate these data. This is also a basic feature of commercial models; see Section 7.

Credit ratings constitute a published ranking of the creditor's ability to meet his obligations. Such ratings are provided by independent agencies, for example Standard & Poor's or Moody's and mostly financed by the gauged companies. The firms

[7] The forward rate is by definition $\bar{f}(t,T) = -\frac{\partial}{\partial T}\ln\bar{B}(t,T)$.

are rated even if they are not willing to pay, but for a fee they get detailed insight in the results of the examinations and might retain fundamental insights in their internal divisions to identify weaknesses.

Each rating company uses a different system of letters to classify the creditworthiness of the rated agencies. Standard & Poor's, for example, describes the highest rated debt (triple-A=AAA) with the words "Capacity to pay interest and repay principal is extremely strong". An obligation with the lowest rating, 'D', is in state of default or is not believed to make payments in time or even during a grace period. The lower the rating, the greater is the risk that interest or principal payments will not be made.

**3.1. Jarrow, Lando and Turnbull (1997).** The model proposed by Jarrow, Lando and Turnbull [**J-L-T**] circumvents some disadvantages of the hitherto introduced models. Especially the use of credit ratings is an attractive feature. The movements between the single rating classes is modeled by a time homogenous Markov chain, the entry into the lowest rating class yielding a default. For example, if a bond is rated AAA, it is a member of the highest rating class (= class 1). If there exist $K-1$ rating classes, denote by $K$ the class of default. Default is assumed to be an absorbing state, restructuring after default is not considered in this model. The generator of the Markov chain is defined as

$$\mathbf{\Lambda} = \begin{pmatrix} -\lambda_1 & \lambda_{12} & \lambda_{13} & \cdots & \lambda_{1K} \\ \lambda_{21} & -\lambda_2 & \lambda_{23} & \cdots & \lambda_{2K} \\ \vdots & \vdots & \ddots & \cdots & \vdots \\ \lambda_{K-1,1} & \lambda_{K-1,2} & \cdots & -\lambda_{K-1} & \lambda_{K-1,K} \\ 0 & 0 & \cdots & \cdots & 0 \end{pmatrix}.$$

The transition rates for the first rating class are in the first row. So $\lambda_1 = \sum_{j\neq 1} \lambda_{1j}$ is the rate for leaving this class, while $\lambda_{12}$ is the rate for downgrading to class 2 and so on. The rate for a default directly from class one is $\lambda_{1K}$.

We denote

$$q_{ij}(0,t) := \mathbb{P}(\text{Rating is in class } i \text{ at } 0 \text{ and in class } j \text{ at } t),$$

and by $\mathbf{Q}(t)$ the matrix of the transition probabilities $q_{ij}(0,t)$. The transition probabilities can be computed from the intensity matrix via[8]

$$\mathbf{Q}(t) = \exp(t\mathbf{\Lambda}) := \mathrm{id}_n + t\mathbf{\Lambda} + \frac{1}{2!}(t\mathbf{\Lambda})^2 + \frac{1}{3!}(t\mathbf{\Lambda})^3 + \ldots,$$

where $\mathrm{id}_n$ is the $n \times n$ identity-Matrix.

[8]See, for example, [**I-R-W**].

Under the recovery of treasury assumption[9] we obtain for the price of a zero coupon bond under default risk

$$\begin{aligned}
\bar{B}(t,T) &= 1_{\{\tau>t\}}\mathbb{E}_t\Big[e^{-\int_t^\tau r_s\,ds}\cdot\delta B(\tau,T)1_{\{\tau\le T\}}+e^{-\int_t^T r_s\,ds}\cdot 1_{\{\tau>T\}}\Big]\\
&= 1_{\{\tau>t\}}\mathbb{E}_t\Big[\delta 1_{\{\tau\le T\}}e^{-\int_t^T r_s\,ds}+1_{\{\tau>T\}}e^{-\int_t^T r_s\,ds}\Big]\\
&= 1_{\{\tau>t\}}\Big[\delta B(t,T)+\mathbb{E}_t\Big((1-\delta)e^{-\int_t^T r_s\,ds}1_{\{\tau>T\}}\Big)\Big]\\
&= 1_{\{\tau>t\}}B(t,T)\Big[\delta+(1-\delta)Q_t^T(\tau>T)\Big]. \qquad (3.1)
\end{aligned}$$

$Q^T$ is the $T$-forward measure[10]. It is therefore crucial to have a model which determines the transition probabilities under this measure. While rating agencies estimate the transition probabilities using historical observations, i.e., under the objective measure $P$, [**J-L-T**] propose a method which uses the defaultable bond prices and calculates transition probabilities under the the risk-neutral measure $Q$.

Consider the bond with rating "i" and set $Q_t^{T,i}(\tau > T)$ the probability that the bond will not default until $T$ given it is rated "i" at $t$. As it makes no sense to talk about bond prices after default, we further on just consider the bond price on $\{\tau > t\}$ and get

$$\bar{B}_i(t,T) = B(t,T)\left(\delta + (1-\delta)Q_t^{T,i}(\tau > T|\tau > t)\right). \qquad (3.2)$$

Jarrow, Lando and Turnbull [**J-L-T**] split the intensity matrices into an empirical part (under $P$) and a risk adjustment like a market price of risk: They assume that the intensities under $Q^T$ have the form $\mathbf{U\Lambda}$ and $\mathbf{U}$ denotes a diagonal matrix where the entries are the risk adjusting factors $\mu_i$. For the transition probabilities this yields that $q_{ij}(t,T)$ is the $ij$'th entry of the matrix $\exp(\mathbf{U\Lambda})$. Time homogeneity of $\mu$ would entail exact calibration being impossible.

For the discrete time approximation, $[0,T]$ is divided into steps of length 1. Starting with (3.2) one obtains

$$\begin{aligned}
Q_0^{T,i}(\tau > T) &= \frac{B(0,T)(1-\delta) - \bar{B}_i(0,T) + \delta B(0,T)}{B(0,T)(1-\delta)}\\
&= \frac{B(0,T) - \bar{B}_i(0,T)}{B(0,T)(1-\delta)}. \qquad (3.3)
\end{aligned}$$

Denote the empirical probabilities from the rating agency by $p_{ij}(t,T)$. This leads to $Q_0^{T,i}(\tau \le 1) = \mu_i(0)p_{iK}(0,1)$, and we obtain

$$\mu_i(0) = \frac{Q_0^{T,i}(\tau > 1)}{p_{iK}(0,1)} = \frac{B(0,1) - \bar{B}_i(0,1)}{p_{iK}(0,1)\cdot B(0,1)(1-\delta)}.$$

By this one obtains $(\mu_1, \ldots, \mu_{K-1})'$ and consequently $q_{ij}(0,1)$. For the step from $t$ to $t+1$ use

$$Q_0^{T,i}(\tau \le t+1) = Q_0^{T,i}(\tau \le t+1|\tau > t)\cdot Q_0^{T,i}(\tau > t)$$

---

[9]The bond holder receives $\delta$ equivalent and riskless bonds in case of default. See Section 1.2.

[10]The $T$-forward measure is the risk neutral measure which has the risk-free bond with maturity $T$ as numeraire. For details see [**Bj**].

to get

$$\begin{aligned} Q_0^{T,i}(\tau \leq t+1) &= \mu_i(t) P^i(\tau \leq t+1 | \tau > t) \cdot \sum_{j=1}^{K-1} q_{ij}(0,t) \\ &= \mu_i(t) p_{iK}(t,t+1) \cdot \sum_{j=1}^{K-1} q_{ij}(0,t). \end{aligned}$$

This leads to

$$\begin{aligned} \mu_i(t) &= \frac{Q_0^{T,i}(\tau \leq t+1)}{\sum_{j=1}^{K-1} q_{ij}(0,t) \cdot p_{iK}(t,t+1)} \\ &\overset{(3.3)}{=} \frac{B(0,t+1) - \bar{B}_i(0,t+1)}{B(0,t+1)(1-\delta) \left( \sum_{j=1}^{K-1} q_{ij}(0,t) \right) p_{iK}(t,t+1)}, \end{aligned}$$

and, via $q_{ij}(0,t+1) = \mu_i(t) p_{ij}(0,t+1)$, the required probabilities are obtained.

This model extends [**J-T**] using time dependent intensities but still working with constant recovery rates. Das and Tufano [**D-T**] propose a model which also allows for correlation between interest rates and default intensities.

It seems problematic that all bonds with the same rating automatically have the same default probability. In reality this is definitely not the case. Naturally different credit spreads occur for bonds with the same rating.

A further restrictive assumption is the time independence of the intensities. The yield of a bond in this model may only change if the rating changes. Usually the market price precedes the ratings with informations on a possible rating change which is an important insight of the KMV model; see Section 7.1.

**3.2. Lando (1998).** The work of Lando [**La1**] uses a conditional Markov chain[11] to describe the rating transitions of the bond under consideration. All available market information like interest rates, asset values or other company specific information is modeled as a stochastic process $(X_t)_{t \geq 0}$. This is analogous to the case without ratings, where Lando used $\lambda_t = \lambda(X_t)$.

Assume that a risk-neutral martingale measure $Q$ is already chosen. Then the arbitrage-free price of a contingent claim is the conditional expectation under this measure $Q$. The author lays out the framework for rating transitions where all probabilities are already under the risk-neutral measure and calibrates them to available market prices. As no historical information is used the probability distribution under the objective measure is not needed. If one wants to consider risk-measures like Value-at-Risk, note that the objective measure is still required.

[11]See also Section 11.3 in [**B-R**].

We denote the generator of the conditional Markov chain $C_t$ by

$$\mathbf{\Lambda}(s) = \begin{pmatrix} -\lambda_1(s) & \lambda_{12}(s) & \lambda_{13}(s) & \cdots & \lambda_{1K}(s) \\ \lambda_{21}(s) & -\lambda_2(s) & \lambda_{23}(s) & \cdots & \lambda_{2K}(s) \\ \vdots & \vdots & \ddots & \cdots & \vdots \\ \lambda_{K-1,1}(s) & \lambda_{K-1,2}(s) & \cdots & -\lambda_{K-1}(s) & \lambda_{K-1,K}(s) \\ 0 & 0 & \cdots & \cdots & 0 \end{pmatrix}.$$

We assume $\lambda_{ij}(t)$ to be adapted processes and nonnegative for $i \neq j$. Furthermore, for all $s$

$$\lambda_i(s) = \sum_{j \neq i}^{K} \lambda_{ij}(s), \quad i = 1, \dots, K-1.$$

It is important for the intensities to depend on both time and interest rates. Especially for low rated companies the default rates vary considerably over time[12]. It was observed by Duffee [**Du**], e.g., that default rates significantly depend on the term structure of interest rates. It is certainly bad news for companies with high debt when interest rates increase whereas for other companies it might be good news.

Consider a series of independent exponential(1)-distributed random variables $E_{11}$, $\dots, E_{1K}, E_{21}, \dots, E_{2K}, \dots$ which are also independent of $\sigma(\mathbf{\Lambda}(s) : s \geq 0)$ and denote the rating class of the company at the beginning of the observation by $\eta_0$. Define

$$\tau_{\eta_0,i} := \inf\{t : \int_0^t \lambda_{\eta_0,i}(X_s)\, ds \geq E_{1i}\}, \quad i = 1, \dots, K$$

and

$$\tau_0 := \min_{i \neq \eta_0} \tau_{\eta_0,i}, \qquad \eta_1 := \arg\min_{i \neq \eta_0} \tau_{\eta_0,i}.$$

The $\tau_{\eta_0,i}$ model the possible transitions to other rating classes starting from rating $\eta_0$. The first transition to happen determines the transition that really takes place. The reached rating class is denoted by $\eta_1$ while $\tau_0$ denotes the time at which this occurs. Analogously, the next change in rating starting in $\eta_1$ is defined, and similarly for $\eta_i$ and $\tau_i$.

Default is assumed to be an absorbing state of the Markov chain and we denote the overall-time to default by $\tau$. This is the first time when $\eta_i = K$.
The transition probabilities $P(s,t)$ for the time interval $(s,t)$ satisfy Kolmogorov's backward differential equation[13]

$$\frac{\partial P_X(s,t)}{\partial s} = -\mathbf{\Lambda}(s)\, P_X(s,t).$$

Consider the price of a defaultable zero recovery bond at time t, $\bar{B}^i(t,T)$, which has maturity $T$ and is rated in class $i$ at time $t$. Then we obtain the following Theorem.

---

[12]Cf. Chapter 15 in [**C-A-N**].

[13]For non-commutative $\mathbf{\Lambda}$ the solution is in general not of the form $P_X(s,t) = \exp \int_s^t \mathbf{\Lambda}(u)\, du$. See [**Gi-J**] for solutions using product integrals.

THEOREM 3.1. *Under the above assumptions the price of the defaultable bond equals*

$$\bar{B}^i(t,T) = \mathbb{E}\Big( \exp\big(-\int_t^T r_s\, ds\big)(1 - P_X(t,T)_{i,K}\big)\Big| \mathcal{F}_t\Big).$$

*Here $P_X(t,T)_{i,K}$ is the (i,K)-th element of the matrix of transition probabilities for the time interval $(t,T)$, $P_X(t,T)$.*

PROOF. As already mentioned the Markov chain is modeled under $Q$ so that the arbitrage-free price of the bond is the following conditional expectation:

$$\bar{B}^i(t,T) = \mathbb{E}\Big( \exp\big(-\int_t^T r_s\, ds\big) 1_{\{\tau > T\}}\Big| \mathcal{F}_t\Big).$$

Using conditional expectations and the independence of $E_{1K}$ and $(\mathbf{\Lambda}(s))$ one concludes

$$\begin{aligned} \bar{B}^i(t,T) &= 1_{\{C_t = i\}} \mathbb{E}\Big( \exp\big(-\int_t^T r_s\, ds\big) \mathbb{P}\big(\tau > T \big| \sigma(\mathbf{\Lambda}_s : 0 \le s \le T) \vee \mathcal{F}_t\big)\Big| \mathcal{F}_t\Big) \\ &= \mathbb{E}\Big( \exp\big(-\int_t^T r_s\, ds\big)(1 - P_X(t,T)_{i,K}\big)\Big| \mathcal{F}_t\Big). \end{aligned}$$

□

For the calibration to observed credit spreads explicit formulas are needed and therefore further assumptions will be necessary. Lando chooses an Eigenvalue-representation of the generator.

Denote with $\mathbf{A}(s)$ the matrix with entries $\lambda_1(s), \dots, \lambda_{K-1}(s), 0$ on the diagonal and zero otherwise. Assume that $\mathbf{\Lambda}(s)$ admits the representation

$$\mathbf{\Lambda}(s) = \mathbf{B}\, \mathbf{A}(s)\, \mathbf{B}^{-1},$$

where $\mathbf{B}$ is the $K \times K$-matrix of the Eigenvectors of $\mathbf{\Lambda}(s)$.

We conclude $P_X(s,t) = \mathbf{B}\, \mathbf{C}(s,t)\, \mathbf{B}^{-1}$ with

$$\mathbf{C}(s,t) = \begin{pmatrix} \exp\int_s^t \lambda_1(u)du & 0 & \cdots & 0 \\ 0 & \ddots & \cdots & \vdots \\ \vdots & \cdots & \exp\int_s^t \lambda_{K-1}(u)du & 0 \\ 0 & \cdots & 0 & 1 \end{pmatrix}.$$

It is easy to see that $P_X(s,t)$ satisfies the Kolmogorov-backward differential equation. For uniqueness, see [**Gi-J**].

Under these additional assumptions the price of the defaultable bond in Theorem 3.1 simplifies considerably.

PROPOSITION 3.2. *Denoting by $b_{ij}$ the entries of $\mathbf{B}$, the price of the defaultable bond equals*

$$\bar{B}^i(t,T) = \sum_{j=1}^{K-1} -\frac{b_{ij}}{b_{jK}} \mathbb{E}\Big[ \exp\Big( \int_t^T (\lambda_j(u) - r_u)\, du\Big)\Big| \mathcal{F}_t\Big].$$

PROOF. In this setup the conditional probability for a default when the bond is in rating class $i$ equals

$$\mathbb{P}_X(t,T)_{i,K} = 1_{\{\tau>t\}} \sum_{j=1}^{K} b_{ij} \exp(\int_t^T \lambda_j(u)du) b_{jK}^{-1}.$$

With $b_{iK}b_{KK}^{-1} = 1$ we obtain

$$1 - \mathbb{P}_X(t,T)_{i,K} = \sum_{j=1}^{K-1} -\frac{b_{ij}}{b_{jK}} \exp(\int_t^T \lambda_j(u)du)$$

and the conclusion follows as in 3.1. □

Using the readily available tools for hazard rate models it is now easy to consider options which explicitly depend on the credit rating or credit derivatives with a credit trigger.

3.2.1. *Calibration.* Assuming a Vasiček model[14] for the interest rate we are in the position to use the model laid out above for calibration to observed credit spreads. There are no economic factors considered other than the interest rate and, as a consequence, $\lambda_t$ must be adapted to $\mathcal{G}_t = \sigma(r_s : 0 \le s \le t)$.

Furthermore, we assume

$$\lambda_j(s) = \gamma_j + \kappa_j r_s, \quad j = 1, \dots, K-1,$$

with constants $\gamma_j, \kappa_j$.

The dynamics of the generator matrix is $\mathbf{\Lambda}(s) = \mathbf{B}\,\mathbf{A}(s)\,\mathbf{B}^{-1}$ and $\mathbf{B}$ has to be estimated from historical data while $\gamma_j, \kappa_j$ are calibrated.

The credit spread is the difference of the offered yield to the spot rate. By Theorem 3.1 the bond price satisfies

$$\bar{B}^i(t,T) = -\sum_{j=1}^{K-1} -\frac{b_{ij}}{b_{jK}} \mathbb{E}\Big[\exp\Big(\int_t^T \gamma_j - (1-\kappa_j) r_u \, du\Big)\Big| \mathcal{F}_t\Big].$$

Therefore, we obtain for the bond's yield

$$\begin{aligned} -\frac{\partial}{\partial T}\Big|_{T=t} \log \bar{B}^i(t,T) &= -\frac{\partial}{\partial T}\Big|_{T=t} \sum_{j=1}^{K-1} \beta_{ij} \mathbb{E}\Big[\exp\Big(\int_t^T \gamma_j - (1-\kappa_j) r_u \, du\Big)\Big| \mathcal{F}_t\Big] \\ &= -\sum_{j=1}^{K-1} \beta_{ij} \lim_{T\to t} \mathbb{E}\Big[(\gamma_j + (\kappa_j - 1)r_T) \exp\Big(\int_t^T \gamma_j + \kappa_j r_s - r_s ds\Big)\Big| \mathcal{F}_t\Big] \\ &= -\sum_{j=1}^{K-1} \beta_{ij}(\gamma_j + (\kappa_j - 1) r_t). \end{aligned}$$

[14] see equation (1.2).

Hence the credit spread equals

$$s^i(t) = -\sum_{j=1}^{K-1} \beta_{ij}(\gamma_j + \kappa_j r_t).$$

For calibration a second relation is needed. Lando uses the sensitivity of the credit spreads w.r.t. the spot rate:

$$\frac{\partial}{\partial r_t} s^i(t) = -\sum_{j=1}^{K-1} \beta_{ij}\kappa_j.$$

Denote by $\hat{s}_0, d\hat{s}_0$ the observed credit spreads and their estimated sensitivities. One finally has to solve the following equation to calibrate the model:

$$\begin{aligned} -\beta(\gamma + \kappa r_0) &= \hat{s}_0 \\ -\beta\kappa &= d\hat{s}_0. \end{aligned}$$

It turns out to be problematic that observed credit spreads are not always monotone with respect to the ratings. The author argues that in practice this would occur rather seldom.

## 4. Basket Models

Usually there is a whole portfolio under consideration instead of just one single asset. Therefore the so far presented models were extended to models which may handle the behavior of a larger number of individual assets with default risk, a so-called *portfolio* or *basket.*

There are several approaches in the literature and they can be grouped into models which use a conditional independence concept and others which are based on copulas.

From the first class we present the methods of Kijima and Muromachi [**K-M**], which provide a pricing formula for a credit derivative on baskets with a first- or second-to-default feature. An example is the first-to-default put, which covers the loss of the first defaulted asset in the considered portfolio, see also Section 8.6. From the second class we discuss an implementation based on the normal copula in Section 4.2.

Besides that, Jarrow and Yu [**J-Y**] model a kind of direct interaction between default intensities of different companies. In their model the default of a *primary* company has some impact on the hazard rate of a *secondary* company, whose income significantly depends on the primary company.

**4.1. Kijima and Muromachi (2000).** Consider a portfolio of $n$ defaultable bonds and denote by $\tau_i$ the default time of the $i$-th bond. Let $(\mathcal{G}_t)_{t\geq 0}$ represent the general market information and assume that for any $t_1, \ldots, t_n \leq T$

$$Q(\tau_1 > t_1, \ldots, \tau_n > t_n | \mathcal{G}_T) = Q(\tau_1 > t_1 | \mathcal{G}_T) \cdot \cdots \cdot Q(\tau_n > t_n | \mathcal{G}_T), \tag{4.1}$$

where $Q$ is assumed to be the unique risk neutral measure. Using the representation via Cox processes, this yields

$$(4.1) = \exp(-\sum_{i=1}^{n} \int_0^{t_i} \lambda_i(s)\, ds).$$

In the recovery of treasury model, the loss of bond $i$ upon default equals the prespecified constant $w_i := (1 - \delta_i)$. So the first-to-default put is the option which pays $w_i$ if the $i$th asset is the first one to default before $T$ and zero if there is no default. Denote the event that the first defaulted bond is number $i$ by

$$D_i := \{\tau_i \leq T, \tau_j > \tau_i, \forall j \neq i\}.$$

Then, using the risk neutral valuation principle, the price of the bond can be computed as the expectation w.r.t. the risk-neutral measure $Q$ and equals

$$\begin{aligned} \bar{S}_F &= \mathbb{E}\Big[\exp(-\int_0^T r_u\, du) \sum_{i=1}^{n} w_i 1_{A_i}\Big] \\ &= \sum_{i=1}^{n} w_i \mathbb{E}\big[\exp(-\int_0^T r_u\, du) Q(A_i|\mathcal{G}_T)\big]. \end{aligned}$$

We obtain this probability using the factorization

$$\begin{aligned} &\mathbb{P}(\tau_i \leq T, \tau_k > \tau_i, \forall k \neq i | \mathcal{G}_T \vee \{\tau_i = x\}) \\ &\qquad = 1_{\{x \leq T\}} \mathbb{P}(\tau_k > x, \forall k \neq i | \mathcal{G}_T \vee \{\tau_i = x\}) \\ &\qquad = 1_{\{x \leq T\}} \exp(-\sum_{k \neq i} \int_0^x \lambda_k(s)\, ds). \end{aligned}$$

We therefore obtain[15]

$$\begin{aligned} &\mathbb{P}(\tau_i \leq T, \tau_k > \tau_i, \forall k \neq i | \mathcal{G}_T) \\ &\quad = \mathbb{E}\Big[1_{\{\tau_i \leq T\}} \exp(-\sum_{k \neq i} \int_0^{\tau_i} \lambda_k(s)\, ds) | \mathcal{G}_T\Big] \\ &\quad = \mathbb{E}\Big[\int_0^T \lambda_i(u) \exp(-\int_0^u \lambda_i(s)\, ds) \exp(-\sum_{k \neq i} \int_0^u \lambda_k(s)\, ds)\, du\Big] \\ &\quad = \int_0^T \mathbb{E}\big[\lambda_i(u) \exp(-\int_0^u \sum_{k=1}^{n} \lambda_k(s)\, ds)\big]\, du. \end{aligned}$$

We conclude for the price of the first-to-default put:

$$\bar{S}_F = \sum_{i=1}^{n} \delta_i \int_0^T \mathbb{E}\Big[\lambda_i(u) \exp(-\int_0^T r_s\, ds - \sum_{k=1}^{n} \int_0^u \lambda_k(s)\, ds)\Big]\, du.$$

[15]See [**B-R**], Proposition 5.1.1.

This formula simplifies considerably if $w_i \equiv w$, as in that case

$$\begin{aligned}\bar{S}_F &= w\mathbb{E}\Big[\int_0^T \sum_{i=1}^n \lambda_i(u)\exp(-\int_0^u \sum_{k=1}^n \lambda_k(s)\,ds)\,du\exp(-\int_0^T r_s\,ds)\Big]\\ &= w\mathbb{E}\Big[\big(-\exp(-\sum_{i=1}^n \int_0^T \lambda_i(u)\,du)\big)\Big|_0^T \cdot \exp(-\int_0^T r_s\,ds)\Big]\\ &= (1-\delta)B(0,T)\big[1-\mathbb{E}^T\big(\exp(-\int_0^T \sum_{i=1}^n \lambda_i(u)\,du)\big)\big].\end{aligned}$$

Using similar methods, we determine the swap-price, if $w_i$ is paid immediately at default to the swap-holder. Set

$$\bar{S}_F^* = \mathbb{E}\Big[\exp(-\int_0^\tau r_u\,du)\cdot\sum_{i=1}^n w_i 1_{A_i}\Big].$$

Certainly, $\int_0^\tau r_u\,du$ is not $\mathcal{G}_T$-measurable, so that a slight modification of the previously used method is necessary. We obtain for the factorization

$$\begin{aligned}&\mathbb{E}\big[\exp(-\int_0^x r_u\,du)1_{\{x\le T\}}1_{\{\tau_k > x, \forall k\neq i\}}\big|\mathcal{G}_T \vee \{\tau_i = x\}\big]\\ &\qquad = 1_{\{x\le T\}}\exp(-\int_0^x r_u + \sum_{k\neq i}\lambda_k(u)\,du)\end{aligned}$$

and conclude

$$\bar{S}_F^* = \sum_{i=1}^n w_i \int_0^T \mathbb{E}\big[\lambda_i(u)\exp(-\int_0^u r_s + \sum_{k=1}^n \lambda_k(s)\,ds)\big]\,du.$$

Similarly, the authors provide the following price of a (first and) second-to-default swap, which protects the holder against the first two defaults in the portfolio:

$$\begin{aligned}\bar{S}_S\ -\ & \sum_{i=1}^n \delta_i\mathbb{E}\big[\exp(\ \int_0^T \lambda_i(u)\,du)\big]\quad B(0,T)\sum_{i=1}^n \delta_i\\ &+\sum_{i\neq j}(\delta_i+\delta_j)\int_0^T \mathbb{E}\Big[\lambda_k(u)\exp(-\int_0^T r_s\,ds - \sum_{j=1}^n\int_0^u \lambda_j(u)\,du)\big]\\ &-(n-2)\sum_{i=1}^n \delta_i \int_0^T \mathbb{E}\big[\lambda_i(u)\exp(-\int_0^T r_s\,ds - \int_0^u \sum_{j=1}^n \lambda_j(s)\,ds)\big]\end{aligned}$$

4.1.1. *Extended Vasiček implementation.* Kijima and Muromachi [**K-M**] discuss a special case of the above implementation. The main idea is to perform a calibration similar to the one of Hull and White [**H-W**] for credit risk models. Assume for the dynamics of the hazard rates

$$d\lambda_i(t) = \big(\phi_i(t) - a_i\lambda_i(t)\big)\,dt + \sigma_i\,dw_i(t), \qquad i = 1,\dots,n, \tag{4.2}$$

where $w_i$ are standard Brownian motions with correlation $\rho_{ij}$, which is sometimes stated as $dw_i dw_j = \rho_{ij}\,dt$. Furthermore, assume for the short rate $r_t$

$$dr_t = \big(\phi_0(t) - a_0 r_t\big)\,dt + \sigma_0\,dw_0(t).$$

Note that equations of the type (4.2) admit explicit solutions, see [**Sc**]. From this, we get

$$\lambda_i(t) = \lambda_i(0)e^{-a_i t} + \int_0^t \phi_i(s)e^{-a_i(t-s)}\,ds + \sigma_i \int_0^t e^{-a_i(t-s)}\,dw_i(s).$$

Using the recovery of treasure assumption the bond price equals

$$\bar{B}_i(0,t) = \delta_i B(0,t) + (1-\delta_i)\mathbb{E}\big[\exp(-\int_0^t (r_u + \lambda_i(u))\,du)\big].$$

Note that $\int (r_u + \lambda_i(u))\,du$ is normally distributed and therefore the expectation equals the Laplace transform of a normal random variable with mean

$$\begin{aligned}\mathbb{E}\big[-\int_0^t (r_u + \lambda_i(u))\,du\big] &= -\int_0^t \big(r_0 e^{-a_0 u} + \int_0^u \phi_0(s)e^{-a_0(u-s)}\,ds\big) \\ &\quad -\int_0^t \big(\lambda_i(0)e^{-a_i u} + \int_0^u \phi_i(s)e^{-a_i(u-s)}\,ds\big)\,du\end{aligned}$$

and variance

$$\begin{aligned}&\mathrm{Var}\Big[\int_0^t (r_u + \lambda_i(u))\,du\Big] \\ &\qquad = \mathrm{Var}\Big[\int_0^t \sigma_0 \int_0^u e^{-a_0(u-s)}dz_0(s)\,du + \int_0^t \sigma_i \int_0^u e^{-a_i(u-s)}dw_i(s)\,du\Big].\end{aligned}$$

To compute the variances it is sufficient to calculate the variances of all summands and the covariances. Setting $\rho_{ii} = 1$, we have

$$\begin{aligned}&\mathbb{E}\Big[\int_0^t\int_0^t \sigma_i\sigma_j \int_0^{u_1}\int_0^{u_2} \exp(-a_i(u_1-s_1) - a_j(u_2-s_2))\,dw_j(s_2)\,dw_i(s_1)\,du_2\,du_1\Big] \\ &= \sigma_i\sigma_j\mathbb{E}\Big[\int_0^t\int_0^t\int_0^{s_1}\int_0^{s_2} \exp(-a_i(u_1-s_1) - a_j(u_2-s_2))\,du_2\,du_1\,dw_j(s_2)\,dw_i(s_1)\Big] \\ &= \sigma_i\sigma_j\mathbb{E}\Big[\int_0^t\int_0^t e^{a_i s_1 + a_j s_2}\frac{1}{a_i a_j}(1-e^{-a_i s_1})(1-e^{-a_j s_2})\,dw_j(s_2)\,dw_i(s_1)\Big] \\ &= \sigma_i\sigma_j\rho_{ij}\int_0^t e^{a_i s + a_j s}\frac{1}{a_i a_j}(1-e^{-a_i s})(1-e^{-a_j s})\,ds \\ &= \frac{\sigma_i\sigma_j\rho_{ij}}{a_i a_j}\Big[t + \frac{1}{a_i}(e^{-a_i t}-1) + \frac{1}{a_j}(e^{-a_j t}-1) + \frac{1}{a_i+a_j}(1-e^{-(a_i+a_j)t})\Big] \\ &=: c_{ij}(t)\end{aligned}$$

Therefore,

$$\begin{aligned}&\mathrm{Var}\Big[\int_0^t \sigma_i \int_0^u e^{-a_i(u-s)}dw_i(s)\,du\Big] \\ &\qquad = \frac{\sigma_i^2}{a_i^2}\Big[t + \frac{2}{a_i}(e^{-a_i t}-1) + \frac{1}{2a_i}(1-e^{-2a_i t})\Big] =: v_2(t).\end{aligned}$$

Recall that we want to calibrate the model to the bond prices, which means calculating $\phi_i(s)$. $\phi_0(s)$ is computed as in the risk neutral case, see [**H-W**]. Consider

$$\frac{1}{B(0,t)}\mathbb{E}\big[\exp(-\int_0^t (r_u + \lambda_i(u))\,du)\big] = \frac{1}{1-\delta_i}\Big[\frac{B_i(0,t)}{B(0,t)} - \delta_i\Big] =: \gamma_i(t),$$

which can be obtained from available prices, since $\delta_i$ is assumed to be known. Note that $\gamma_i(t)$ does not involve $\phi_0(s)$ as

$$\begin{aligned}\gamma_i(t) &= \exp\Big[-\int_0^t \big(\lambda_i(0)e^{-a_i u} + \int_0^u \phi_i(s)e^{-a_i(u-s)}\,ds\big)\,du \\ &\quad + \frac{1}{2}\big(c_{0i}(t) + v_2(t)\big)\Big].\end{aligned}$$

As we want to solve this expression for $\phi_i$, we consider the following derivatives:

$$\begin{aligned}-\frac{\partial}{\partial t}\ln\gamma_i(t) &= \lambda_i(0)e^{-a_i t} + \int_0^t \phi_i(s)e^{-a_i(t-s)}\,ds - \frac{1}{2}\Big[c_{0i}(t) + v_2(t)\Big]' \\ &=: g_i(t)\end{aligned}$$

With

$$\frac{\partial}{\partial t}g_i(t) = -a_i\lambda_i(0)e^{-a_i t} + \phi_i(t) - a_i e^{-a_i t}\int_0^t \phi_i(s)e^{a_i s}\,ds - \frac{1}{2}\Big[c_{0i}(t) + v_2(t)\Big]''$$

we conclude

$$\phi_i(t) = \frac{\partial}{\partial t}g_i(t) + a_i g_i(t) + a_i\frac{1}{2}\Big[c_{0i}(t) + v_2(t)\Big]' + \frac{1}{2}\Big[c_{0i}(t) + v_2(t)\Big]''.$$

Hence

$$\begin{aligned}a_i c_{0i}(t)' + c_{0i}(t)'' &= \sigma_0\sigma_i\rho_{0i}\Big[\frac{1}{a_0} - \frac{1}{a_0}e^{-a_0 t} - \frac{1}{a_0}e^{-a_i t} + \frac{1}{a_0}e^{-(a_0+a_i)t}\Big] \\ &\quad + \sigma_0\sigma_i\rho_{0i}\Big[\frac{1}{a_i}e^{-a_0 t} + \frac{1}{a_0}e^{-a_i t} - \frac{a_0+a_i}{a_0 a_i}e^{-(a_0+a_i)t}\Big] \\ &= \sigma_0\sigma_i\rho_{0i}\Big[\frac{1-e^{-a_0 t}}{a_0} + e^{-a_0 t}\frac{1-e^{-a_i t}}{a_i}\Big]\end{aligned}$$

and

$$\begin{aligned}a_i v_2(t)' + v_2(t)'' &= \sigma_i^2\Big[\frac{1}{a_i} - \frac{2}{a_i}e^{-a_i t} - \frac{1}{a_i}e^{-2a_i t} + \frac{2}{a_i}e^{-a_i t} + \frac{2}{a_i}e^{-2a_i t}\Big] \\ &= \frac{\sigma_i^2}{a_i}\Big[1 + e^{-2a_i t}\Big]\end{aligned}$$

which finally leads to

$$\begin{aligned}\phi_i(t) &= \frac{\partial}{\partial t}g_i(t) + a_i g_i(t) + \frac{\sigma_i^2}{2a_i}(1 - e^{-2a_i t}) \\ &\quad + \frac{1}{2}\sigma_0\sigma_i\rho_{0i}\Big[\frac{1-e^{-a_0 t}}{a_0} + e^{-a_0 t}\frac{1-e^{-a_i t}}{a_i}\Big].\end{aligned}$$

Using similar methods Kijima and Muromachi [**K-M**] obtain an explicit formula for the first-to-default swap. In Kijima [**K**] these methods are extended to pricing a credit swap on a basket, which might incorporate a first-to-default feature.

**4.2. Copula Models.** The concept of copulas is well known in statistics and probability theory, and has been applied to finance quite recently. Modeling dependent defaults using copulas can be found, for example, in [**Li**] or [**F-M**]. We give an outline of Schmidt and Ward [**S-W**], who apply a special copula, the normal copula, to the pricing of basket derivatives.

Fix $t = 0$. The goal of the model is to present a calibration method. Consider the default times $\tau_1, \ldots, \tau_n$ and assume for the beginning that $t = 0$. The link between the marginals $Q_i(t) := Q(\tau_i \leq t)$ and the joint distribution is the so-called copula $C(t_1, \ldots, t_n)$. Assuming continuous marginals, $U_i := Q_i(\tau_i)$ is uniformly distributed. The joint distribution of the transformed random times is the copula

$$C(u_1, \ldots, u_n) := Q(U_1 \leq u_1, \ldots, U_n \leq u_n)$$

and defines the joint distribution of the $\tau_i$'s via

$$Q(\tau_1 \leq t_1, \ldots, \tau_n \leq t_n) = C\big(Q_1(t_1), \ldots, Q_n(t_n)\big).$$

For more detailed information on copulas see [**N**].

The choice of the copula certainly depends on the application. Schmidt and Ward [**S-W**] choose the normal copula because in a Merton framework with correlated firm value processes such a dependence is obtained, and secondary the normal copula is determined by correlation coefficients which can be estimated from data.

Assume that $(Y_1, \ldots, Y_n)$ follows an $n-$dimensional normal distribution with correlation matrix $\mathbf{\Sigma} = (\rho_{ij})$, where $\rho_{ii} = 1$ for all $i$. Denoting their joint distribution function by $\Phi_n(y_1, \ldots, y_n, \mathbf{\Sigma})$ yields the *normal copula*

$$C(u_1, \ldots, u_n) = \Phi_n\big(\Phi^{-1}(u_1), \ldots, \Phi^{-1}(u_n)\big).$$

For modeling purposes it is useful to note that setting

$$\tau_i := Q_i^{-1}(\Phi(Y_i)),$$

results in $\{\tau_1, \ldots, \tau_n\}$ having a normal copula with correlation matrix $\mathbf{\Sigma}$.

The above methods enable us to calculate the joint distribution of $n$ default times, and the required correlations can be estimated using historical data. Thus, a value at risk can be determined.

For the pricing of a derivative with first-to-default feature, note that

$$Q(\tau^{1\text{st}} \leq T) = 1 - Q(\tau_1 > T, \ldots, \tau_n > T) \tag{4.3}$$

which can be calculated from the copula and the marginals. A more involved, but also explicit formula can be obtained for a $k$th-to-default option.

For example, consider a *first-to-default swap*, which is also discussed in Section 8.6. This is a derivative which offers default protection against the first defaulted asset in a specified portfolio. Under the assumption, that all credits have the same recovery rate $\delta_i \equiv \delta$, the swap pays $(1 - \delta)$ at $\tau^{1\text{st}}$ if $\tau^{1\text{st}} \leq T$. In exchange to this, the swap holder pays the premium $S$ at times $T_1, \ldots, T_m$, but at most until $\tau^{1\text{st}}$.

As explained in Section 8.3, calculating expectations of the discounted cash flows yields the first-to-default swap premium. Thus, using Equation (8.1), we obtain

$$S_{1\text{st}} = \frac{(1-\delta)\mathbb{E}\big[\exp(-\int_t^{\tau^{1\text{st}}} r_u\,du)1_{\{\tau^{1\text{st}}\le T\}}\big]}{\sum_{i=1}^m \mathbb{E}\big[\exp(-\int_0^{T_i} r_u\,du)1_{\{\tau^{1\text{st}}>T_i\}}\big]}.$$

To calculate the expectations, the distribution of $\tau^{1\text{st}}$ under any forward measure is needed. Assuming, for simplicity, independence of the default intensity and the risk-free interest rate, one obtains

$$\mathbb{E}\big[\exp(-\int_0^{T_i} r_u\,du)1_{\{\tau^{1\text{st}}>T_i\}}\big] = B(0,T_i)Q(\tau^{1\text{st}} > T_i).$$

The bond prices are readily available and the probability can be calculated via (4.3), once the copula is determined.

For the second expectation, use

$$\begin{aligned}\mathbb{E}\big[\exp(-\int_t^{\tau^{1\text{st}}} & r_u\,du)1_{\{\tau^{1\text{st}}\le T\}}\big] \\ &= \int_0^T B(0,s)\mathbb{E}\big[\exp(-\int_t^s \lambda_u^{1\text{st}}\,du)\lambda^{1\text{st}}(s)].\end{aligned}$$

Note that this expectation can be obtained via

$$\begin{aligned}\frac{\partial}{\partial s}Q(\tau^{1\text{st}} > s) &= \frac{\partial}{\partial s}\mathbb{E}\big[\exp(-\int_t^s \lambda_u^{1\text{st}}\,du)\big] \\ &= \mathbb{E}\big[\exp(-\int_t^s \lambda_u^{1\text{st}}\,du)\lambda^{1\text{st}}(s)\big].\end{aligned}$$

Further on, Schmidt and Ward [**S-W**] derive interesting results on spread widening, once a default occurred. For example, if one of two strongly related companies defaults, it might be likely that the remaining one gets into difficulties, and therefore credit spreads increase. It seems interesting that traders have a good intuition on this amount of spread widening, which also could be used as an input parameter to the model, which determines the copulas.

## 5. Hybrid models

Hybrid models incorporate both preceding models, for example the firm value is modeled, and a hazard rate framework is derived within this model.
The approach of Madan and Unal [**M-U**] mimics the behavior of the Merton model in a hazard rate framework. They assume the following structure for the default intensity:

$$\lambda(t) = \frac{c}{\left(\ln\frac{V(t)}{F\cdot B(t)}\right)^2}.$$

Here $V(t)$ denotes the firm value which as in Merton's model is assumed to follow a geometric Brownian motion. $B(t)$ is the discounting factor $\exp(-\int_0^t r_u\,du)$ and $F$ is the amount of outstanding liabilities. If the firm value approaches $F$ the default intensity increases sharply and it is very likely that the bond defaults. As defaults can happen at any time this model is much more flexible than the Merton model.

Unlike in Longstaff and Schwartz's model, the default can even happen when the firm value is far above $F$, though with low probability.

The authors also consider parameter estimation in their model. A closed form solution for the bond price is not available and for calculating the prices of derivatives numerical methods need to be used.

Further hybrid models of this type can be found in [**Am**] or [**B-R**].

The approach of Duffie and Lando [**D-L**] accounts for the fact that bond holders only obtain imperfect information on the firm value. Thus, starting in a structural framework, this leads to a hazard rate model.

## 6. Market Models with Credit Risk

Schönbucher [**Scho**] discusses the framework for a defaultable *market model*. The difference between the market models and the continuous time models is that market models rely only on a finite number of bonds, whereas continuous time models assume a continuity of bonds traded in the market. As a matter of fact, many important variables are not available in these models as, for example, the short rate or continuously derived forward rates, which form the basis for the setting in [**H-J-M**]. Introductions to market models without default risk can be found for example in [**B-G-M**], [**R**] or [**B-M**].

Assume we are given a collection of settlement dates $T_1 < \cdots < T_K$, the tenor structure, which denotes the maturities of all traded bonds.

Denote by $B_k(t) := B(t, T_k)$ the riskless bonds traded in the market. The discrete forward rate for the interval $[T_k, T_{k+1}]$ is defined as

$$F(t, T_k, T_{k+1}) =: F_k(t) = \frac{1}{T_{k+1} - T_k}\Big(\frac{B_k(t)}{B_{k+1}(t)} - 1\Big).$$

The defaultable zero coupon bond is denoted by $\bar{B}(t, T_k)$. As a starting point for modeling, it is assumed that this is a zero recovery bond, i.e., at default the value of the bond falls to zero. Put $\bar{B}_k(t) = \bar{B}(t, T_k) = 1_{\{\tau > t\}}\bar{B}(t, T_k)$. The *default risk factor* is denoted by

$$D_k(t) := \frac{\bar{B}_k(t)}{B_k(t)}.$$

If there exists an equivalent martingale measure $Q$ we have

$$\begin{aligned} D_k(t) &= \frac{1}{B_k(t)}\mathbb{E}^Q\Big[\exp(-\int_0^{T_k} r_u\,du)1_{\{\tau > T_k\}}\Big|\mathcal{F}_t\Big] \\ &= \frac{B_k(t)}{B_k(t)}\mathbb{E}^{T_k}\big[1_{\{\tau > T_k\}}\big|\mathcal{F}_t\big] \\ &= Q^{T_k}\big(\tau > T_k\big|\mathcal{F}_t\big) \end{aligned}$$

where $Q^{T_k}$ denotes the $T_k$-forward measure[16] and $\mathbb{E}^{T_k}$ the expectation w.r.t. this measure. So $D_k(t)$ denotes the probability that, under the forward measure, the bond survives time $T_k$.

Define

$$H(t, T_k, T_{k+1}) := H_k(t) = \frac{1}{T_{k+1} - T_k}\Big(\frac{D_k(t)}{D_{k+1}(t)} - 1\Big).$$

To simplify the notation we write $B_1$ for $B_1(t)$ (similarly for $F, D, H$) and $T_{j+1} - T_j = \delta_j$.

This leads to the following decomposition

$$\begin{aligned}
\bar{B}_k &= \bar{B}_1 \prod_{j=1}^{k-1} \frac{\bar{B}_{j+1}}{\bar{B}_j} \\
&= \bar{B}_1 \prod_{j=1}^{k-1} \frac{\bar{B}_{j+1}}{B_{j+1}} \frac{B_j}{\bar{B}_j} \frac{B_{j+1}}{B_j} \\
&= D_1 \prod_{j=1}^{k-1} \frac{D_{j+1}}{D_j}\, B_1 \prod_{j=1}^{k-1} \frac{B_{j+1}}{B_j} \\
&= D_1 B_1 \prod_{j=1}^{k-1} \big(1 + \delta_j H_j\big)^{-1} \cdot \big(1 + \delta_j F_j\big)^{-1}.
\end{aligned}$$

The discrete forward rates of the defaultable bond are split into a risk-free part and a risky part which is represented by the "discrete-tenor hazard rate" $H$.

Defining the *credit spread*

$$S_k(t) = S(t, T_k, T_{k+1}) := \bar{F}_k(t) - F_k(t),$$

we immediately obtain

$$\begin{aligned}
S_k(t) &= \frac{1}{\delta_k}\Big(\frac{\bar{B}_{k+1}}{\bar{B}_k} - 1\Big) - \frac{1}{\delta_k}\Big(\frac{B_{k+1}}{B_k - 1}\Big) \\
&= \frac{B_k}{B_{k+1}} \frac{1}{\delta_j}\Big(\frac{\bar{B}_{k+1}}{\bar{B}_k} \frac{B_k}{B_{k+1}} - 1\Big) \\
&= (1 + \delta_k F_k)\, H_k.
\end{aligned}$$

The main motivation for market models was to reproduce Black-like formulas for prices of caps and swaptions. This was particularly possible in the so-called LIBOR-market models. The basic assumption in these models is that the discrete forward rate has a log-normal distribution. There are also other models, see, for example, [**A-A**].

Schönbucher [**Scho**] concentrates on LIBOR-like models and assumes

---

[16]The $T_k$-forward measure is the risk neutral measure which has the risk-free bond with maturity $T_k$ as numeraire. For details see [**Bj**].

$$\frac{dF_k(t)}{F_k(t)} = \mu_k^F(t)\,dt + \boldsymbol{\sigma}_k^F \cdot d\mathbf{W}(t)$$
$$\frac{dS_k(t)}{S_k(t)} = \mu_k^S(t)\,dt + \boldsymbol{\sigma}_k^S \cdot d\mathbf{W}(t).$$

Here $\mathbf{W}$ denotes a $N$-dimensional standard Brownian motion, whereas $\boldsymbol{\sigma}_k$ are constant vectors and $\mu_k$ are adapted processes.

Alternatively, also the dynamics of $H$ could be specified and the dynamics of $S$ derived.

Since $H_k = S_k/(1+\delta_k F_k)$, we obtain

$$\begin{aligned} dH_k(t) &= \frac{1}{(1+\delta_k F_k)^2}\Big[(1+\delta_k F_k)S_k(\mu_k^S(t)\,dt + \boldsymbol{\sigma}_k^S \cdot d\mathbf{W}_t) \\ &\qquad - S_k\delta_k F_k(\mu_k^F(t)\,dt + \boldsymbol{\sigma}_k^F \cdot d\mathbf{W}_t) - S_k\delta_k F_k \boldsymbol{\sigma}_k^S \cdot \boldsymbol{\sigma}_k^F\,dt\Big] \\ &\quad - \frac{S_k}{(1+\delta_k F_k)^3}\delta_k^2 F_k^2 \sigma_k^F \cdot \sigma_k^F\,dt \\ &= \ldots dt + \frac{S_k}{1+\delta_k F_k}\Big[\boldsymbol{\sigma}_k^S - \frac{\delta_k F_k}{1+\delta_k F_k}\boldsymbol{\sigma}_k^F\Big] \cdot d\mathbf{W}_t \\ &=: H_k(t)\Big[\mu_k^H(t)\,dt + \boldsymbol{\sigma}_k^H(t) \cdot d\mathbf{W}_t\Big]. \end{aligned}$$

Note that $\boldsymbol{\sigma}_k^H$ is not a constant, but an adapted process with

$$\boldsymbol{\sigma}_k^H(t) = \boldsymbol{\sigma}_k^S - \frac{\delta_k F_k(t)}{1+\delta_k F_k(t)}\boldsymbol{\sigma}_k^F.$$

Using Itô's formula we obtain for the dynamics of the defaultable forward rates

$$\begin{aligned} d\bar{F}_k(t) &= dS_k(t) + dF_k(t) + d < S_k, F_k >_t \\ &= \Big[S_k\mu_k^S + F_k\mu_k^F + S_k F_k \boldsymbol{\sigma}_k^S \cdot \boldsymbol{\sigma}_k^F\Big]\,dt + \big(S_k\boldsymbol{\sigma}_k^S + F_k\boldsymbol{\sigma}_k^F\big) \cdot d\mathbf{W}_t \\ &=: \bar{F}_k(t)\Big[\mu_k^{\bar{F}}(t)\,dt + \boldsymbol{\sigma}_k^{\bar{F}}(t) \cdot d\mathbf{W}_t\Big]. \end{aligned}$$

The main reason for the popularity of the market models lies in the agreement between the model and well-established market formulas for basic derivative products. Therefore the model is usually calibrated to actual market data and afterwards used, for example, to price more complicated derivatives. For this reason the dynamics are directly modeled under the risk-neutral measure, or even more conveniently, under the $T_k$-forward measures. In search of something analogous for market models with credit risk, the $T_k$-survival measure turns up naturally. It is the measure under which the defaultable bond $\bar{B}_k(t)$ becomes a numeraire.

The $T_k$-*survival measure* $\bar{Q}_k$ is defined by the density

$$\bar{L}_k := \frac{\beta(t)1_{\{\tau > T_k\}}}{\bar{B}_k(0)} = \frac{d\bar{Q}_k}{dQ}.$$

Note that the density has $Q$-expectation 1 but becomes zero at default. In view of this, $\bar{Q}_k$ is not equivalent to $Q$ but only absolutely continuous w.r.t. $Q$.

At this point different changes of measures can be obtained. Changes from the survival to the forward measure and the analogy of the spot LIBOR measure in a credit risk context are also discussed in [**Scho**].

Finally, consider an $\mathcal{F}_T$-measurable claim $X_T$, which is paid only when $\tau > T$. Assuming zero recovery, then this claim can be valued by the following result, see [**B-R**]:

$$S_t = \bar{B}(t,T)\bar{E}_k\big(X_T\big|\mathcal{F}_t\big).$$

Here $\bar{E}_k$ denotes the expectation with respect to $\bar{Q}_k$.

## 7. Commercial Models

The models presented in this section, the so-called commercial models, are quite different from the models presented up to now. These models were developed by several companies and are widely accepted in practice. They all offer an implemented software, but the complete procedure of this implementation is published only for some models.

**7.1. The KMV Model (1995) - CreditMonitor.** The procedure of KMV is based on Merton's approach (see Section 1.1) and combines it with historical information via a statistical procedure.

KMV do not publish the exact procedure implemented in their software but the following illustrative example may be considered to be very close to their approach.

In Merton's model the firm value of the company was assumed to be observable. In reality this is unfortunately not the case. Usually shares of a company are traded but the real firm value is even difficult to estimate for internals. Using the traded shares as an estimate of the unknown firm value dates back to Modigliani and Miller, see [**C A N**], p. 142, for more information. The share is viewed as a call option on the firm value, where the exercise price is the level of the company's debt.

With the dynamic chosen as in Merton's model and denoting by $D$ the debt level at time $T$, the value of the shares $E$ corresponds to the Black-Scholes formula

$$E = V\,\Phi(d_1) - De^{-r(T-t)}\Phi(d_2),$$

where the constants $d_1, d_2$ are

$$\begin{aligned} d_1 &= \frac{\ln\frac{V}{De^{-r(T-t)}} + \frac{1}{2}\sigma^2(T-t)}{\sigma\sqrt{T-t}} \\ d_2 &= d_1 - \sigma\sqrt{T-t}. \end{aligned}$$

Inverting this relation results in the firm value. Also an estimate for the volatility of the share results in an estimate of the firm's value.

KMV found that in general firms do not default when their asset value reaches the book value of their total liabilities. This is due to the long-term nature of some of their liabilities which provides some breathing space. The *default point* therefore lies somewhere in between the total liabilities and the short-term (or current) liabilities. For this reason set

$$\text{default point} := \text{short-term debt} + 50\% \text{ long-term debt}.$$

In the next step they calculate the *distance-to-default*

$$DD = \frac{\text{firm value} - \text{default point}}{\text{firm value} \times \text{vola of firm value}}.$$

Finally KMV obtains the default probability from data on historical default and bankruptcy frequencies including over 250,000 company-years of data and over 4,700 incidents of bankruptcy[17].

**7.2. Moody's.** Besides Merton's approach, which is often stated as contingent claims analysis (CCA), there are statistical approaches, pioneered by Altman [**Al**], which predict default events using market information and accounting variables via econometric methods. Moody's public firm risk model bridges between these models and is therefore named a 'hybrid' model. The procedure, as described in [**S-K**], uses a variant of Merton's CCA as well as rating information (if available), certain reported accounting information and some macroeconomic variables to represent the state of the economy and of specific industries through logistic regression. On this basis they provide a one-year estimated default probability (EDP).

**7.3. CreditMetrics.** CreditMetrics was originally developed by J.P. Morgan and belongs to RiskMetrics Group since 1998. The procedure is totally published to clarify the model and the used data are provided in the Internet.

The target of CreditMetrics is a full valuation of a whole portfolio. This includes different assets and derivatives like loans, bonds, commitments to lend, financial letters-of-credit, receivables and market driven instruments like swaps, forwards and options.

The determination of the actual price of the portfolio proceeds in three steps. First the probability of a default is determined, second the probability of changes in rating (which directly results in a different price) and third the determination of the changes in value which are evoked by either a default or a change in rating.

For the three steps certain inputs are needed. They can be obtained by historical estimation or are observable in the market[18]:

- Transition matrices - transition probabilities for changes in rating,
- Recovery rates in default - ordered by seniority, countries and sectors,
- Risk-free yield curve,
- Credit spreads - for all maturities and ratings.

The transition matrices are also provided by Moody's and Standard & Poor's and therefore have to be listed separately (Moody's rates in eight and Standard & Poor's in 18 classes). In our example we consider the Table 1.

---

[17]See [**C-B**] for further information.

[18]See www.riskmetrics.com/products/data/datasets/creditmetrics.

TABLE 1

| Rating (now) | Rating in 1 year - Prob. in % | | | | | | | |
|---|---|---|---|---|---|---|---|---|
| | AAA | AA | A | BBB | BB | B | CCC | D |
| AAA | 90.81 | 8.33 | 0.68 | 0.06 | 0.12 | 0 | 0 | 0 |
| AA | 0.7 | 90.65 | 7.79 | 0.64 | 0.06 | 0.14 | 0.02 | 0 |
| A | 0.09 | 2.27 | 91.05 | 5.52 | 0.74 | 0.26 | 0.01 | 0.06 |
| BBB | 0.02 | 0.33 | 5.95 | 86.93 | 5.3 | 1.17 | 0.12 | 0.18 |
| BB | 0.03 | 0.14 | 0.67 | 7.73 | 80.53 | 8.84 | 1 | 1.06 |
| B | 0 | 0.11 | 0.24 | 0.43 | 6.48 | 83.46 | 4.07 | 5.2 |
| CCC | 0.22 | 0 | 0.22 | 1.3 | 2.38 | 11.24 | 64.86 | 19.79 |

Observe that there are some unusual figures in this table. For example, the probability that a company rated CCC is rated AAA after one year equals 0.22 %. This seems to be unusually high in comparison to the other entries. As there are few CCC ratings this seems to be a consequence of an exceptional event. Also critical is that the probability to default for a company rated AAA or AA equals zero. For sure there is a small but positive probability that such an event may happen. At this point smoothing algorithms are recommended to obtain a transition-matrix which is well suited for further calculations; see [**G-F-B**], pp.66-67.

For the second set of data, recovery rates are estimated on a historical basis. Usually this information is provided by rating agencies. There are some studies on recovery rates, and we discuss an example of Asarnov and Edwards [**A-E**]. CreditMetrics though uses just mean and standard deviation. The use of a beta distribution is discussed but not implemented.

The seniority of the bond certainly has a significant influence on the recovery rate. Table 2 illustrates this.

TABLE 2

| Seniority | mean (%) | SD (%) |
|---|---|---|
| Senior Secured | 53.80 | 26.86 |
| Senior Unsecured | 51.13 | 25.45 |
| Senior Subordinated | 38.52 | 23.81 |
| Subordinated | 32.74 | 20.18 |
| Junior Subordinated | 17.09 | 10.90 |

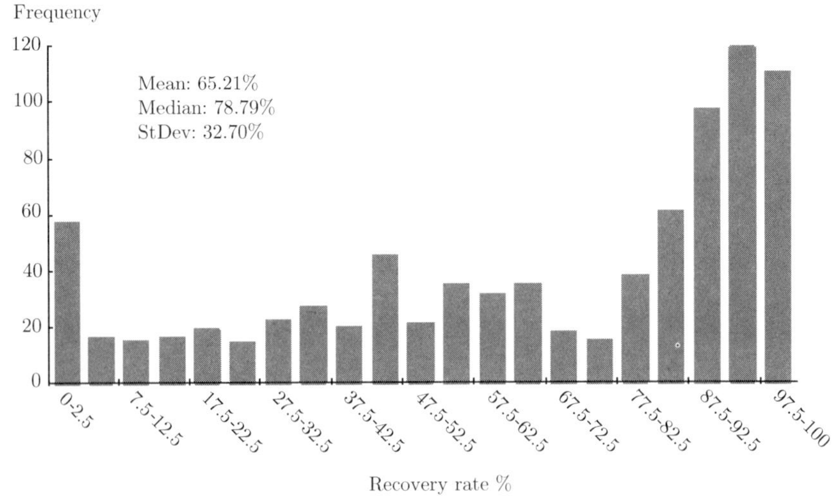


FIGURE 2. Recovery Rates

CreditMetrics also uses the actual term structure of interest rates and observable credit spreads. As the target is the valuation of bonds in a year's horizon not only default information should be used but also price changes due to rating changes. One needs to answer the question "What will be the value of a bond rated XXX in a year?". This is done by calculating stripped forward rates with respect to the rating. Stripping is the procedure to calculate zero coupon prices from a set of bonds offering coupons.

Assume for now that the current credit spreads do not change. The risk-free term structure provides forward rates and the current credit spreads are added to obtain the future (defaultable) forward-rates.

We show the full procedure in the context of an example. We face the problem to price a BBB-rated senior unsecured bond with maturity 5Y and annual coupons of 6%. Face value is 100 USD.

As described above one strips the bond prices to obtain the defaultable forward zero coupon curve. We want to explain this procedure in greater detail using the figures in Table 3.

Assume the bond has rating A at the end of the year. The forward value then becomes

$$FV = 6 + \frac{6}{1+3,72\%} + \frac{6}{(1+4.32\%)^2} + \frac{6}{(1+4.93\%)^3} + \frac{106}{(1+5.32\%)^4} = 108.64.$$

The other forward values are

TABLE 3

| Category | 1Y | 2Y | 3Y | 4Y (in %) |
|---|---|---|---|---|
| AAA | 3.60 | 4.17 | 4.73 | 5.12 |
| AA | 3.65 | 4.22 | 4.78 | 5.17 |
| A | 3.72 | 4.32 | 4.93 | 5.32 |
| BBB | 4.10 | 4.67 | 5.25 | 5.63 |
| BB | 5.55 | 6.02 | 6.78 | 7.27 |
| B | 6.05 | 7.02 | 8.03 | 8.52 |
| CCC | 15.05 | 15.02 | 14.03 | 13.52 |

| Rating | AAA | AA | A | BBB | BB | B | CCC |
|---|---|---|---|---|---|---|---|
| Forward Value($) | 109.35 | 109.17 | 108.64 | 107.53 | 102.01 | 98.09 | 83.63 |

The results may be found in Table 4.

TABLE 4

| State in 1Y | Prob. (%) | Forward Value | $(FV - \bar{FV})^2$ |
|---|---|---|---|
| AAA | 0.02 | 109.35 | 5.21 |
| AA | 0.33 | 109.17 | 4.42 |
| A | 5.95 | 108.64 | 2.48 |
| BBB | 86.93 | 107.53 | 0.21 |
| BB | 5.3 | 102.01 | 25.63 |
| B | 1.17 | 98.09 | 80.70 |
| CCC | 0.12 | 83.63 | 549.60 |
| Default | 0.18 | 51.13 | 3129.21 |
| | mean/ SD: | **107.07** | **8.94** |

The value at default is assumed to be the mean of historical recovery values for senior unsecured debt. In the above calculation we followed the CreditMetrics Technical Document. For the standard deviation they do not include the estimated standard deviation of the recovery rates. If this is incorporated (SD for senior unsecured debt = 25.45%, see the table on the previous page) one obtains a standard deviation of 10.11 which is considerably higher.

## 8. Credit Derivatives

In this section we introduce several types of derivatives that relate to credit risk. Unless explicitly mentioned, we assume that the protection seller has no default risk. In reality, strong correlations between protection seller and underlying prove to be quite dangerous. The protection seller might default shortly after the underlying and the protection becomes worthless.

Additionally to the derivatives presented in this section, there exist so-called *vulnerable options*. These are derivatives whose writer may default, thus facing a counterparty risk. They are considered, for example, in [**Am**] or [**B-R**]. We do not consider derivatives on large baskets like *collateralized debt obligations* or others. See [**B-O-W**] for more information.

**8.1. Credit Default Swaps and Options.** A *credit default swap* or a *credit default option* is an exchange of a fee for a contingent payment if a credit default event occurs. The fee is usually called *default swap premium*. The difference between swap and option is determined by the way the fee is paid. If the fee is paid up-front, the agreement is called option, while if the fee is paid over time, it is called swap[19].

The "default event" is not a precise notion. Quite contrary, the event, which triggers the payment, is negotiable. It could be a certain level of spread widening, occurrence of publicly available information of failure to pay or an event, that the partners can agree upon. See [**Da**] for examples of credit derivatives and the underlying contracts. Not surprisingly, terms of documentation risk or legal risk arise in the context of credit risk.

If the payoff is some predetermined constant, the derivative is called *digital*, for example default digital put or default digital swap.

There are also options on a basket which have specific features. For example, a first-to-default swap is based on a basket of underlyings, where the protection seller agrees to cover the exposure of the first entity triggering a default event. The first-to-default structure is similar to a collateralized bond or loan obligation. Usually there are bonds or loans with similar credit ratings in the basket, because otherwise the weakest credit would dominate the derivative's behavior.

Like in the interest rate case, there are options with early exercise possibility, called American, credit derivatives with knock-in/out features, options directly on the credit spreads or leveraged credit default structures, see [**T**]. Also reduced loss credit default options are mentioned therein, which yields a way to reduce the cost of default protection. In this contract the protection buyer still takes a fixed percentage of the loss on a default event, while the further loss is covered by the protection seller.

**8.2. Digital Options.** In the case of a digital swap or option the payment, which is exchanged if the default event occurs within the lifetime of the option, is

[19]See, for example, [**T**], p. 61.

fixed. Assume, for simplicity, that the payoff equals 1. There are two possibilities for the time, when the payoff is exchanged, either at maturity $T$ of the option or directly at default $\tau$:

1. If the payoff takes place at maturity, the price of the option (usually called put) at time $t$, if there was no default before $t$, equals[20]

$$\begin{aligned} 1_{\{\tau>t\}}P_d(t,T) &= 1_{\{\tau>t\}}\mathbb{E}_t\Big[\exp(-\int_t^T r_u\,du)1_{\{\tau\le T\}}\Big] \\ &= 1_{\{\tau>t\}}B(t,T)\,Q_t^T\big[\tau\le T\big]. \end{aligned}$$

This default digital put is closely related to a zero recovery bond, as

$$P_d(t,T)+B^0(t,T)=B(t,T), \qquad \forall t.$$

2. If the payoff is done at default, we obtain[21]

$$\begin{aligned} 1_{\{\tau>t\}}P_d(t,T) &= 1_{\{\tau>t\}}\mathbb{E}_t\Big[\exp(-\int_t^\tau r_u\,du)1_{\{\tau\le T\}}\Big] \\ &= 1_{\{\tau>t\}}\mathbb{E}_t\Big[\int_t^T \exp(-\int_t^s (r_u+\lambda_u)\,du)\lambda_s\,ds\Big] \\ &= 1_{\{\tau>t\}}\int_t^T B(t,s)\,\mathbb{E}_t^s\big[\exp(-\int_t^s \lambda_u\,du)\lambda_s\big]\,ds. \end{aligned}$$

REMARK 8.1. The payoff of the digital default put is similar to the payoff of the zero recovery bond. In fact, if we denote the defaultable bond with zero recovery and maturity $T$ by $B^0(\cdot,T)$, we obtain

$$\begin{aligned} 1_{\{\tau>t\}}P_d(t,T) &= 1_{\{\tau>t\}}\mathbb{E}_t\big[\exp(-\int_t^T r_u\,du)(1-1_{\{\tau>T\}})\big] \\ &= 1_{\{\tau>t\}}[B(t,T)-B^0(t,T)]. \end{aligned}$$

So, once the price of the zero recovery bond is known, the price of the default put can be easily calculated. Economically spoken, as a defaultable put and a zero recovery bond with same maturities guarantee the payoff 1, their price must be equal to the price of a risk-free bond, which is $B(t,T)$.

**8.3. Default Option and Default Swap.** To clarify the payments taking place for a *default option* or a *default swap*, consider figures 3 and 4. In the case of the default option, the protection buyer pays a fee up-front, which equals the price of the option. For the default swap the premium $S$ is paid at time points $t_1,\dots,t_n$ until either maturity of the contract or default.

There are several structural options for the default payment[22]:
*Difference to par*: If a default event occurs, the protection seller has either to pay the par value (which we always assume to be 1) in exchange for the defaulted bond,

---

[20]For convenience we write $\mathbb{E}_t(\cdot)$ for $\mathbb{E}^Q(\cdot|\mathcal{F}_t)$ and $\mathbb{E}_t^T(\cdot)$ for $E^{Q^T}(\cdot|\mathcal{F}_t)$, when $Q^T$ is the T-forward measure.

[21]See [**B-R**] Proposition 5.1.1.

[22]See, for example, [**Da**], p.63.

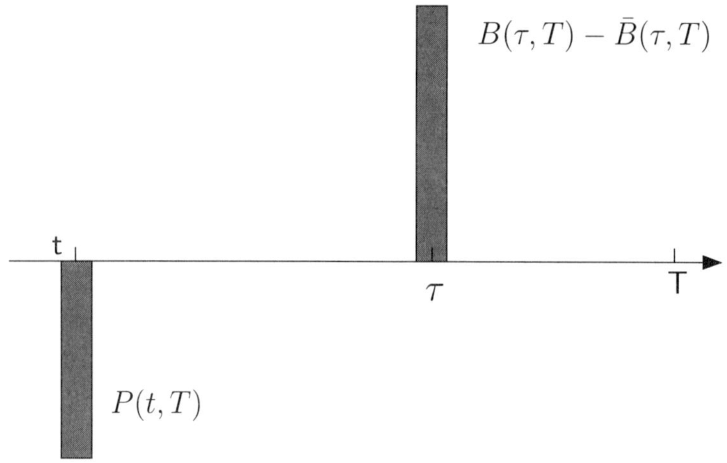


FIGURE 3. Cash flows for a default put. Default occurs at $\tau$ before the option expires. The payoff is agreed to be the "difference to an equivalent default-free bond", which is denoted by $B(\tau, T) - \bar{B}(\tau, T)$. The price of the default put is denoted by $P(t, T)$ and is paid initially at $t$.

or pay the par value minus the post-default price of the underlying bond. The payoff is equivalent to

$$1 - \bar{B}(\tau, T), \qquad \text{if } \tau \leq T.$$

*Difference to an equivalent bond*: The payoff in the case that a default event occurs is the value of an equivalent, default-free bond minus the market value of the defaulted bond. In this case the payoff equals

$$B(\tau, T) - \bar{B}(\tau, T), \qquad \text{if } \tau \leq T.$$

In the case of a coupon bond, there is usually a protection of the principal, and possibly of the accrued interest.

The first step in pricing the defaultable swap is the pricing of the defaultable option with the same payoff. The price of the option, denoted by $P(t, T)$, yields the discounted value of the payoff at time $t$. The premium $S$ is paid at times $t_1, \ldots, t_n$ until a default event occurs. Denoting the price of a zero recovery bond by $B^0(t, T)$, this yields

$$P(t, T) = \sum_{i=1}^{n} S \cdot B^0(t, T_i).$$

Consequently, the swap premium can be obtained, once the price of the defaultable option and the zero recovery bond prices are known, as

$$S(t) = \frac{P(t, T)}{\sum_{i=1}^{n} B^0(t, T_i)}. \tag{8.1}$$

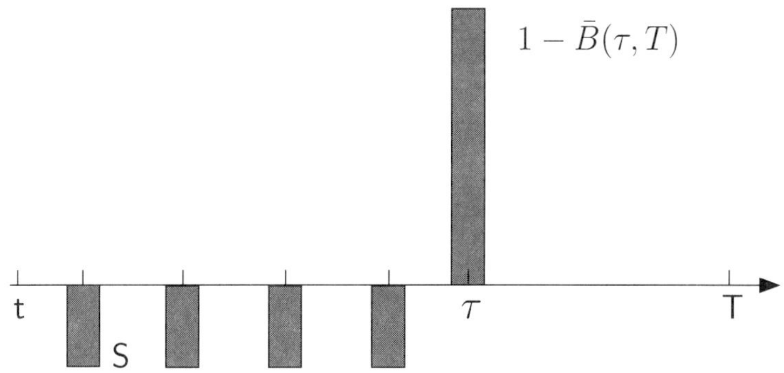


FIGURE 4. Cash flows for a credit default swap. Default occurs at $\tau$ before the option expires. The payoff is agreed to be the "difference to par", $1-\bar{B}(\tau,T)$. The default swap spread, $S$, is paid regularly at times $t_1,\ldots,t_4$ until default.

For example, if we assume recovery of treasury for the defaultable bond, we have

$$P(t,T)=\mathbb{E}_t\big[\exp(-\int_t^\tau r_u\,du)\,(1-\delta)1_{\{t<\tau\le T\}}\big],$$

which can be expressed using the default digital put as

$$P(t,T)=\mathbb{E}_t(1-\delta)P_d(t,T).$$

As already mentioned, this gets slightly more difficult if the underlying is a coupon bond, see [**S**] for details.

**8.4. Default Swaptions.** A *credit default swaption* offers the right, but not the obligation, to buy or sell a credit default swap at a future time point $T$ for a pre-specified swap premium $K$. The contract is knocked out if a default of the reference entity occurs before $T$. We refer to a *credit default swap call* (CDS call) if the assigned right is to buy a credit default swap and otherwise to a credit default put (CDS put). Credit default swaptions are not yet standard instruments which are liquidly traded, but, for example, Hull and White [**H-W1**] report that a market for such contracts is developing.

Denoting the tenor structure of the underlying swap by $\mathcal{T}=\{T_1,\ldots,T_n\}$ and the price of the CDS call at time $t$ by $C_S(t,T,\mathcal{T})$, we obtain for the payoff of the CDS call at maturity $T\le T_1$

$$C_S(T,T,\mathcal{T})=\big[\bar{S}(T)-K\big]^+\sum_{i=1}^n B^0(T,T_i)1_{\{\tau>T\}}.$$

$\bar{S}(T)$ is the swap rate at time $T$. For simplicity we set the day-count fraction to one[23].

[23]For a discussion on the different day-count fractions, see [**J-W**], p.51. With arbitrary day-count fraction $\Delta_i$ we would have to consider $\sum_{i=1}^n \Delta_i B^0(T,T_i)$.

If the swap offers the replacement of the difference to an equivalent default-free bond in the case of a default, the swap rate equals

$$\bar{S}(T) = \frac{B(T,T_n) - \bar{B}(T,T_n)}{\sum_{i=1}^{n} B^0(T,T_i)}.$$

We conclude for the price of the CDS call

$$C_S(0,T,\mathcal{T}) = \\ \mathbb{E}\Big[\exp(-\int_0^T r_u\,du)\Big(B(T,T_n) - \bar{B}(T,T_n) - K\sum_{i=1}^{n} B^0(T,T_i)\Big)^+ 1_{\{\tau > T\}}\Big].$$

Otherwise, if difference to par is considered, the swap price depends on the recovery. In a recovery of treasury model, the swap rate, as shown in the previous section, equals

$$\bar{S}(T) = \frac{(1-\delta)\,P_d(T,T_n)}{\sum_{i=1}^{n} B^0(T,T_i)}.$$

This yields that the price of the CDS call can be computed via

$$C_S(0,T,\mathcal{T}) \quad = \quad \mathbb{E}\Big[\exp(-\int_0^T r_u\,du)\Big((1-\delta)\,P_d(T,T_n) - K\sum_{i=1}^{n} B^0(T,T_i)\Big)^+\Big].$$

**8.5. Credit Spread Options.** A *credit spread option* is an option which depends on the credit spread, that is the difference between the yield of the underlying defaultable bond and the yield of a reference bond, which is usually assumed to be default-free. For example, a credit spread call with strike (yield) $K$ at maturity $T$ has the payoff

$$\Big(\bar{B}(T,T') - e^{-K(T'-T)}B(T,T')\Big)^+,$$

where $T' > T$ is the maturity of the underlying defaultable bond.

Thus the call is in the money if the yield of the defaultable bond is higher than the yield of the riskless bond plus the strike (yield) $K$. We use continuous compounding[24] of the yield rate, and note that this represents an annual yield, if the time scale is denoted in entities of 1 year.

Schmid [**S**] discusses credit spread options with a *knock-out* feature. In this case a credit spread call option with maturity $T$ on an underlying defaultable bond with maturity $T'$ and strike $K$, knocked out at default, has the payoff

$$1_{\{\tau > T\}}\Big(\bar{B}(T,T') - e^{-K(T'-T)}B(T,T')\Big)^+.$$

[24]The relation to the discrete time value of money concept is the following. The discounting factor for a time period of $T$ years are

$$\frac{1}{(1+y)^{nT}} = e^{-K\cdot T},$$

if the yield $y$ is paid $n$ times a year. This yields the relation

$$y = (\ln K)^{\frac{1}{n}}.$$

In contrast to the option-specific payoff, a credit spread swap with strike $K$ and maturity $T$ has the payoff

$$\bar{B}(T,T') - e^{-K(T'-T)}B(T,T').$$

To replicate the payoff of the credit spread swap, the seller buys a portfolio at time $t$, which consists of the defaultable bond with maturity $T'$ and sells $(1+K\cdot B(t,T))$ risk free bonds with maturity $T'$. A replicating argument yields the value at time $t$ of the above payoff to be $\bar{B}(t,T') - B(t,T')\exp[-K(T'-T)]$. Consequently, the credit spread swap premium, which has to be paid at times $t_1,\ldots,t_n$, equals

$$S = \frac{\bar{B}(t,T') - e^{-K(T'-T)}B(t,T')}{\sum_{i=1}^{n} B(t_i,T)}.$$

If the credit spread swap is knocked out at default of the underlying, the premium relates to zero recovery bonds $B^0(\cdot,T')$, which promise the par value, 1, if the reference bond $\bar{B}(\cdot,T')$ did not default until its maturity $T'$ and zero otherwise. Then the premium equals

$$S = \frac{\bar{B}(t,T') - e^{-K(T'-T)}B(t,T')}{\sum_{i=1}^{n} B^0(t_i,T)}.$$

**8.6. *k*th-to-default Options.** Derivatives with a *kth-to-default* feature are quite common in the market. For example, a first-to-default put covers the loss of the first defaulted asset in a considered portfolio. These types of products offer a cheaper protection against losses, if one considers more than $k$ assets to default in a certain time interval as unlikely, and therefore offer tailor-made credit risk profiles, which may be used to redistribute credit risk or release regulatory capital.

Once a price for a $k$th-to-default put is obtained, the premium of a $k$th-to-default swap can be calculated via formula (8.1). See Section 4 for applications, where we already obtained the following formula for the premium of a first-to-default swap

$$S_{1\text{st}} \quad = \quad \frac{(1-\delta)\mathbb{E}\big[\exp(-\int_t^{\tau^{1\text{st}}} r_u\,du)1_{\{\tau^{1\text{st}}<T\}}\big]}{\sum_{i=1}^{m}\mathbb{E}\big[\exp(-\int_0^{T_i} r_u\,du)1_{\{\tau^{1\text{st}}>T_i\}}\big]}.$$

## References


[Al] E. Altman, *Financial ratios, discriminant analysis and the prediction of corporate bankruptcy*, J. of Finance **23** (1968), pp. 589-609.

[Am] M. Ammann, *Pricing Derivative Credit Risk*, Lecture Notes in Econom. and Math. Systems vol. 470, Springer-Verlag, Berlin-Heidelberg-New York (1999).

[A-A] L. Andersen and J. Andreasen, *Volatility skews and extensions of the LIBOR market model*, Appl. Math. Finance **7** (2000) pp. 1-32.

[A-E] E. Asarnow and D. Edwards, *Measuring loss on defaulted bank loans: A 24-year study*, J. of Commer. Lending **77** (1995), pp. 11-23.

[B-R] T. Bielecki and M. Rutkowski, *Credit Risk: Modeling, Valuation and Hedging*, Springer-Verlag, Berlin-Heidelberg-New York (2002).

[Bj] T. Björk, *Interest Rate Theory*, Lecture Notes in Math., vol. 1656, Springer-Verlag, Berlin-Heidelberg-New York (1997).

[B-C] F. Black and J.C. Cox, *Valuing corporate securities: Some effects of bond indenture provisions*, J. of Finance **31** (1976), pp. 351-367.

[B-S] F. Black and M. Scholes, *The pricing of options and corporate liabilities*, J. of Political Economy **81** (1973), pp. 637-653.

[B-O-W] C. Blum, L. Overbeck and C. Wagner, *Credit Risk Modeling.*, Chapman & Hall/CRC, New-York-London, (2003).

[B-G-M] A. Brace, D. Gatarek and M. Musiela, *The market model of interest rate dynamics*, Math. Finance **7** (1995) pp. 127-155.

[Br] P. Brémaud, *Point Processes and Queues*, Springer-Verlag, Berlin-Heidelberg-New York, (1981).

[B-M] D. Brigo and F. Mercurio, *Interest Rate Models - Theory and Practice*, Springer-Verlag, Berlin-Heidelberg-New York, (2001).

[Bu] E. Buffett, *Credit risk: The structural approach revisited*, Progr. Probab. **52** (2002), pp. 45-53.

[C-A-N] J.B. Caouette, E.I. Altmann and P. Narayanan, *Managing Credit Risk*, John Wiley & Sons, New York, (1998).

[C-I-R] J.C. Cox, J.W. Ingersoll and S.A. Ross, *A theory of the term structure of interest rates*, Econometrica **54** (1985), pp. 385-407.

[C-R-R] J.C. Cox., S.A. Ross and M. Rubinstein, *Option pricing: A simplified approach*, J. of Financial Econometrics **7** (1979), pp. 229-265.

[C-B] P. Crosbie and J.R. Bohn, *Modeling Default Risk*, KMV Corporation [http://www.kmv.kom/insight/index.html], (2001).

[Da] S. Das, *Credit Derivatives*, John Wiley & Sons, New York, (1998).

[D-T] S. Das and P. Tufano, *Pricing credit-sensitive debt when interest rates, credit ratings and credit spreads are stochastic*, J. of Financial Eng. **5** (1996), pp. 161-198.

[Du] G. Duffee, *Estimating the price of default*, Rev. of Financial Stud. **12** (1999), pp. 187-226.

[D-L] D. Duffie and D. Lando, *Term structures of credit spreads with incomplete accounting information*, Econometrica **69** (2001), pp. 633-664.

[D-S] D. Duffie and K. Singleton, *Modeling term structures of defaultable bonds*, Rev. of Financial Stud. **12** (1999), pp. 687-720.

[F-M] R. Frey and A.J. McNeil, *Modelling dependent defaults*, Working paper, (2001).

[Ge-J] R. Geske and H.E. Johnson, *The valuation of corporate liabilities as compound options: A correction*, J. of Financial and Quant. Anal. **19** (1984), pp. 231-232.

[Gi-J] R. Gill and S. Johannsen, *A survey of product-integration with a view towards applications in survival analysis*, Ann. Statist. **18** (1990), pp. 1501-1555.

[Go] R. Goldstein, *The term structure of interest rates as a random field*, Working paper, (1997).

[Gr] J. Grandell, *Mixed Poisson Processes*, Monogr. Statist. Appl. Probab. vol. 77, Chapman & Hall, London (1997).

[G-F-B] G.M. Gupton, C. Finger and M. Bahtia, *CreditMetrics: Technical document*, J.P.Morgan & Incorporated, New York [http://www.riskmetrics.com/ research], (1997).

[H-P] J.M. Harrison and S.R. Pliska, *Martingales and stochastic integrals in the theory of continuous trading*, Stochastic Process. Appl. **11** (1981), pp. 215-260.

[H-J-M] D. Heath, R.A. Jarrow and A.J. Morton, *Bond pricing and the term structure of interest rates*, Econometrica **60** (1992), pp. 77-105.

[H-W] J.C. Hull and A. White, *Pricing interest-rate derivative securities*, Rev. of Financial Stud. **3** (1990), pp. 573-592.

[H-W1] J.C. Hull and A. White, *The valuation of credit default swap options*, Working paper, (2002).

[I-R-W] R.B. Israel, J.S. Rosenthal and J.Z. Wei, *Finding generators for markov chains via empirical transition matrices, with applications to credit ratings*, Math. Finance **11** (2001), pp. 245-265.

[J-W] J. James and N. Webber, *Interest Rate Modelling*, John Wiley & Sons, New York, (2000).

[Ja] F. Jamshidian, *Libor and swap market models and measures*, Finance and Stoch. **1** (1997), pp. 293-330.

[J-L-T] R. Jarrow, D. Lando and S. Turnbull, *A Markov model for the term structure of credit risk spreads*, Rev. of Financial Stud. **10** (1997), pp. 481-523.

[J-T] R. Jarrow and S. Turnbull, *Pricing options on financial securities subject to default risk*, J. of Finance **5** (1995), pp. 53-86.

[J-T1] R. Jarrow and S. Turnbull, *The intersection of market and credit risk*, J. of Banking and Finance **24** (2000), pp. 271-299.

[J-Y] R. Jarrow and F. Yu, *Counterparty risk and the pricing of defaultable securities*, J. of Finance **56** (2001), pp. 1765-1799.

[Je] M. Jeanblanc, *Credit risk*, Credit Risk - Munich Spring School, (2002).

[J-M-R] E. Jones, S. Mason and E. Rosenfeld, *Contingent claim analysis of corporate capital structures: An empirical investigation*, J. of Finance **39** (1984), pp. 611-625.

[K] M. Kijima, *Valuation of a credit swap of the basket type*, Rev. Deriv. Res. **4** (2000), pp. 81-97.

[K-M] M. Kijima and Y. Muromachi, *Credit events and the valuation of credit derivatives of basket type*, Rev. Deriv. Res. **4** (2000), pp. 55-79.

[K-R-S] I.J. Kim, K. Ramaswamy and S. Sundaresan, *The valuation of corporate fixed income securities*, Working paper, Wharton School, University of Pennsylvania, (1993).

[La] D. Lando, *Three Essays on Contingent Claim Pricing*, Ph. D. thesis, Cornell University, (1994).

[La1] D. Lando, *On Cox processes and credit risky securities*, Rev. Deriv. Res. **2** (1998), pp. 99-120.

[Le] F. Lehrbass, *Defaulters get intense*, Risk - Credit Risk Supplement July (1997), pp. 56-59.

[Li] D. Li, *On default correlation: a copula function approach*, J. of Fixed Income **9** (2000), pp. 43-54.

[L-S] F. Longstaff and E. Schwartz, *A simple approach to valuing risky fixed and floating rate debt*, J. of Finance **50** (1995), pp. 789-819.

[M-U] D.B. Madan and H. Unal, *Pricing the risks of default*, Rev. Deriv. Res. **2** (1998), pp. 121-160.

[M-B] S.P. Mason and S. Bhattacharya, *Risky debt, jump processes, and safety covenants*, J. of Financial Economics **9** (1981), pp. 281-307.

[Me] R. Merton, *On the pricing of corporate debt: the risk structure of interest rates*, J. of Finance **29** (1974), pp. 449-470.

[N] R.B. Nelsen, *An introduction to copulas*, Lecture Notes in Statist., vol. 139, Springer-Verlag, Berlin-Heidelberg-New York (1999).

[N-S-S] T.N. Nielsen, J. Saà-Requejo and P. Santa-Clara, *Default risk and interest rate risk: The term structure of default spreads*, Working paper, (1993).

[Pr] P. Protter, *Stochastic Integration and Differential Equations*, Springer-Verlag, Berlin-Heidelberg-New York, (1992).

[Py] G. Pye, *Gauging the default premium*, Financial Anal. J. **30** (1974), pp. 49-52.

[R] R. Rebonato, *Interest-rate option models*, John Wiley & Sons, New York (1996).

[S] B. Schmid, *Pricing Credit Linked Financial Instruments*, Lecture Notes in Econom. and Math. Systems vol. 516, Springer-Verlag, Berlin-Heidelberg-New York (2002).

[Sc] W.M. Schmidt, *On a general class of one-factor models for the term structure of interest rates*, Finance and Stochastics **1** (1997), pp. 3-24.

[S-W] W.M. Schmidt and I. Ward, *Pricing default baskets*, Risk January (2002), pp. 111-114.

[Scho] P. Schönbucher, *A LIBOR market model with default risk*, Working paper (2000).

[S-T-D] D. Shimko, N. Tejima and D. van Deventer, *The pricing of risky debt when interest rates are stochastic*, J. of Fixed Income **3** (1993), pp. 58-66.

[S-K] J. Sobehart and R. Klein, *Moody's public firm risk model: A hybrid approach to modeling short term default risk*, Moody's investors service, [http://riskcalc.moodysrms.com/us/research/crm/53853.asp], (2000).

[T] J. Tavakoli, *Credit Derivatives*, John Wiley & Sons, New York (1998).

[V] O. Vasiček, *An equilibrium characterization of the term structure*, J. of Financial Economics **5** (1977), pp. 177-188.

[V1] O. Vasiček, *Credit valuation*, Working paper, KMV Corporation, (1984).

[W-G] D.G Wei and D. Guo, *Pricing risky debt: An empirical comparison of the Longstaff and Schwartz and Merton models*, J. of Fixed Income **7** (1991), pp. 8-28.

[Z] C. Zhou, *A jump-diffusion approach to modeling credit risk and valuing defaultable securities*, Finance and Economics Discussion Paper Series 1997/15, Board of Governors of the Federal Reserve System., (1997).

Mathematical Institute, University of Giessen, Arndtstr. 2, 35392 Giessen, Germany.
*E-mail address*: Thorsten.Schmidt@math.uni-giessen.de

Mathematical Institute, University of Giessen, Arndtstr. 2, 35392 Giessen, Germany.
*E-mail address*: Winfried.Stute@math.uni-giessen.de

# Research Papers

Contemporary Mathematics
Volume **336**, 2003

# Optimal Investment in Incomplete Financial Markets with Stochastic Volatility

Netzahualcóyotl Castañeda–Leyva and Daniel Hernández-Hernández

ABSTRACT. Our goal is to solve an optimal investment problem in the context of an incomplete financial market. The market model is a generalization of the Black and Scholes one, consisting of a bank account, a risky asset, and a stochastic correlated external factor. The coefficients of the model: interest rate, rate return and volatility, depend on the external factor. The problem is solved using both, the martingale method and stochastic control techniques. The associated *dual* problem is presented and the relationship between the optimal solutions to the primal and dual problems is studied. Finally, explicit solutions are given when the utility function is logarithmic and HARA.

## 1. Introduction

Since the fundamental work of Black and Scholes to valuate European options, their model became a cornerstone in the development and study of many other problems in mathematical finance. In recent years, different generalizations of this classical model have been studied, trying to model more precisely the dynamics of the asset prices. In this paper we model the asset price as a diffusion process, with volatility and drift coefficients depending on correlated economic factors. Empirical works (see [**F-P-S**]) have verified that this model captures some of the qualitative effects on the distributions of the stock prices, like asymmetric heavy tails, while the economic factor have typically an ergodic behavior. In particular, the case when the economic factor is modelled as an Ornstein-Uhlenbeck process is covered in this paper. Another point of view to model the volatility is given in [**B-S**], in a framework of Levy processes.

The goal of this work is to solve an optimal investment portfolio problem on a finite time horizon. More precisely, consider a single investor trying to maximize the expected *utility* of *terminal wealth* over the set of admissible portfolio processes, as well as to find an optimal strategy. We assume that the financial market is

2000 *Mathematics Subject Classification.* Primary: 60J70, 91B28, 93E20.

*Key words and phrases.* Stochastic volatility, incomplete markets, optimal final wealth, martingale method, Black & Scholes model.

The research of the first author was supported by Universidad Autónoma de Aguascalientes and PROMEP, under Ph.D. grant UAAGS-69. The research of the second author was supported by Conacyt grant 37643-E.

composed by a bank account, a risky asset, and an external correlated factor. The dynamics of the risky asset price and the external factor are modelled as diffusion processes. The external factor affects the coefficients of the model, in particular the volatility. Since the external factor is not traded, this problem has the particular difficulty that the market is incomplete.

This problem will be solved using the martingale approach. The original primal problem is solved once the solution to the associated *dual* problem is obtained. In this case, the dual problem turns out to be equivalent to an optimal stochastic control problem, where the control processes are identified with a subset of the *equivalent local martingale measures*. When the utility function is HARA, this stochastic optimal control problem is solved using dynamic programming arguments, characterizing the corresponding value function as the unique solution, in a suitable class, of the associated Hamilton-Jacobi equation, obtaining also an optimal control. Using a power transformation, previously introduced by Zariphopoulou [**Z**], a Feynman-Kac representation of the value function is obtained.

Another way to study this problem is to use stochastic control techniques to solve the optimal investment portfolio problem directly. Zariphopoulou [**Z**] elaborates on this direction, while in Fleming and Hernández-Hernández [**F-H**] an optimal consumption problem is solved using this approach.

The martingale method goes back to the fundamental contribution by Harrison and Pliska [**H-P**], and it has became a standard approach to study optimal wealth and/or consumption problems. This method is especially powerful when the financial market is incomplete. Karatzas and Shreve [**Ka-S2**] studied these kind of problems using this approach for diffusion models. Kramkov and Schachermayer [**Kr-S**] solved similar problems when the prices are driven by semimartingales. In these works it is shown, under suitable conditions, that there is no duality *gap* between the primal and dual problems. However, explicit optimal solutions are not presented in general, except when the utility function is logarithmic or the coefficients are deterministic.

The paper is organized as follows. In Section 2 the model is established, while the primal problem and its characterization are given in Section 3. These results turn out to be important to write down the original problem as a convex optimization one. In Section 4, the martingale method is explained, and we give a practical condition to show that there is no duality gap. Furthermore, a relevant relationship between the optimal solutions of both problems is obtained. Finally, in Section 5 closed form solutions, when the utility function is logarithmic and HARA, are presented.

NOTATION. For $k \geq 0$, $C^k(\mathbf{R})$ denotes the class of real functions with continuous $k$-th derivative. Whereas $C_b^k(\mathbf{R})$ $[C_p^k(\mathbf{R})]$ is the class of real functions in $C^k(\mathbf{R})$, such that $f^{(j)}$ is bounded [polynomial growing]; $0 \leq j \leq n$. Similar notation is applied for other Euclidean spaces. For $f \in C_b(\mathbf{R})$, its supremum norm is denoted by $|f|_\infty$. The supremum [infimum] of $D \subset \mathbf{R}$ is denoted by $\max D$ [$\min D$]. While, given a collection of nonnegative random variables $\mathcal{X}$, $\sup \mathcal{X}$ denotes its *essential* supremum.

## 2. The Model

Let $\{(W_{1t}, W_{2t}), \mathcal{F}_t\}_{0 \leq t \leq T}$ be a standard two-dimensional Brownian motion $(BM)$ in a complete probability space $(\Omega, \mathcal{F}, P)$, where $\mathcal{F} \doteq \mathcal{F}_T$ and $\{\mathcal{F}_t\}_{0 \leq t \leq T}$ is the augmentation of the filtration $\{\mathcal{F}_t^{(W_1,W_2)}\}_{0\leq t\leq T}$. Consider a financial market governed by this BM, consisting of a bank account, a *risky* asset, and a correlated external factor, such that, for $t\in[0,T]$:

(1) The bank account process is given by $S_t^0 \triangleq \exp\left(\int_0^t r(Y_u)\,du\right)$.

(2) The asset price process $S$ is assumed to satisfy the following stochastic differential equation ($SDE$)

$$dS_t = S_t\left[\mu(Y_t)\,dt + \sigma(Y_t)\,dW_{1t}\right] \quad \text{with} \quad S_0 = 1. \tag{2.1}$$

(3) The dynamics of the *external factor* $Y$ is modelled as a diffusion process solving the SDE

$$dY_t = g(Y_t)\,dt + \beta\left[\rho dW_{1t} + \varepsilon dW_{2t}\right] \quad \text{with} \quad Y_0 = y \in \mathbf{R}, \tag{2.2}$$

where $|\rho|\leq 1$, $\varepsilon \triangleq \sqrt{1-\rho^2}$, and $\beta\neq 0$.

This financial market is *incomplete*, except when $\rho=\pm 1$, since the external factor cannot be traded. On the other hand, without loss of generality, we take $\beta=1$.

ASSUMPTION A. 1. $\mu(\cdot), r(\cdot), \theta(\cdot)\in C_b^2(\mathbf{R})$, where

$$\theta(y) \triangleq \frac{\mu(y)-r(y)}{\sigma(y)}; \quad y\in\mathbf{R}.$$

2. $0<\sigma(\cdot)\in C^1(\mathbf{R})$ and $g(\cdot)\in C^1(\mathbf{R})$, with $g'(\cdot)\in C_b(\mathbf{R})$.

In this context, a single investor generates a *wealth process* $X$ with *initial capital* $x$, spliting at each time $t\in[0,T]$ his capital $X_t$ between $\pi_t$ and $X_t-\pi_t$, where $\pi_t$ is the net amount allocated in the risky asset. Then, the fluctuations of the wealth process are described by the following difference equation

$$\triangle X_t = (X_t-\pi_t)\frac{\triangle S_t^0}{S_t^0} + \pi_t\frac{\triangle S_t}{S_t}; \quad t\in[0,T], \quad \text{with} \quad X_0 = x>0.$$

These concepts are formalized in the following definition.

DEFINITION 2.1. The real process $\{\pi_t,\mathcal{F}_t\}_{0\leq t\leq T}$ is a **portfolio** if it is progressively measurable and $\int_0^T \pi_u^2 du<\infty$ almost sure ($a.s.$). Its associated **wealth** process $X^\pi \triangleq X^{x,y,\pi}$ is the solution of the integral equation

$$X_t^\pi \triangleq x + \int_0^t \left[r(Y_u)X_u^\pi + (\mu(Y_u)-r(Y_u))\pi_u\right]du + \int_0^t \pi_u\sigma(Y_u)\,dW_{1u}. \tag{2.3}$$

We say that $\pi$ is **admissible** if $X^\pi$ satisfies the *state constraint* $X^\pi\geq 0$, and the set of such processes is denoted by $\mathcal{A}(x,y)$.

Finally, the investor's objective is to

$$\text{maximize} \quad \{EU(X_T^\pi)\} \quad \text{over} \quad \pi\in\mathcal{A}(x,y),$$

and provide the *optimal* portfolio $\hat{\pi}$, where $U:\mathbf{R}_+\to\mathbf{R}$ is a *utility* function; which captures the investor's attitude about risk. This problem will be referred as the **primal** problem. Throughout the paper the initial values $x\in\mathbf{R}_+\triangleq(0,\infty)$, $y\in\mathbf{R}$, and the terminal time $T>0$ are fixed, unless the opposite is stated.

## 3. Primal Problem

In this section the optimal investment problem is represented as a convex optimization problem. The first step will be to obtain a characterization of the convex family of admissible processes $\mathcal{A}(x,y)$, in terms of another convex family of nonnegative random variables.

Let $\mathcal{M}(y)$ be the set of progressively measurable processes $\{\nu_t, \mathcal{F}_t\}_{t\in[0,T]}$, with $E\int_0^T \nu_u^2 du < \infty$, such that

$$Z_t^\nu \mathrel{\hat=} \exp\left(-\int_0^t [\theta(Y_u)\, dW_{1u} + \nu_u dW_{2u}] - \frac{1}{2}\int_0^t [\theta^2(Y_u) + \nu_u^2]\, du\right), \tag{3.1}$$

is a $\mathcal{F}_t$-martingale. Note that the bounded processes $\nu$ belong to $\mathcal{M}(y)$, since the function $\theta(\cdot)$ is bounded. For each $\nu \in \mathcal{M}(y)$ the following probability measure can be associated

$$dP^\nu \mathrel{\hat=} Z_T^\nu dP; \quad \text{in} \quad \mathcal{F}_T. \tag{3.2}$$

Note that the measures $P^\nu$ and $P$ are equivalent, and $Z_t^\nu \mathrel{\hat=} dP^\nu/dP|_{\mathcal{F}_t}$; $t \in [0,T]$. Under the measure $P^\nu$ the two-dimensional process $\{(W_{1t}^\nu, W_{2t}^\nu), \mathcal{F}_t\}_{0\le t\le T}$, defined as

$$W_{1t}^\nu \mathrel{\hat=} W_{1t} + \int_0^t \theta(Y_u)\, du \quad \text{and} \quad W_{2t}^\nu \mathrel{\hat=} W_{2t} + \int_0^t \nu_u du, \tag{3.3}$$

is also a BM. Moreover, the dynamics of the processes defined above can be written as

$$dZ_t^\nu = Z_t^\nu \left([\theta^2(Y_t) + \nu_t^2]\, dt - \theta(Y_t)\, dW_{1t}^\nu - \nu_t dW_{2t}^\nu\right), \tag{3.4}$$

$$dY_t = [g(Y_t) - \rho\theta(Y_t) - \varepsilon\nu_t]\, dt + \rho dW_{1t}^\nu + \varepsilon dW_{2t}^\nu, \tag{3.5}$$

while the *discounted* price and wealth processes satisfy

$$\begin{aligned} d\frac{S_t}{S_t^0} &= \frac{S_t}{S_t^0}\sigma(Y_t)\, dW_{1t}^\nu, \\ d\frac{X_t^\pi}{S_t^0} &= \frac{\pi_t}{S_t^0}\sigma(Y_t)\, dW_{1t}^\nu; \quad \pi \in \mathcal{A}(x,y). \end{aligned} \tag{3.6}$$

REMARK 3.1. The above imply the following: 1. For $\nu \in \mathcal{M}(y)$, the discounted price process $S/S^0$ is a $P^\nu$-continuous local martingale.
2. The discounted wealth process $X^\pi/S^0$ is a $P^\nu$-continuous local martingale and, by Fatou's lemma, it is also a $P^\nu$-supermartingale.

The following proposition will be useful to solve the dual optimization problem. It is similar to Theorem 5.6.2 in [**Ka-S2**] and, in fact, some parts of their proof are quoted. An analogous version for a consumption problem can be found in [**C**]. Condition (3.7) below is usually referred as *budget* constraint.

PROPOSITION 3.2. *For each $\mathcal{F}_T$-measurable random variable $B \ge 0$ with*

$$\max_{\nu\in\mathcal{M}(y)} E^\nu \frac{B}{S_T^0} \le x, \tag{3.7}$$

*there exists $\pi \in \mathcal{A}(x,y)$ such that $X_T^\pi \ge B$ a.s.*
*Conversely, if $\pi \in \mathcal{A}(x,y)$, then $B \mathrel{\hat=} X_T^\pi$ satisfies the budget constraint (3.7).*

PROOF. The last part of the theorem is straightforward, since, from Remark 3.1, when $\pi \in \mathcal{A}(x,y)$ and $\nu \in \mathcal{M}(y)$, the discounted process $X^\pi/S^0$ is a $P^\nu$-supermartingale. Hence $E^\nu X_T^\pi/S_T^0 \leq E^\nu X_0^\pi/S_0^0 = x$.

Now, to show the first part, define the following nonnegative process:

$$\frac{\check{X}_t}{S_t^0} \triangleq \sup_{\nu\in\mathcal{M}(y)} E^\nu\left[\frac{B}{S_T^0} \mid \mathcal{F}_t\right]; \quad t\in[0,T]. \tag{3.8}$$

Note that

$$\check{X}_0 = \max_{\nu\in\mathcal{M}(y)} E^\nu \frac{B}{S_T^0} \leq x \quad \text{and} \quad \check{X}_T \equiv B. \tag{3.9}$$

It will be shown that the process in (3.8) induces an admissible portfolio $\pi$ such that the associated final wealth $X_T^\pi$ is greater than or equal to $B$. First, it will be verified that $\check{X}$ satisfies the following *dynamic programming equation* (*DPE*):

$$\frac{\check{X}_s}{S_s^0} = \sup_{\nu\in\mathcal{M}(y)} E^\nu\left[\frac{\check{X}_t}{S_t^0} \mid \mathcal{F}_s\right]; \quad 0\leq s\leq t\leq T. \tag{3.10}$$

Since

$$E^\nu\left[\frac{B}{S_T^0} \mid \mathcal{F}_s\right] = E^\nu\left[E^\nu\left(\frac{B}{S_T^0} \mid \mathcal{F}_t\right) \mid \mathcal{F}_s\right] \leq E^\nu\left[\frac{\check{X}_t}{S_t^0} \mid \mathcal{F}_s\right],$$

then

$$\frac{\check{X}_s}{S_s^0} = \sup_{\nu\in\mathcal{M}(y)} E^\nu\left[\frac{B}{S_T^0} \mid \mathcal{F}_s\right] \leq \sup_{\nu\in\mathcal{M}(y)} E^\nu\left[\frac{\check{X}_t}{S_t^0} \mid \mathcal{F}_s\right].$$

To obtain the reverse inequality, we shall verify that

$$\frac{\check{X}_s}{S_s^0} \geq E^\nu\left[\frac{\check{X}_t}{S_t^0} \mid \mathcal{F}_s\right]; \quad \text{for all} \quad \nu\in\mathcal{M}(y). \tag{3.11}$$

Given $\nu\in\mathcal{M}(y)$ and $t\in[0,T]$, define $\mathcal{M}^\nu(t,y) \triangleq \{\eta\in\mathcal{M}(y) : \eta\equiv\nu \text{ in } [0,t]\}$ and

$$J_t^\eta \triangleq E^\eta\left[\frac{B}{S_T^0} \mid \mathcal{F}_t\right] = E\left[\frac{Z_T^\eta}{Z_t^\eta}\frac{B}{S_T^0} \mid \mathcal{F}_t\right]; \quad \eta\in\mathcal{M}^\nu(t,y).$$

The second equality in the last expression is due to Bayes' formula for conditional expectations (see equation III.3.9 in [**J-S**] or Lemma 3.5.3 in [**Ka-S1**]). On the other hand, since $Z_T^\eta/Z_t^\eta$ depends only on the values of $\nu$ in $[t,T]$, then

$$\frac{\check{X}_t}{S_t^0} = \sup_{\eta\in\mathcal{M}^\nu(t,y)} J_t^\eta.$$

Furthermore, it is not difficult to check that

$$\frac{\check{X}_t}{S_t^0} = \lim_{n\to\infty} J_t^{\eta_n},$$

for some increasing sequence $\{J_t^{\eta_n}\}_{n\geq1}$ with $\eta_n\in\mathcal{M}^\nu(t,y)$. See equation (5.6.13) in [**Ka-S2**]. Therefore, from the conditional monotone convergence theorem, inequality (3.11) holds if

$$\frac{\check{X}_s}{S_s^0} \geq E^\nu\left[J_t^{\eta_n} \mid \mathcal{F}_s\right] = E^\nu\left[E^{\eta_n}\left(\frac{B}{S_T^0} \mid \mathcal{F}_t\right) \mid \mathcal{F}_s\right]; \quad n\geq 1. \tag{3.12}$$

However, we have

$$\frac{\check{X}_s}{S_s^0} \geq E^{\eta_n}\left[\frac{B}{S_T^0} \mid \mathcal{F}_s\right] = E^{\eta_n}\left[E^{\eta_n}\left(\frac{B}{S_T^0} \mid \mathcal{F}_t\right) \mid \mathcal{F}_s\right] = E^{\nu}\left[E^{\eta_n}\left(\frac{B}{S_T^0} \mid \mathcal{F}_t\right) \mid \mathcal{F}_s\right].$$

In consequence, from the DPE (3.10), the process $\check{X}/S^0$ is a $P^\nu$-supermartingale; for each $\nu \in \mathcal{M}(y)$. Now, using the Doob-Meyer supermartingale decomposition theorem and the local martingale representation theorem, the above discounted process can be written as

$$\frac{\check{X}_t}{S_t^0} = \check{X}_0 + \int_0^t [\psi_{1s}^\nu dW_{1s}^\nu + \psi_{2s}^\nu dW_{2s}^\nu] - A_t^\nu, \tag{3.13}$$

where $\psi_1^\nu$, $\psi_2^\nu$, and $A^\nu$ are progressively measurable real processes, such that a.s. $\int_0^T ([\psi_{1s}^\nu]^2 + [\psi_{2s}^\nu]^2) ds < \infty$ and $A^\nu$ is predictable increasing with $A_0^\nu \equiv 0$. See Theorem 3.3.9 in [**L-S**] and Problem 3.4.16 in [**Ka-S1**]. Thus, from (3.13), the following identity holds

$$\int_0^t [\psi_{1s}^\nu dW_{1s}^\nu + \psi_{2s}^\nu dW_{2s}^\nu] - A_t^\nu = \int_0^t [\psi_{1s}^0 dW_{1s}^0 + \psi_{2s}^0 dW_{2s}^0] - A_t^0.$$

According to (3.3), we get

$$\begin{aligned} 0 &= \int_0^t \left[\left(\psi_{1s}^\nu - \psi_{1s}^0\right) dW_{1s} + \left(\psi_{2s}^\nu - \psi_{2s}^0\right) dW_{2s}\right] + A_t^0 - A_t^\nu \\ &\quad + \int_0^t \left[\left(\psi_{1s}^\nu - \psi_{1s}^0\right) \theta(Y_s) + \psi_{2s}^\nu \nu_s\right] ds. \end{aligned}$$

This equation has the implicit form $M + V + \phi \equiv 0$, where $M$ is a continuous local martingale, $V$ is a predictable finite variation process, and $\phi$ is a continuous process with zero quadratic variation, such that $M_0 \equiv V_0 \equiv \phi_0 \equiv 0$. The above suggests that all those terms should be the zero process. In fact, by Proposition I.4.49.d in [**J-S**], the *covariation* $\langle M, V\rangle$ is identically zero. Thus, $0 = \langle \phi, \phi \rangle = \langle M + V, M + V\rangle = \langle M\rangle + \langle V\rangle + 2\langle M, V\rangle = \langle M\rangle + \langle V\rangle$. Hence $\langle M\rangle = \langle V\rangle = 0$. That means $\psi_1^\nu \equiv \psi_1^0$, $\psi_2^\nu \equiv \psi_2^0$, and $A^\nu \equiv A^0 + \int_0^\cdot \psi_{2s}^0 \nu_s ds \geq 0$ a.s. In particular, for constant processes $\nu \equiv v \in \mathbf{R}$, $A_t^v = A_t^0 + v \int_0^t \psi_{2s}^0 ds$ with probability one. Now, define the events in $\mathcal{F}_t$

$$\Psi_t^- \doteq \left\{\int_0^t \psi_{2s}^0 ds < 0\right\} \quad \text{and} \quad \Psi_t^+ \doteq \left\{\int_0^t \psi_{2s}^0 ds > 0\right\}.$$

Noting that, when $v \to \infty$ [$v \to -\infty$], then $A_t^v \to -\infty$ in $\Psi_t^-$ [$A_t^v \to -\infty$ in $\Psi_t^+$], one concludes that $\Psi_t^- \cup \Psi_t^+ = \emptyset$. That is, $A_t^v = A_t^0$ a.s. Hence, $A^v \equiv A^0$ and $\psi_2^0 \equiv 0$; since $A^v$ is *cadlag*. Summarizing:

$$\psi \doteq \psi_1^0 \equiv \psi_1^\nu, \quad \psi_2^\nu \equiv 0, \quad \text{and} \quad A^\nu \equiv A^0; \quad \nu \in \mathcal{M}(y).$$

Thus, (3.13) results

$$\frac{\check{X}_t}{S_t^0} = \check{X}_0 + \int_0^t \psi_s dW_{1s}^\nu - A_t^0; \quad \nu \in \mathcal{M}(y). \tag{3.14}$$

Now, assume for a moment that the budget constraint (3.7) holds with equality, *i.e.*

$$\check{X}_0 = \max_{\nu \in \mathcal{M}(y)} E^\nu \frac{B}{S_T^0} = x. \tag{3.15}$$

Next, define the portfolio

$$\pi_t \triangleq \frac{S_t^0}{\sigma(Y_t)}\psi_t; \quad t \in [0,T]. \tag{3.16}$$

Then, by (3.6), (3.14), and (3.15), the associated wealth process $X^{x,y,\pi}$ satisfies

$$\begin{aligned}\frac{X_t^{x,y,\pi}}{S_t^0} &= x + \int_0^t \frac{\pi_s}{S_s^0}\sigma(Y_s)\,dW_{1s}^{\nu} = x + \int_0^t \psi_s dW_{1s}^{\nu} \\ &= \frac{\check{X}_t}{S_t^0} + A_t^0 \geq \frac{\check{X}_t}{S_t^0} \geq 0 \quad \text{a.s.}\end{aligned}$$

In particular, $X_T^{x,y,\pi} \geq \check{X}_T = B$. Otherwise, when $\check{X}_0 < x$, substitute $\check{X}_0$ by $x$ and apply the above arguments to the portfolio $\pi$, but also investing in the bank account the exceeding initial capital $x - \check{X}_0$. Thus

$$X^{x,y,\pi} \geq X^{\check{X}_0,y,\pi} \geq 0 \quad \text{and} \quad X_T^{x,y,\pi} \geq B \quad \text{a.s.}$$

□

THEOREM 3.3. *Given $\check{\nu} \in \mathcal{M}(y)$, the following statements are equivalent:*
*(i)*

$$B \in \mathcal{B}(x,y) \quad \textit{and} \quad E^{\check{\nu}}\frac{B}{S_T^0} = x.$$

*(ii) There exists an admissible portfolio $\pi$ such that $X_T^{\pi} \equiv B$ and $X^{\pi}/S^0$ is a $P^{\check{\nu}}$-martingale with representation*

$$\frac{X_t^{\pi}}{S_t^0} = x + \int_0^t \psi_s dW_{1s}^{\check{\nu}}; \quad t \in [0,T], \tag{3.17}$$

*where $\psi$ is a progressively measurable process with $\int_0^T \psi_u^2 du < \infty$ a.s.*

PROOF. (i) implies (ii): We check that $X^{\pi} \equiv \check{X}$, where $\check{X}$ is as in (3.8) and $\pi$ is the portfolio given in (3.16). From (3.9) and (i), we have

$$E^{\check{\nu}}\frac{\check{X}_T}{S_T^0} = E^{\check{\nu}}\frac{B}{S_T^0} = x = \max_{\nu \in \mathcal{M}(y)} E^{\nu}\frac{B}{S_T^0} = \check{X}_0.$$

Then $\check{X}/S^0$ is a $P^{\check{\nu}}$-martingale, since it is a $P^{\check{\nu}}$-supermartingale with constant mean. Thus $A^0 \equiv 0$ and hence, $X^{\pi} \equiv \check{X}$. The rest follows from martingale representation theorem.

(ii) implies (i): From Remark 3.1, the discounted wealth process $X^{\pi}/S^0$ is a $P^{\nu}$-supermartingale, for each $\nu \in \mathcal{M}(y)$. Hence, $B \equiv X_T^{\pi}$ belongs to $\mathcal{B}(x,y)$. Moreover, $E^{\check{\nu}}B/S_T^0 = x$, since $X^{\pi}/S^0$ is a $P^{\check{\nu}}$-martingale. □

REMARK 3.4. In Theorem 3.3 the portfolio $\pi$ is given by

$$\pi_t \triangleq \frac{S_t^0}{\sigma(Y_t)}\psi_t. \tag{3.18}$$

The next assertions are also equivalent to parts (i) and (ii) in Theorem 3.3.
(a) There exists $\pi \in \mathcal{A}(x,y)$ such that $X_T^{\pi} \equiv B$ and the following DPE holds

$$\frac{X_s^{\pi}}{S_s^0} = \sup_{\nu \in \mathcal{M}(y)} E^{\nu}\left[\frac{X_t^{\pi}}{S_t^0} \mid \mathcal{F}_s\right] = E^{\check{\nu}}\left[\frac{X_t^{\pi}}{S_t^0} \mid \mathcal{F}_s\right]; \quad 0 \leq s \leq t \leq T.$$

(b) There exists $\pi \in \mathcal{A}(x,y)$ such that

$$\frac{X_t^\pi}{S_t^0} = \sup_{\nu\in\mathcal{M}(y)} E^\nu\left[\frac{B}{S_T^0} \mid \mathcal{F}_t\right] = E^{\tilde{\nu}}\left[\frac{B}{S_T^0} \mid \mathcal{F}_t\right]; \quad 0 \le t \le T.$$

Thanks to Proposition 3.2, the optimal investment problem is equivalent to

$$\text{(P)} \qquad \text{maximize } E\{U(B)\} \quad \text{over} \quad B \in \mathcal{B}(x,y),$$

where

$$\mathcal{B}(x,y) \triangleq \left\{B \ge 0 \mid B \text{ is } \mathcal{F}_T\text{-measurable and } \max_{\nu\in\mathcal{M}(y)} E^\nu \frac{B}{S_T^0} \le x\right\}.$$

## 4. Convex Optimization

In this section we will formulate the dual problem using techniques from convex analysis. The relationship between the optimal solutions to the primal and dual problems will also be discussed. The section is self-contained, taking some basic concepts from [**L**]. See also Section 3.4 in [**Ka-S2**].

Given a **utility** function $U : \mathbf{R}_+ \to \mathbf{R}$, we assume the following:

Assumption B. 1. $U(\cdot)$ is strictly increasing, strictly concave, and differentiable. This implies in particular that $U'(\cdot)$ is strictly decreasing.
2. $U'(\infty) \triangleq \lim_{b\to\infty} U'(b) = 0$ and $U'(0+) \triangleq \lim_{b\downarrow 0} U'(b) = \infty$.

The *conjugate* convex function of the utility function $U(\cdot)$ is defined as

$$\tilde{U}(z) \triangleq \max_{b>0} \{U(b) - zb\}; \quad z > 0. \tag{4.1}$$

The function $-\tilde{U}(\cdot)$ is the concave conjugate of $U(\cdot)$ (see Section 7.11 in [**L**]). Further, from (4.1) and elementary calculus, it follows that

$$\tilde{U}(z) = U(I(z)) - zI(z); \quad z > 0, \tag{4.2}$$

where $I(\cdot)$ is the inverse function of $U'(\cdot)$.

The associated *dual functional* to the primal problem (P) is defined as

$$L(\nu,\lambda) \triangleq L(\nu,\lambda;x,y) \triangleq \max_{B\ge 0}\left\{EU(B) + \lambda\left(x - E^\nu \frac{B}{S_T^0}\right)\right\}; \quad \nu \in \mathcal{M}(y), \quad \lambda \ge 0.$$

Here the argument "$B \ge 0$" means that $B$ is a $\mathcal{F}_T$-measurable and nonnegative random variable. This dual functional with two *Lagrange parameters*: $\lambda$ and $Z^\nu$ (or well $\lambda$ and $\nu$), is analogous to equation (22) in [**C**], where it was used to study a consumption problem. On the other side, note that the present definition is different to the classical dual functional[1]. See equation (8.6.2) in [**L**].

From (4.1), observe that

$$EU(B) - \lambda E^\nu \frac{B}{S_T^0} = E\left[U(B) - \lambda \frac{Z_T^\nu}{S_T^0} B\right] \le E\tilde{U}\left(\lambda \frac{Z_T^\nu}{S_T^0}\right); \quad B \ge 0. \tag{4.3}$$

[1]In this case it would be $L(\lambda) \triangleq \max_{B\ge 0}\left\{EU(B) + \lambda\left(x - \max_{\nu\in\mathcal{M}(y)} E^\nu B/S^0(T)\right)\right\}$.

This inequality, together with (4.2), imply that

$$\begin{aligned} L(\nu,\lambda) &\leq E\tilde{U}\left(\lambda\frac{Z_T^\nu}{S_T^0}\right)+\lambda x =: E\left[U\left(B^{\nu,\lambda}\right)-\lambda\frac{Z_T^\nu}{S_T^0}B^{\nu,\lambda}+\lambda x\right] \\ &= E\left[U\left(B^{\nu,\lambda}\right)+\lambda\left(x-E^\nu\frac{B^{\nu,\lambda}}{S_T^0}\right)\right] \\ &\leq L(\nu,\lambda), \end{aligned}$$

where $B^{\nu,\lambda} \triangleq I\left(\lambda Z_T^\nu / S_T^0\right)$; which is $\mathcal{F}_T$-measurable and nonnegative. Hence, the dual functional can be written as

$$L(\nu,\lambda) = E\tilde{U}\left(\lambda\frac{Z_T^\nu}{S_T^0}\right)+\lambda x; \quad \nu\in\mathcal{M}(y), \quad \lambda\geq 0. \tag{4.4}$$

Now, we can state the **dual** problem, which consists in

$$\text{(D)} \qquad \text{minimize} \quad \left\{E\tilde{U}\left(\lambda\frac{Z_T^\nu}{S_T^0}\right)+\lambda x\right\} \quad \text{over} \quad \nu\in\mathcal{M}(y) \quad \text{and} \quad \lambda>0.$$

This dual representation is inspired as a natural extension, from complete to incomplete markets, of the results presented in Section 3.6 in [**Ka-S2**]. In this book, the martingale method is implemented building a family of auxiliary complete markets, indexed by $\nu\in\mathcal{P}(y)$. In this sense, the problem is reduced to find the *optimal market.*

Now, observe that, in general

$$\max_{B\in\mathcal{B}(x,y)} EU(B) \leq \min_{\nu\in\mathcal{M}(y),\lambda>0} L(\nu,\lambda). \tag{4.5}$$

This follows from the budget constraint (3.7), since

$$\begin{aligned} \max_{B\in\mathcal{B}(x,y)} EU(B) &\leq \max_{B\in\mathcal{B}(x,y)}\left\{EU(B)+\lambda\left(x-E^\nu\frac{B}{S_T^0}\right)\right\} \\ &\leq \max_{B\geq 0}\left\{EU(B)+\lambda\left(x-E^\nu\frac{B}{S_T^0}\right)\right\} \\ &= L(\nu,\lambda); \quad \nu\in\mathcal{M}(y), \quad \lambda\geq 0. \end{aligned}$$

When the equality holds in (4.5), we say there is no duality *gap*. In the next section we will check that this is the case for logarithmic and HARA utility functions.

The following proposition shows the relationship between the optimal solutions to the primal (P) and dual (D) problems.

PROPOSITION 4.1. *Assume, for some* $(\hat{\nu},\hat{\lambda})\in\mathcal{M}(y)\times\mathbf{R}_+$, *that*

$$\hat{B} \triangleq I\left(\hat{\lambda}\frac{Z_T^{\hat{\nu}}}{S_T^0}\right)\in\mathcal{B}(x,y) \quad \text{and} \quad E^{\hat{\nu}}\frac{\hat{B}}{S_T^0}=x. \tag{4.6}$$

*Then,* $\hat{B}$ *is an optimal solution to the primal problem (P), whereas* $(\hat{\nu},\hat{\lambda})$ *is the optimal solution of the dual problem (D). Further, there is no duality gap.*

PROOF. From (4.4) and (4.2), it follows that

$$\begin{aligned}
\min_{\nu\in\mathcal{M}(y),\lambda>0} L(\nu,\lambda) &= \min_{\nu\in\mathcal{M}(y),\lambda>0}\left\{E\tilde{U}\left(\lambda\frac{Z_T^{\nu}}{S_T^0}\right)+\lambda x\right\} \leq E\tilde{U}\left(\hat{\lambda}\frac{Z_T^{\hat{\nu}}}{S_T^0}\right)+\hat{\lambda}x \\
&= EU\left(I\left(\hat{\lambda}\frac{Z_T^{\hat{\nu}}}{S_T^0}\right)\right)+\hat{\lambda}\left[x-E\frac{Z_T^{\hat{\nu}}}{S_T^0}I\left(\hat{\lambda}\frac{Z_T^{\hat{\nu}}}{S_T^0}\right)\right] \\
&= EU(\hat{B})+\hat{\lambda}\left[x-E^{\hat{\nu}}\frac{\hat{B}}{S_T^0}\right]=EU(\hat{B}) \\
&\leq \max_{B\in\mathcal{B}(x,y)} EU(B).
\end{aligned}$$

Finally, from (4.5), $\hat{B}$ is the optimal final wealth for the primal problem (P) and $(\hat{\nu},\hat{\lambda})$ is the optimal solution to the dual problem (D). □

REMARK 4.2. The optimal final wealth (4.6) is similar to the one obtained in (6.3.16) by Karatzas and Shreve [**Ka-S2**] (including its Remark 6.5.8), with $\hat{\lambda}=\mathcal{Y}_{\hat{\nu}}(x)$, where $\mathcal{Y}_{\hat{\nu}}(\cdot)$ is the inverse function of $\mathcal{X}_{\hat{\nu}}(\cdot)$; defined in (6.3.15). An existence result and a characterization of the solution to the dual problem (D) is also given in Section 6.5 in [**Ka-S2**]. However, except when the coefficients are deterministic (Section 6.6) and the logarithmic case (Example 6.7.2), they do not give the explicit form of the optimal process $\hat{\nu}$.

## 5. Examples

In this section the optimal solution of the investor's problem shall be given when the utility function is logarithmic or HARA. Based on the results obtained in the previous sections, we can formulate the following methodology:

(1) Given a utility function $U(\cdot)$ pose and solve the dual problem. That is, get the optimal solution $(\hat{\nu},\hat{\lambda})\in\mathcal{M}(y)\times\mathbf{R}_+$.
(2) Check that the random variable $\hat{B}\triangleq I(\hat{\lambda}Z_T^{\hat{\nu}}/S_T^0)$ belongs to $\mathcal{B}(x,y)$ and satisfies $E^{\hat{\nu}}\hat{B}/S_T^0=x$. Then, Proposition 4.1 and Theorem 3.3 can be applied.
(3) Finally, from (3.16) and representation (3.17), get the optimal portfolio $\hat{\pi}$.

### 5.1. Logarithmic Case.

If $U(b)=\log b;\ b>0$, then

$$I(z)=\frac{1}{z}\quad\text{and}\quad\tilde{U}(z)=U(I(z))-zI(z)=-(1+\log z);\qquad z>0.$$

The dual functional (4.4) gets the form

$$L(\nu,\lambda)=E\int_0^T r(Y_u)\,du+\lambda x-(1+\log\lambda)-E\log Z_T^{\nu}. \tag{5.1}$$

The optimal value of the parameter $\lambda$ is given by $\hat{\lambda}=1/x$, which does not depend on $\nu$. On the other hand, from (3.1), we get

$$-\log Z_T^{\nu}=\int_0^T[\theta(Y_t)\,dW_{1t}+\nu_t dW_{2t}]+\frac{1}{2}\int_0^T[\theta^2(Y_t)+\nu_t^2]\,dt.$$

Thus, the dual problem is equivalent to

$$\text{minimize} \quad \left\{ E\int_0^T \nu_t^2 dt \right\} \quad \text{over} \quad \nu \in \mathcal{M}(y).$$

Clearly, the optimal solution for the dual problem is $(\hat{\nu}, \hat{\lambda}) \equiv (0, 1/x)$.

Now, as suggests (4.6), define

$$\hat{B} \mathrel{\hat{=}} I\left(\frac{1}{x}\frac{Z_T^0}{S_T^0}\right) = x\frac{S_T^0}{Z_T^0}.$$

Next, define the nonnegative $P^0$-martingale:

$$\frac{\hat{X}_t}{S_t^0} \mathrel{\hat{=}} E^0\left[\frac{\hat{B}}{S_T^0} \mid \mathcal{F}_t\right]; \quad t \in [0,T].$$

Thus

$$\frac{\hat{X}_t}{S_t^0} = E\left[\frac{Z_T^0}{Z_t^0}\frac{\hat{B}}{S_T^0} \mid \mathcal{F}_t\right] = E\left[\frac{x}{Z_T^0} \mid \mathcal{F}_t\right] = \frac{x}{Z_t^0}.$$

From Ito's formula, and expressions (3.4) and (3.1), note that

$$d\left[Z_t^0\right]^{-1} = -\left[Z_t^0\right]^{-2} dZ_t^0 + \left[Z_t^0\right]^{-3}\left[dZ_t^0\right]^2 = \frac{\theta(Y_t)}{Z_t^0} dW_{1t}^0$$

with

$$\int_0^T \left[Z_t^0\right]^{-2} dt = \int_0^T e^{2\int_0^t \theta(Y_u)dW_{1u} + \int_0^t \theta^2(Y_u)du} dt < \infty.$$

Then, if define

$$\psi_t \mathrel{\hat{=}} x\frac{\theta(Y_t)}{Z_t^0} \quad \text{and} \quad \hat{\pi}_t \mathrel{\hat{=}} \psi_t \frac{S_t^0}{\sigma(Y_t)} = \frac{\theta(Y_t)}{\sigma(Y_t)}\hat{X}_t,$$

we have $X^{\pi} \equiv \hat{X} \geq 0$. Furthermore, note that $\hat{B} \in \mathcal{B}(x,y)$, since it holds condition (ii) of Theorem 3.3 with $\check{\nu} \equiv 0$.

This result is similar to the solution of a consumption-investment problem presented in Example 6.7.2 in [**Ka-S2**].

**5.2. HARA Case.** If $U(b) = b^{\gamma}/\gamma$; $b > 0$, with $\gamma < 1$ and $\gamma \neq 0$, then

$$I(z) = z^{-\frac{1}{1-\gamma}} \quad \text{and} \quad \tilde{U}(z) = U(I(z)) - zI(z) = -\frac{1}{\alpha}z^{\alpha}; \quad z > 0,$$

where $\alpha \mathrel{\hat{=}} -\gamma/(1-\gamma)$. Observe that $\gamma = -\alpha/(1-\alpha)$, $\alpha < 1$, and $\alpha \neq 0$. In this example, the dual functional (4.4) is given by

$$L(\nu,\lambda) = \lambda x - \frac{1}{\alpha}\lambda^{\alpha}\Lambda_{\nu}, \quad \text{where} \quad \Lambda_{\nu} \mathrel{\hat{=}} E\left(\frac{Z_T^{\nu}}{S_T^0}\right)^{\alpha}. \tag{5.2}$$

The first and second derivatives of $L$ with respect to $\lambda$ are $L_{\lambda}(\nu,\lambda) = x - \lambda^{-(1-\alpha)}\Lambda_{\nu}$ and $L_{\lambda\lambda}(\nu,\lambda) = (1-\alpha)\lambda^{-(2-\alpha)}\Lambda_{\nu} > 0$. Hence, the optimal value for the parameter $\lambda$ is $\hat{\lambda}(\nu) \mathrel{\hat{=}} [\Lambda_{\nu}/x]^{1/(1-\alpha)}$. Substituting this value in (5.2), we get

$$L(\nu, \hat{\lambda}(\nu)) = x^{\gamma}\Lambda_{\nu}^{1-\gamma} - \frac{1}{\alpha}x^{\gamma}\Lambda_{\nu}^{1+\frac{\alpha}{1-\alpha}} = \frac{1}{\gamma}x^{\gamma}\Lambda_{\nu}^{1-\gamma}. \tag{5.3}$$

When $\gamma < 0$ $[0 < \gamma < 1]$, minimizing $L(\nu, \hat{\lambda}(\nu))$ is equivalent to maximize [minimize] $\{\Lambda_\nu\}$ over $\nu \in \mathcal{M}(y)$. This is an optimal control problem, and will be identified hereafter as the *auxiliary problem.*

The solution for the negative HARA parameter $\gamma < 0$ is provided in the rest of this work. See Remark 5.5 for the case $0 < \gamma < 1$. The auxiliary problem will be solved using stochastic control methods.

Note that

$$\begin{aligned} [Z_t^\nu]^\alpha &= e^{-\int_0^t \alpha[\theta(Y_u)dW_{1u}+\nu_u dW_{2u}]-\frac{1}{2}\alpha\int_0^t\left[\theta^2(Y_u)+\nu_u^2\right]du} \\ &= e^{-\int_0^t \alpha[\theta(Y_u)dW_{1u}+\nu_u dW_{2u}]-\frac{1}{2}\alpha^2\int_0^t\left[\theta^2(Y_u)+\nu_u^2\right]du-\frac{1}{2}\alpha(1-\alpha)\int_0^t\left[\theta^2(Y_u)+\nu_u^2\right]du} \\ &=: Z_t^{\alpha,\nu} e^{-\frac{1}{2}\alpha(1-\alpha)\int_0^t\left[\theta^2(Y_u)+\nu_u^2\right]du}, \end{aligned}$$

where $Z^{\alpha,\nu}$ is defined as $Z^\nu$, substituting $\alpha\theta(\cdot)$ by $\theta(\cdot)$ and $\alpha\nu$ by $\nu$ in (3.1). Proceeding as in (3.2) and (3.3), we can define respectively the measure $P^{\alpha,\nu}$ in $\mathcal{F}_T$ and the process $(W_1^{\alpha,\nu}, W_2^{\alpha,\nu})$.

Now we will define a family of auxiliary problems indexed by the temporal parameter $s$ in $[0,T]$. With this in mind, the set of control processes $\nu$ defined in $[s,T]$ will be denoted by $\mathcal{M}(s,y)$ and the process $Z^{s,\alpha,\nu}$ is given by

$$Z_t^{s,\alpha,\nu} \doteq z^\alpha e^{-\alpha\int_s^t[\theta(Y_u^s)dW_{1u}+\nu_u dW_{2u}]-\frac{1}{2}\alpha^2\int_s^t[\theta^2(Y_u^s)+\nu_u^2]du}; \quad t \in [s,T], \quad z > 0.$$

Again, this martingale induces a probability measure $P^{s,\alpha,\nu}$. Under this new measure, the dynamics of the external factor satisfies the following SDE in $[s,T]$

$$dY_t^s = [g(Y_t^s) - \alpha\rho\theta(Y_t^s) - \alpha\varepsilon\nu_t]dt + \rho dW_{1t}^{\alpha,\nu} + \varepsilon dW_{2t}^{\alpha,\nu} \quad \text{and} \quad Y_s^s = y \in \mathbf{R}. \tag{5.4}$$

Now, for $(s,y) \in [0,T] \times \mathbf{R}$, $\nu \in \mathcal{M}(s,y)$, and $z > 0$, define

$$\begin{aligned} J(s,y,\nu) &\doteq E^{s,\alpha,\nu} \exp \int_s^T q(Y_u^s, \nu_u)\,du \\ &\doteq E^{s,\alpha,\nu} \exp\left(-\alpha\int_s^T \left[r(Y_u) + \frac{1}{2}(1-\alpha)\left(\theta^2(Y_u^s) + \nu_u^2\right)\right] du\right), \end{aligned}$$

where $q(y,v) \doteq -\alpha\left[r(y) + \frac{1}{2}(1-\alpha)\left(\theta^2(y) + v^2\right)\right]$; for $(y,v) \in \mathbf{R}^2$. This function is bounded above. The *value* function associated with the auxiliary problem is

$$W(s,y) \doteq \max_{\nu\in\mathcal{M}(s,y)} J(s,y,\nu) \quad \text{with} \quad W(T,y) = 1. \tag{5.5}$$

To solve this optimal control problem some basic properties of $W(s,y)$ are needed. Since

$$e^{-\alpha\int_s^T\left[|r|_\infty+\frac{1}{2}(1-\alpha)\left(|\theta|_\infty^2+\nu_u^2\right)\right]du} \le J(s,y,\nu) \le 1,$$

then

$$0 < K_1 \le W(s,y) \le 1, \tag{5.6}$$

where $K_1 \doteq e^{-\alpha\left(|r|_\infty+\frac{1}{2}(1-\alpha)|\theta|_\infty^2\right)T}$; which does not depend on $\nu$ and $y$.

Now, let us check that $W(s,\cdot)$ is a Lipschitz function. From dominated convergence theorem, $J(s,y,\nu)$ is $y$-differentiable and

$$J_y(s,y,\nu) = E^{s,\alpha,\nu}\int_s^T q_y(Y_u^s,\nu_u)\frac{\partial}{\partial y}Y_u^s du \times \exp\int_s^T q(Y_u^s,\nu_u)\,du,$$

where $q_y(y,v) = -\alpha[r'(y) + (1-\alpha)\theta(y)\theta'(y)]$ and $\frac{\partial}{\partial y}Y_u^s$ is the solution of the ordinary differential equation

$$d\frac{\partial}{\partial y}Y_u^s = [g'(Y_u^s) - \alpha\rho\theta'(Y_u^s)]\frac{\partial}{\partial y}Y_u^s du \quad \text{with} \quad \frac{\partial}{\partial y}Y_s^s = 1,$$

or equivalently

$$\frac{\partial}{\partial y}Y_u^s = \exp\left(\int_s^u [g'(Y_t^s) - \alpha\rho\theta'(Y_t^s)]\,dt\right); \quad \text{in} \quad [s,T].$$

Since

$$\left|\int_s^T q_y(Y_u^s, \nu_u)\frac{\partial}{\partial y}Y_u^s du\right| \le K_2 \doteq \alpha\left(|r'|_\infty + (1-\alpha)|\theta|_\infty|\theta'|_\infty\right)Te^{\left(|g'|_\infty + \alpha|\theta'|_\infty\right)T},$$

where $K_2$ does not depend on $\nu$ and $y$, then

$$|J_y(s,y,\nu)| \le K_2.$$

Hence, $W(s,\cdot)$ is Lipschitz, with Lipschitz constant $K_2$. Finally, from (5.6), we get

$$\text{(5.7)} \qquad \frac{|W_y(s,y)|}{W(s,y)} \le K \doteq \frac{K_2}{K_1};$$

for all $(s,y)$ where $W_y(s,y)$ is defined.

To solve the auxiliary problem (5.5), temporarily we assume that the control space is bounded, that is, $\nu \in \mathcal{M}(y) \cap [[-M,M]]$, where $[[-M,M]]$ is the class of progressively measurable real processes taking values in $[-M,M]$, for some constant $M > 0$. We will show below that the value function $W(s,y)$ does not depend on $M$, for $M$ large enough.

We will show that

$$w(s,y) = W(s,y);$$

where $w(s,y)$ is the unique smooth function in $C^{1,2}([0,T]\times\mathbf{R}) \cap C_p([0,T]\times\mathbf{R})$, satisfying the associated *Hamilton-Jacobi-Bellman* (*HJB*) equation:

$$\begin{aligned}\text{(5.8)} \quad 0 \quad = \quad & w_t + \frac{1}{2}w_{yy} + (g - \alpha\rho\theta)w_y - \alpha\left[r + \frac{1}{2}(1-\alpha)\theta^2\right]w \\ & + \alpha \max_{v\in[-M,M]}\left\{-\varepsilon w_y v - \frac{1}{2}(1-\alpha)wv^2\right\} \quad \text{with} \quad w(T,y) = 1.\end{aligned}$$

The existence and uniqueness of $w$ follows from Theorem IV.4.3 and Remark IV.4.1 in [**F-S**].

Now, define the Markov *policy,* for $(t,y) \in [s,T]\times\mathbf{R}$

$$\begin{aligned}\text{(5.9)} \quad v^*(t,y) \quad & \doteq \quad \arg\max_{v\in[-M,M]}\left\{-\varepsilon w_y(t,y)v - \frac{1}{2}(1-\alpha)w(t,y)v^2\right\} \\ & = \quad \begin{cases} -\dfrac{\varepsilon}{1-\alpha}\dfrac{w_y}{w} & \text{if} \quad \dfrac{\varepsilon}{1-\alpha}\dfrac{|w_y|}{w} \le M \quad \text{and} \quad w \ne 0 \\ -M\operatorname{sgn} w_y & \text{otherwise} \end{cases}.\end{aligned}$$

THEOREM 5.1 (Verification). *For $M > 0$, let $w$ be the unique solution to (5.8). Then:*

*(i)* $w(s,y) \ge J(s,y,\nu)$; *for $(s,y) \in [0,T]\times\mathbf{R}$ and $\nu \in \mathcal{M}(s,y) \cap [[-M,M]]$.*

*(ii)* $w(s,y) = W(s,y) = J(s,y,\hat{\nu})$, *where* $\hat{\nu}$ *is a Markov control process in* $\mathcal{M}(T,y) \cap [[-M,M]]$, *defined as*

$$\hat{\nu}_t \mathrel{\hat{=}} v^*(t, Y_t^s); \quad t \in [s,T]. \tag{5.10}$$

*Hence,* $\hat{\nu}$ *is optimal for the auxiliary problem relative to (5.5).*

PROOF. (i) For $v \in [-M,M]$, let $\mathcal{L}^v$ be the functional defined by

$$\mathcal{L}^v f \mathrel{\hat{=}} f_t + \frac{1}{2} f_{yy} + (g - \alpha\rho\theta - \varepsilon f_y v) f_y; \quad f \in C^{1,2}([0,T] \times \mathbf{R}).$$

In particular, for $f(t,y) \mathrel{\hat{=}} w(t,y)$ results

$$[\mathcal{L}^v + q(y,v)] w(t,y) = w_t + \frac{1}{2} w_{yy} + (g - \alpha\rho\theta) w_y - \alpha \left[ r + \frac{1}{2}(1-\alpha)\theta^2 \right] w$$
$$+ \alpha \left[ -\varepsilon w_y v - \frac{1}{2}(1-\alpha) w v^2 \right].$$

From the HJB equation (5.8), for $t \in [s,T]$ and $\nu \in \mathcal{M}(s,y) \cap [[-M,M]]$

$$[\mathcal{L}^{\nu_t} + q(Y_t^s, \nu_t)] w(t, Y_t^s) \le 0. \tag{5.11}$$

This inequality, together with the Feynman Kac formula (see equation (D.13) in [**F-S**]) and a localization argument, imply that

$$w(s,y) = E^{s,\alpha,\nu} \left[ e^{\int_s^T q(Y_u^s,\nu_u)du} w(T, Y_T^s) \right.$$
$$\left. - \int_s^T e^{\int_s^t q(Y_u^s,\nu_u)du} [\mathcal{L}^{\nu_t} + q(Y_t^s,\nu_t)] w(t,Y_t^s) dt \right]$$
$$\ge E^{s,\alpha,\nu} \left[ e^{\int_s^T q(Y_u^s,\nu_u)du} w(T, Y_T^s) \right] = E^{s,\alpha,\nu} \left[ e^{\int_s^T q(Y_u^s,\nu_u)du} \right] \tag{5.12}$$
$$= J(s,y,\nu);$$

for $(s,y) \in [0,T] \times \mathbf{R}$ and $\nu \in \mathcal{M}(s,y) \cap [[-M,M]]$.

(ii) Since $w_y(t,y)$ is continuous, then the Markov policy $v^*(t,y)$ is bounded, continuous and $y$-locally Lipschitz. Then, $\hat{\nu}$ is a Markov control process in $\mathcal{M}(s,y) \cap [[-M,M]]$. Moreover, for $\nu \mathrel{\hat{=}} \hat{\nu}$, inequalities (5.11) and (5.12) become equalities. Hence, $w(s,y) = W(s,y) = J(s,y,\hat{\nu})$. □

The following result is a direct consequence of the previous theorem.

COROLLARY 5.2. *Let* $w$ *be the unique solution to (5.8) with* $M > K\varepsilon/(1-\alpha)$. *Then,* $w = W$, *where* $W$ *is the unconstrained value function, and*

$$\hat{\nu}_t = v^*(t, Y_t^s) = -\frac{\varepsilon}{1-\alpha} \frac{W_y(t, Y_t^s)}{W(t, Y_t^s)}; \quad t \in [s,T], \tag{5.13}$$

*is the optimal control.*

Consequently, $W \in C^{1,2}([0,T] \times \mathbf{R}) \cap C_b^{0,1}([0,T] \times \mathbf{R})$ and the HJB equation (5.8) can be written as

$$0 = W_t + \frac{1}{2} W_{yy} + (g - \alpha\rho\theta) W_y - \alpha \left[ r + \frac{1}{2}(1-\alpha)\theta^2 \right] W - \frac{1}{2}\gamma\varepsilon^2 \frac{W_y^2}{W}, \tag{5.14}$$

with $W(T,y) = 1$.

On the other hand, assuming that $W(s,y) =: [h(s,y)]^{\delta}$ for some $\delta > 0$, we get

$$\begin{aligned} W_t &= \delta h^{\delta-1} h_t, \quad W_y = \delta h^{\delta-1} h_y, \quad \frac{W_y^2}{W} = \delta^2 h^{\delta-2} h_y^2, \\ W_{yy} &= \delta h^{\delta-1} h_{yy} + \delta(\delta-1) h^{\delta-2} h_y^2. \end{aligned}$$

Thus, the HJB equation (5.14) results: $h(T,y) = 1$ and

$$0 = h_t + \frac{1}{2} h_{yy} + (g - \alpha\rho\theta) h_y - \frac{\alpha}{\delta}\left[r + \frac{1}{2}(1-\alpha)\theta^2\right] h + \frac{1}{2}\left(\delta - 1 - \gamma\delta\varepsilon^2\right)\frac{h_y^2}{h}.$$

If $\delta - 1 - \gamma\delta\varepsilon^2 = 0$, then $\delta \mathrel{\hat{=}} 1/\left(1-\gamma\varepsilon^2\right)$ is the unique number in $(0,1]$ such that the last term in the previous equation vanishes. That means, the HJB equation (5.14) is equivalent to

$$(5.15)\quad 0 = h_t + \frac{1}{2} h_{yy} + (g - \alpha\rho\theta) h_y - \frac{\alpha}{\delta}\left[r + \frac{1}{2}(1-\alpha)\theta^2\right] h \quad \text{with} \quad h(T,y) = 1.$$

The previous power transformation was used in [**Z**], for a similar optimal wealth problem. In fact, equation (5.15) agree with its equation (3.11).

On the other side, since the function $h(s,y)$ belongs to $C^{1,2}([0,T]\times \mathbf{R})$, the Feynman-Kac formula can be applied to get the following representation:

$$(5.16)\qquad h(s,y) = E\exp\left(-\frac{\alpha}{\delta}\int_s^T \left[r(\check{Y}_t^s) + \frac{1}{2}(1-\alpha)\theta^2(\check{Y}_t^s)\right] dt\right);$$

for $(s,y) \in [0,T]\times\mathbf{R}$, where $\{\check{Y}_t^s\}_{t\in[s,T]}$ is the solution to the SDE

$$d\check{Y}_t^s = \left[g(\check{Y}_t^s) - \alpha\rho\theta(\check{Y}_t^s)\right] dt + d\check{W}_t \quad \text{with} \quad \check{Y}_s^s = y,$$

and $\check{W} \mathrel{\hat{=}} \rho W_1 + \varepsilon W_2$; which is a BM relative to $(\Omega, \{\mathcal{F}_t\}_{t\in[0,T]}, P)$. See Theorem 5.7.6 in [**Ka-S1**]. Finally, from (5.16), it is easy to see that $h_t$ is bounded.

The following remark resumes the above results. Before, recall that $\alpha = -\gamma/(1-\gamma)$, $\delta = 1/(1-\gamma\varepsilon^2) = (1-\alpha)/(1-\alpha\rho^2)$, and $\hat{\nu}_t = -[\varepsilon/(1-\alpha)]W_y(t,Y_t^s)/W(t,Y_t^s) = \frac{-\varepsilon\delta}{1-\alpha} h_y(t,Y_t^s)/h(t,Y_t^s)$.

REMARK 5.3. The value function $W$ belongs to $C^{1,2}([0,T]\times\mathbf{R}) \cap C_b^{1,1}([0,T]\times\mathbf{R})$. Further, the unique smooth function solving (5.15), is given by $h(s,y) = [W(s,y)]^{1/\delta}$.

REMARK 5.4. Letting $\alpha \to 0$, then $\delta \to 1$ and, from Feynman Kac's formula (5.16), $h(s,y) \to 1$ and $h_y(s,y) \to 0$. Hence, from (5.13), the limit process is $\hat{\nu} \equiv 0$, which agrees with the optimal process for the logarithmic utility case.

Now, going back to the original problem, consider the initial time $s = 0$. According to the methodology described at the beginning of this section, we take the optimal process $\hat{\nu}$ from (5.13), and consider

$$\hat{B} \mathrel{\hat{=}} I\left(\hat{\lambda}\frac{Z_T^{\hat{\nu}}}{S_T^0}\right) = \frac{x}{\Lambda_{\hat{\nu}}}\left(\frac{Z_T^{\hat{\nu}}}{S_T^0}\right)^{-(1-\alpha)}, \quad \text{where} \quad \Lambda_{\hat{\nu}} = E\left(\frac{Z_T^{\hat{\nu}}}{S_T^0}\right)^{\alpha} = W(0,y).$$

Next we shall identify the optimal portfolio $\hat{\pi}$. Write down the system of SDE

$$\left.\begin{aligned} dY_t &= g(Y_t)\,dt + \rho dW_{1t} + \varepsilon dW_{2t} & Y_0 &= y \\ d\frac{Z_t^{\hat{\nu}}}{S_t^0} &= -\frac{Z_t^{\hat{\nu}}}{S_t^0}\left(r dt + \theta(Y_t)\,dW_{1t} + \hat{\nu}_t dW_{2t}\right) & \frac{Z_0^{\hat{\nu}}}{S_0^0} &= z = 1 \end{aligned}\right\}; \quad \text{in} \quad [0,T].$$

For this system define the operator

$$\mathcal{L}f \stackrel{\circ}{=} f_t + gf_y - rzf_z - z(\rho\theta + \varepsilon v^*)f_{yz} + \frac{1}{2}f_{yy} + \frac{1}{2}z^2(\theta^2 + v^{*2})f_{zz}, \quad f \in C^{1,2}(\mathbf{R}).$$

In particular, when $f(t,y,z) \stackrel{\circ}{=} W(t,y)\, z^\alpha$, from (5.14), we get

$$\begin{aligned}
&\mathcal{L}f(t,y,z) \\
= \ & z^\alpha \left( W_t + \frac{1}{2}W_{yy} + [g - \alpha(\rho\theta + \varepsilon v^*)]\, W_y - \alpha\left[r + \frac{1}{2}(1-\alpha)\left(\theta^2 + v^{*2}\right)\right] W \right) \\
= \ & z^\alpha \left( W_t + \frac{1}{2}W_{yy} + \left[g - \alpha\left(\rho\theta - \frac{\varepsilon^2}{1-\alpha}\frac{W_y}{W}\right)\right] W_y \right. \\
& \left. -\alpha\left[r + \frac{1}{2}(1-\alpha)\left(\theta^2 + \frac{\varepsilon^2}{(1-\alpha)^2}\frac{W_y^2}{W^2}\right)\right] W \right) \\
= \ & z^\alpha \left( W_t + \frac{1}{2}W_{yy} + (g - \alpha\rho\theta)\, W_y - \alpha\left[r + \frac{1}{2}(1-\alpha)\,\theta^2\right] W - \frac{1}{2}\gamma\varepsilon^2\frac{W_y^2}{W} \right) \\
= \ & 0.
\end{aligned}$$

Hence, from Proposition 5.4.2 in [**Ka-S1**], the next process is a local martingale:

$$M_t \stackrel{\circ}{=} f\left(t, Y_t, \frac{Z_t^{\hat{\nu}}}{S_t^0}\right) - \int_0^t \mathcal{L}\left(u, Y_u, \frac{Z_u^{\hat{\nu}}}{S_u^0}\right) du = \left(\frac{Z_T^{\hat{\nu}}}{S_T^0}\right)^\alpha W(t, Y_t); \quad t \in [0,T].$$

Further, this process is also a supermartingale with constant mean, since

$$M_0 = W(0,y) = \Lambda_{\hat{\nu}} \quad \text{and} \quad M_T = \left(\frac{Z_T^{\hat{\nu}}}{S_T^0}\right)^\alpha.$$

Then, $M$ is a martingale.

Now, define the following nonnegative process

$$\frac{\hat{X}_t}{S_t^0} \stackrel{\circ}{=} E^{\hat{\nu}}\left[\frac{\hat{B}}{S_T^0} \mid \mathcal{F}_t\right], \quad t \in [0,T].$$

Thus

$$\begin{aligned}
\frac{\hat{X}_t}{S_t^0} &= E^{\hat{\nu}}\left[\frac{\hat{B}}{S_T^0} \mid \mathcal{F}_t\right] = E\left[\frac{Z_T^{\hat{\nu}}}{Z_t^{\hat{\nu}}}\frac{\hat{B}}{S_T^0} \mid \mathcal{F}_t\right] = \frac{x}{W(0,y)\, Z_t^{\hat{\nu}}} E\left[\frac{Z_T^{\hat{\nu}}}{S_T^0}\left(\frac{Z_T^{\hat{\nu}}}{S_T^0}\right)^{\alpha-1} \mid \mathcal{F}_t\right] \\
&= \frac{x}{W(0,y)\, Z_t^{\hat{\nu}}} E\left[\left(\frac{Z_T^{\hat{\nu}}}{S_T^0}\right)^\alpha \mid \mathcal{F}_t\right] = \frac{x}{W(0,y)\, Z_t^{\hat{\nu}}} E\left[M_T \mid \mathcal{F}_t\right] = \frac{x}{W(0,y)}\frac{M_t}{Z_t^{\hat{\nu}}} \\
&= \frac{x}{W(0,y)}\frac{\left[Z_t^{\hat{\nu}}\right]^{\alpha-1}}{\left[S_t^0\right]^\alpha} W(t, Y_t).
\end{aligned}$$

On the other hand, by Ito's formula

$$\begin{aligned}
d\left[Z^{\hat{\nu}}\right]^{\alpha-1} &= (1-\alpha)\left[Z^{\hat{\nu}}\right]^{\alpha-1}\left[\frac{1}{2}(2-\alpha)\left(\theta^2 + \hat{\nu}^2\right) dt + \theta dW_1 + \hat{\nu} dW_2\right], \\
dW &= \left(W_t + \frac{1}{2}W_{yy} + gW_y\right) dt + W_y\left(\rho dW_1 + \varepsilon dW_2\right), \\
d\left[S^0\right]^{-\alpha} &= -\alpha r\left[S^0\right]^{-\alpha} dt.
\end{aligned}$$

Then

$$
\begin{aligned}
& d\left(\frac{\left[Z_t^{\hat{\nu}}\right]^{\alpha-1}}{[S_t^0]^{\alpha}} W\right) \\
= & \frac{\left[Z_t^{\hat{\nu}}\right]^{\alpha-1}}{[S_t^0]^{\alpha}}\left(-\alpha r W dt+(1-\alpha) W\left[\frac{1}{2}(2-\alpha)\left(\theta^2+\hat{\nu}^2\right) dt+\theta dW_1+\hat{\nu} dW_2\right]\right. \\
& \left.+\left[W_t+\frac{1}{2} W_{yy}+g W_y\right] dt+W_y\left[\rho dW_1+\varepsilon dW_2\right]+(1-\alpha)(\rho \theta+\varepsilon \hat{\nu}) W_y dt\right) \\
= & \frac{\left[Z_t^{\hat{\nu}}\right]^{\alpha-1}}{[S_t^0]^{\alpha}}\left(\left[(1-\alpha) \theta^2 W+\rho \theta W_y\right] dt+\left[(1-\alpha) \theta W+\rho W_y\right] dW_1+\right. \\
& \left.\left[W_t+\frac{1}{2} W_{yy}+(g-\alpha \rho \theta) W_y-\alpha\left(r+\frac{1}{2}(1-\alpha) \theta^2\right) W-\frac{1}{2} \gamma \varepsilon^2 \frac{W_y^2}{W}\right] dt\right) \\
= & \frac{\left[Z_t^{\hat{\nu}}\right]^{\alpha-1}}{[S_t^0]^{\alpha}} W\left[(1-\alpha) \theta+\rho \frac{W_y}{W}\right](\theta dt+dW_1) \\
= & \frac{\left[Z_t^{\hat{\nu}}\right]^{\alpha-1}}{[S_t^0]^{\alpha}} W\left[(1-\alpha) \theta+\rho \frac{W_y}{W}\right] dW_1^{\hat{\nu}} .
\end{aligned}
$$

Thus

$$
\begin{aligned}
d \frac{\hat{X}}{S^0} & =\frac{x}{W(0, y)} d\left(\frac{\left[Z^{\hat{\nu}}\right]^{\alpha-1}}{[S^0]^{\alpha}} W\right) \\
& =\frac{x}{W(0, y)} \frac{\left[Z^{\hat{\nu}}\right]^{\alpha-1}}{[S^0]^{\alpha}} W\left[(1-\alpha) \theta+\rho \frac{W_y}{W}\right] dW_1^{\hat{\nu}} \\
& =\frac{\hat{X}}{S^0}\left[(1-\alpha) \theta+\rho \frac{W_y}{W}\right] dW_1^{\hat{\nu}} \\
& =: \frac{\hat{\pi}}{S^0} \sigma dW_1^{\hat{\nu}},
\end{aligned}
$$

where, for $(t, x, y) \in[0, T] \times \mathbf{R}_{+} \times \mathbf{R}$:

$$
\hat{\pi}_t \triangleq p^*(t, \hat{X}_t, Y_t) \quad \text{and} \tag{5.17}
$$

$$
p^*(t, x, y) \triangleq \frac{x}{\sigma(y)}\left[(1-\alpha) \theta(y)+\rho \frac{W_y(t, y)}{W(t, y)}\right] . \tag{5.18}
$$

It follows, from the derivation of $\hat{X}$ together with (3.6), $\hat{\pi} \in \mathcal{A}(x, y)$ and $X^{\hat{\pi}} \equiv \hat{X}$. In addition, since $X_T^{\hat{\pi}} \equiv \hat{B}$ and the discounted process $X^{\hat{\pi}} / S^0$ is a $P^{\hat{\nu}}$-martingale then, we can apply Theorem 3.3 and Remark 3.4 to get $\hat{B} \in \mathcal{B}(x, y)$ and $E^{\hat{\nu}} \hat{B} / S_T^0=x$. Thus, form Proposition 4.1, $\hat{B}$ is the optimal final wealth.

REMARK 5.5. For the case $0<\gamma<1$, the conclusion is the same. That is, the optimal portfolio is given by (5.17) and (5.18), where $W(t, y)$ solves the HJB equation (5.14). In this case, to get an estimation of $W_y$ qualitative properties from the HJB equation (5.14) are needed. See [**C-H**].

## References


[B-S] O. Barndorff-Nielsen and N. Shephard, *Non-Gaussian Ornstein–Uhlenbeck–based models and some of their uses in financial economics,* J. R. Stat. Soc. Ser. B, Stat. Methodol. **63** (2001), pp. 167–241.

[C-H] N. Castañeda-Leyva and D. Hernández–Hernández, *Optimal consumption–investment problems in incomplete financial markets with stochastic coefficients,* preprint, (2003).

[C] D. Cuoco, *Optimal consumption and equilibrium prices with portfolio constraints and stochastic income,* J. Econom. Theory **72** (1997), pp. 33–73.

[F-H] W.H. Fleming and D. Hernández–Hernández, *An optimal consumption model with stochastic volatility,* Finance Stoch. **7** (2003), pp. 245-262.

[F-S] W.H. Fleming and H.M. Soner. *Controlled Markov Processes and Viscosity Solutions,* Springer–Verlag, New York, (1993).

[F-P-S] J.P. Fouque, G. Papanicolaou and K.R. Sircar, *Derivatives in Financial Markets with Stochastic Volatility,* Cambridge University Press, (2000).

[H-P] J.M. Harrison and S.R. Pliska, *Martingales and stochastic integrals in the theory of continuous trading,* Stochastic Process. Appl. **11** (1981), pp. 215–260.

[J-S] J. Jacod and A.N. Shiryaev, *Limit Theorems for Stochastic Processes,* Springer–Verlag, Berlin, (1987).

[Ka-S1] I. Karatzas and S.E. Shreve, *Brownian Motion and Stochastic Calculus,* Springer–Verlag, New York, (1991).

[Ka-S2] I. Karatzas and S.E. Shreve, *Methods of Mathematical Finance,* Springer–Verlag, New York, (1998).

[Kr-S] D. Kramkov and W. Schachermayer, *The Asymptotic elasticity of utility functions and optimal investments in incomplete markets,* Ann. Appl. Probab. **9** (1999), pp. 904-950.

[L-S] R.S. Lipster and A.N. Shiryaev, *Statistics of Random Processes,* Springer–Verlag, Berlin, (2001).

[L] D.G. Luenberger, *Optimization by Vector Space Methods,* Wiley, New York, (1969).

[Z] T. Zariphopoulou, *A solution approach to valuation with unhedgeable risks,* Finance Stoch. **5** (2001), pp. 61–82.



DEPARTAMENTO DE ESTADÍSTICA, CENTRO DE CIENCIAS BÁSICAS, UNIVERSIDAD AUTÓNOMA DE AGUASCALIENTES AND CENTRO DE INVESTIGACIÓN EN MATEMÁTICAS.
AV. UNIVERSIDAD 940 C.U., AGUASCALIENTES, AGS., C.P. 20100 MÉXICO.

*E-mail address*: ncastane@correo.uaa.mx

CENTRO DE INVESTIGACIÓN EN MATEMÁTICAS.
APARTADO POSTAL 402, GUANAJUATO, GTO., C.P. 36000 MÉXICO.

*E-mail address*: dher@cimat.mx

Contemporary Mathematics
Volume **336**, 2003

# Price Calculation for Power Exponential Jump-diffusion Models – A Hermite-series Approach

Manuel Galea, Jin Ma, and Soledad Torres

ABSTRACT. In this paper we study convolution formulae for the independent sum of a normal random variable and several power exponential distributed random variables. This problem is motivated by the numerical simulation for pricing financial derivatives (such as options) when the underlying assets follow a jump-diffusion model in which the logarithm of the jump sizes are assumed to be within the class of power exponential distributions. When the "*kurtosis parameter*" (denoted by $\beta$) of the power exponential distribution equals 1 and $\frac{1}{2}$, the power exponential distribution becomes a standard normal and a double exponential distribution, respectively. Therefore our model contains those of Merton [**M**] and Kuo-Wang [**K-W**] as special cases. We propose a closed form convolution formula, represented in terms of infinite serious expanded using either Hermite polynomials or parabolic cylindrical functions, depending on the value of kurtosis parameter $\beta$. We also analyze the convergence of such series, and perform the numerical experiments to illustrate these formulae.

## 1. Introduction

In this paper we study series forms of convolution formulae for a normal random variable and several i.i.d. power exponential random variables. To be more precise, we represent the convolution formulae of such independent sums as infinite series of either Hermite polynomials or parabolic cylindrical functions. Compared to the usual integral form of convolution formula, the series form has obvious an advantage in numerical simulation because of its explicit nature. This advantage will become more significant when the number of random variables involved increases, as we shall see in this paper.

The study of such convolution formula was motivated by the numerical simulation and/or calculation of asset prices and their derivatives when the dynamics of the asset prices are assumed to follow jump diffusion models. Recall that the jump

2000 *Mathematics Subject Classification.* Primary: 60H10; Secondary: 34F05, 93E03.
*Key words and phrases.* Power exponential distribution, parabolic cylindrical functions, Hermite polinomials, convolution formula, jump-diffusion models.
The first author is supported in part by Fondecyt Grant #1000424.
The second author is supported in part by NSF grant #9971720.
The third author is supported in part by Fondecyt Grant #1000270, #1020211, and DIPUV N. 21/2001.

diffusion model was first considered by Merton (1976) [**M**], in which the logarithm of the jump sizes are assumed to have normal distributions. It is by now well-understood that the Black-Scholes-Merton model does not incorporate the asymmetric leptokurtic natures (heavy tail and skew to the left) of the return distribution and the so-called "volatility smile". One the other hand, it is also understood that while the normal distribution is often used by "default" for any random quantity with unknown statistics, the statistical analysis based on the normal distribution often lacks robustness. In a recent work Kuo and Wang (2001) [**K-W**] considered jump diffusion models assuming that the logarithm of the jump sizes have the "*double exponential distributions*". In that paper the authors noted that by replacing the normal distribution by the double exponential, one captures several fundamental features such as skewness, heavy tails, and kurtosis, as it is observed from the empirical investigations. In that paper the authors also emphasize that one of the main reasons that they choose double exponential (instead of $t$-distribution, another obvious robust class) is the analytical tractability. In fact, using the double exponential distribution the authors were able to derive some closed-form formulae for pricing various types of options, including some path-dependent ones, such as look-back options and barrier options.

At this point we should note that both the normal distribution and double exponential distribution are special cases of the family of *power exponential distributions*, established by Subbotin in 1923 [**S**] as an extension of the normal distribution (see also Box-Tiao (1965) [**B-T**], Hogg (1974) [**Ho**], Rahman (1997) [**R**], Agrò (1995) [**A**], Gokhale-Rahman (1996)[**Go-R**], and Gómez-Gómez-Villegas-Marín (1998) [**G-G-M**], for properties and applications of such distributions). The main purpose of this work is to see whether one can develop a class of jump diffusion models based upon the general power exponential distributions, and compare all the outcomes so as to choose one to "best fit" the empirical data. In fact, it would be rather interesting if one can actually prove or disprove that the double exponential distribution is "optimal" at least among the family of the power exponential distributions to match the fundamental features of Skewness, heavy tail, and kurtosis.

As it turns out, an analysis similar to that of [**K-W**] is possible for power exponential distributions. The main difficulty, however, lies in the explicit formula for the convolution of one normal random variable and a (finite) sequence of i.i.d. power exponential distributions, which leads to this paper. We should note that finding the series form of convolution formula is not only useful for the closed-form solution, but also important in numerical simulation. In fact, it is expected that it is much more efficient than the Monte Carlo simulation for the traditional integral form convolution formulae, especially when the number of power exponential random variables increases, since the main ingredient of the infinite series, the Hermite polynomials or parabolic cylindrical functions, can be computed off line.

We would like to point out that our series representation of the convolution formulae depends on the kurtosis parameter ("$\beta$") of the power exponential distributions. We show that if $\beta > 1$, then the series can be expressed in terms of Hermite polynomials, but if $\beta < 1$ the series would be better expressed in terms of *parabolic cylinder functions* (PCF). Since an integer-indexed PCF can be related to a Hermite polynomial explicitly, we shall call both of them "*Hermite-series*" for simplicity. It is interesting to note that the double exponential distribution actually corresponds to the case when $\beta = 1/2$, for which we would use PCF series which seems to be much more complicated than that of [**K-W**]. We show that such a series can be reduced to the formula of [**K-W**] rather easily, with a little help from the properties of PCF's.

This paper is organized as follows. In section 2 we give all the definitions and basic properties of power exponential distributions, Hermite polynomials, and PCF's. In section 3 we describe the pricing problems for jump diffusion models involving power exponential distributions, and introduce our "convolution problems" $NP(1)$ and $NP(n)$, etc. In section 4 we give the series solutions to the problem $NP(1)$, and discuss their convergence. In section 5 we study two specail cases when a power distribution is degenerated to a normal or a double exponential ($\beta = 1$ and $\frac{1}{2}$, respectively). Finally in section 6 we give a "formal" series solution for $NP(n)$, and in section 7 we show the numerical results.

## 2. Preliminaries

In this section we review some important facts regarding one dimensional power exponential distribution, Hermite Polynomials, and Cylinder Parabolic functions, which will play fundamental roles in the rest of the paper.

### A. Power exponential distribution

A random variable $X$ is said to have *Power exponential distribution* if its probability density function is given by:

$$f(x) = \frac{1}{\phi\Gamma\left(1+\frac{1}{2\beta}\right)2^{1+\frac{1}{2\beta}}} e^{-\frac{1}{2}\left|\frac{x-\mu}{\phi}\right|^{2\beta}}, \qquad x \in \mathbb{R}, \tag{2.1}$$

where $\mu \in \mathbb{R}$, $\phi \in (0, \infty)$, and $\beta \in (0, \infty)$ are traditionally called the *location parameter*, *scale parameter*, and *kurtosis parameter*, respectively. We shall denote $X \sim P(\beta, \mu, \phi)$ to indicate that $X$ has a power exponential distribution with parameters $(\beta, \mu, \phi)$.

We note that the power exponential distribution belongs to the family of symmetrical distributions. Two special cases of such distributions are well-understood:

(i) $\beta = 1$: clearly $P(1, \mu, \phi) = N(\mu, \phi^2)$, the normal distribution;
(ii) $\beta = \frac{1}{2}$: in this case we have the so-called *double-exponential* distribution.

In general, the parameter $\beta$ indicates the disparity of a power exponential from a normal distribution. When $\beta$ decreases, the "tail" of the distribution function gets "heavier". Therefore any $P(\beta, \mu, \phi)$-distribution with $\beta < 1$ will have a heavier tail than a normal distribution.

### B. Hermite polynomials

The family of Hermite polynomials plays an important role in determining the orthogonal basis for an $L^2$-Gaussian space. Let $n \in \mathbb{N}$, the *n-th Hermite polynomial*, denoted by $H_n(x)$, is defined by

$$H_n(x) = (-1)^n \exp(x^2) \frac{d^n}{dx^n} \exp(-x^2), \tag{2.2}$$

and $H_0(x) \equiv 1$. The family of Hermite polynomials comes from the coefficients of the Taylor expansion (in variable $t$) of the function $G(x,t) \triangleq \exp(2xt - t^2)$. In fact,

$$G(x,t) = \exp\{2xt - t^2\} = \sum_{n=0}^{\infty} \frac{t^n H_n(x)}{n!}, \qquad (t,x) \in \mathbb{R}^2. \tag{2.3}$$

Furthermore, using the identities

$$\begin{cases} \dfrac{\partial G}{\partial x} = 2tG, \\ \dfrac{\partial^2 G}{\partial x^2} - 2x\dfrac{\partial G}{\partial x} + 2t\dfrac{\partial G}{\partial t} = 0. \end{cases}$$

one can easily derive the recursive relation on $\{H_n(\cdot)\}$:

$$\begin{cases} H_n'(x) = 2nH_{n-1}(x), & n = 1,2,3,\cdots, \\ H_n''(x) - 2xH_n'(x) + 2nH_n(x) = 0, & n = 0,1,2,\cdots, \\ H_n(x) - 2xH_{n-1}(x) + 2(n-1)H_{n-2}(x) = 0 & n = 2,3,4,\cdots, \end{cases} \tag{2.4}$$

with the "intial values": $H_0 = 1$ and $H_1 = 2x$.

We list some of the properties that will be useful in the future (cf. e.g., Gradshteyn-Ryzhik [**Gr-R**]):

- For each $n \in \mathbb{N}$, one has

$$\begin{cases} H_{2n}(0) = (-1)^n 2^n (2n-1)!!, \\ H_{2n}'(0) = 0, \end{cases} \tag{2.5}$$

and

$$\begin{cases} H_{2n+1}(0) = 0, \\ H_{2n+1}'(0) = (-1)^n \frac{2(2n+1)!}{n!}. \end{cases} \tag{2.6}$$

- The Hermite polynomials $H_n(x)$ are "orthogonal" in the weighted $L^2(\mathbb{R})$:

$$\int_{-\infty}^{\infty} e^{-x^2} H_n(x) H_m(x) dx = \begin{cases} 0, & n \neq m, \\ \sqrt{\pi} 2^n n! & n = m. \end{cases} \tag{2.7}$$

- For any $n \in \mathbb{N}$ and $x \in \mathbb{R}$, it holds that

$$|H_{2n}(x)| \leq 2^n (2n-1)!! e^{\frac{x^2}{2}}. \tag{2.8}$$

**C. Parabolic Cylinder Functions**

Parabolic Cylinder Functions (PCF) have been used in many fields such as Dirichlet problems in parabolic cylinder coordinates (half-integral order) and statistical thermodynamics, crystallography or lattice field theory (integral order). We are interested in them because of their relation to Hermite polynomials. We refer to [**Gr-R**], [**Ha**], and [**T**] for more detailed information on such functions.

By "Parabolic Cylinder Functions" we mean the solutions of the following second order ordinary differential equations with parameter $p$:

$$y'' + \left(p + \frac{1}{2} - \frac{1}{4}x^2\right) y = 0. \tag{2.9}$$

Let $p \in \mathbb{R}$, and we consider PCF's, denoted by $D_p$, that have closed-form formulae. Such closed-form presentation will facilitate the numerical experiment tremendously. To write down the explicit formula let us introduce some auxiliary functions. Let $\Gamma(\alpha)$ be the usual $\Gamma$-function, and denote

$$(a)_n \triangleq \Gamma(a+n)/\Gamma(a), \qquad n = 0, 1, 2, \cdots .$$

The so-called *confluent hypergeometric function* is defined by

$${}_1F_1(a, c, z) = \sum_{n=0}^{\infty} \frac{(a)_n z^n}{(c)_n n!}.$$

Next, we define the following pair of auxiliary functions:

$$\begin{aligned} y_1(a,z) &= e^{-(1/4)z^2} {}_1F_1\left(-\frac{1}{2}a + \frac{1}{4}, \frac{1}{2}; -\frac{1}{2}z^2\right) = e^{(1/4)z^2} F_1\left(\frac{1}{2}a + \frac{1}{4}, \frac{1}{2}; -\frac{1}{2}z^2\right), \\ y_2(a,z) &= ze^{-(1/4)z^2} {}_1F_1\left(\frac{1}{2}a + \frac{3}{4}, \frac{3}{2}; \frac{1}{2}z^2\right) = ze^{(1/4)z^2} {}_1F_1\left(-\frac{1}{2}a + \frac{3}{4}, \frac{3}{2}; -\frac{1}{2}z^2\right). \end{aligned}$$

The Parabolic Cylinder Functions $D_p$ can then be written as follows (see, e.g., Temme [**T**]):

$$(2.10) \quad D_p(z) = \sqrt{\pi} 2^{\frac{p}{2}+\frac{1}{4}} \left\{ \frac{2^{-\frac{1}{4}} y_1(-(p+\frac{1}{2}), z)}{\Gamma(\frac{1}{2} - \frac{p}{2})} - \frac{2^{\frac{1}{4}} y_2(-(p+\frac{1}{2}), z)}{\Gamma(-\frac{p}{2})} \right\}.$$

It is fairly easy to check that the following recursive relations hold for the functions $D_p$ and $D_{-p}$:

$$(2.11) \quad \begin{cases} D_{p+1}(x) = xD_p(x) - pD_{p-1}(x), & p \in \mathbb{R} \\ \frac{d}{dx} D_p(x) = \frac{-x}{2} D_p(x) + pD_{p-1}(x), & p \in \mathbb{R} \\ \frac{d}{dx} D_p(x) = \frac{x}{2} D_p(x) - D_{p+1}(x). & p \in \mathbb{R}. \end{cases}$$

An important case, which is of particular interest to us, is the case when $p = n$, a natural number. In this case the PCFs have the following relations with the Hermite polynomials:

$$D_n(z) = -2^{-\frac{n}{2}} e^{-\frac{z^2}{4}} H_n\left(\frac{z}{\sqrt{2}}\right).$$

Finally, we note that the parabolic cylinder function $D_p(z)$ has the following integral form for $p < 0$:

$$(2.12) \qquad D_p(z) = \frac{e^{-\frac{z^2}{4}}}{\Gamma(-p)} \int_0^\infty t^{-p-1} e^{-tz-\frac{t^2}{2}} dt.$$

## 3. Problem Formulation

In this section we study the jump diffusion models that motivated the convolution formulae that we are interested in. We should note that our framework is almost parallel to the one proposed by Kou-Wang [**K-W**], except for the assumption on the logarithm of the jump sizes. Let $(\Omega, \mathcal{F}, \mathbf{P}; \{\mathcal{F}_t\})$ be a complete filtered probability space on which is defined a Brownian motion $W$ and a compound Poisson process $J$, both adapted to the filtration $\{\mathcal{F}_t\}$. More precisely, we assume that the process $J$ takes the following form:

$$(3.1) \qquad J_t = \sum_{j=1}^{N_t} (V_j - 1), \qquad t \geq 0,$$

where $N = \{N_t\}$ is a standard Poisson process with rate $\lambda$, and $\{V_j\}$ is a sequence of i.i.d. nonnegative random variables. We assume that

(i) for each $j$, $X_j = \log(V_j)$ has a power exponential distribution with density given in (2.1);

(ii) the processes $W$, $N$, and $X_j$'s are independent;

(iii) $\mathcal{F}_t = \sigma\{W_s, J_s : 0 \le s \le t\}$, $t \ge 0$, augmented under $\mathbf{P}$ so that it satisfies the *usual hypotheses* (cf. e.g., [**P**]).

In our jump diffusion model we assume that all the economics have a finite horizon $[0,T]$, and the price of our underlying risky asset is given by the following stochastic differential equation (SDE):

$$\frac{dS_t}{S_t} = \mu dt + \sigma dW_t + dJ_t = \mu dt + \sigma dW_t + d\left(\sum_{i=1}^{N_t}(V_i - 1)\right). \tag{3.2}$$

We assume that the drift $\mu$ and the volatility $\sigma$ are constants. It is then well-known (see, e.g., Protter [**P**]) that the solution to the SDE (3.2) is given by the Doléans-Dade stochastic exponential

$$S_t = S_0 \exp\left\{(\mu - \frac{1}{2}\sigma^2)t + \sigma W_t\right\}\prod_{i=1}^{N_t} V_i. \tag{3.3}$$

Consequently, the return process $Z_t = \log(S_t/S_0)$ is given by

$$Z_t = \left\{\mu - \frac{1}{2}\sigma^2\right\}t + \sigma W_t + \sum_{i=1}^{N_t} V_i, \qquad Z_0 = 0. \tag{3.4}$$

In light of the results of Kuo-Wang [**K-W**], we now consider the discretized version of (3.2)–(3.4), following the well-known Euler method.

Let $\pi : t_0 = 0 \le t_1, \cdots, t_n = T$ be any partition of $[0,T]$. Denote $\triangle_i = t_{i+1} - t_i$, and $|\pi| \stackrel{\triangle}{=} \max_{0\le i\le n-1}\{\triangle_i\}$, the mesh size of the partition. In what follows, for any process $\xi$, we denote $\triangle_i\xi = \xi_{t_{i+1}} - \xi_{t_i}$, $i = 0, 1, \cdots, n-1$. Consider the following discretized version of (3.2): $S_0^\pi = S_0$, and for $i = 0, 1, \cdots, n-1$, define

$$S^\pi_{t_{i+1}} = S^\pi_{t_i} + S^\pi_{t_i}\left\{\mu\triangle_i + \sigma[\triangle_i W] + \triangle_i J\right\}. \tag{3.5}$$

Noting the definition of the process $J$ (see (3.1)), we derive easily that

$$\frac{\triangle_i S^\pi}{S^\pi_{t_i}} = \mu\triangle_i + \sigma[\triangle_i W] + \left\{\sum_{j=N_{t_i}+1}^{N_{t_i+1}}(V_j - 1)\right\}. \tag{3.6}$$

Recall that $X = \log V$ (or $V = e^X$), we can approximate $V$ by $1 + X$ so as to rewrite (3.6) as

$$\frac{\triangle_i S^\pi}{S^\pi_{t_i}} = \mu\triangle_i + \sigma[\triangle_i W] + \left\{\sum_{j=N_{t_i}+1}^{N_{t_i+1}} X_j\right\}. \tag{3.7}$$

for $i = 0, ..., n-1$. In what follows we shall consider only the equi-distant discretization, that is $t_i = \frac{iT}{n}$, $i = 0, 1, \cdots, n$, so that $|\pi| = \triangle_i = \frac{T}{n}$, for all $i$.

Note that for each $1 \le i \le n$ the random variable $\triangle_i W \sim N(0, |\pi|)$, we can write $\triangle_i W = \sqrt{|\pi|}Z_i$, where $Z_i$'s are i.i.d. $N(0,1)$-random variables. The equation

(3.7) then becomes

$$\frac{\triangle_i S^\pi}{S^\pi_{t_i}} = \mu\triangle_i + \sigma\sqrt{|\pi|}Z_i + \left\{ \sum_{j=N_{t_i}+1}^{N_{t_i+1}} X_j \right\}. \tag{3.8}$$

As it was shown in [**K-W**], for $\triangle_i$ small enough we have

$$\sum_{j=N_{t_i}+1}^{N_{t_i+1}} X_j = \begin{cases} X_{N_{t+\triangle_i}} & \text{w.p.} \quad \lambda\triangle_i, \\ 0 & \text{w.p.} \quad 1-\lambda\triangle_i. \end{cases}$$

In other words, if $\delta \triangleq |\pi|$ is sufficiently small, the return can be approximated in distribution by

$$\frac{\triangle S_t}{S_t} = \mu\delta + \sigma Z\sqrt{\delta} + B\cdot X \tag{3.9}$$

where $B$ is a Bernoulli random variable with $\mathbf{P}(B=1)=\lambda\delta$ and $\mathbf{P}(B=0)=1-\lambda\delta$, and $Z \sim N(0,1)$. Note that

$$\begin{aligned} \mathbf{P}\{\sigma\sqrt{\delta}Z + BX \le x\} &= \mathbf{P}\{\sigma\sqrt{\delta}Z + X \le x\}\mathbf{P}\{B=1\} + \mathbf{P}\{\sigma\sqrt{\delta}Z \le x\}\mathbf{P}\{B=0\} \\ &= \mathbf{P}\{\sigma\sqrt{\delta}Z + X \le x\}\lambda\delta + \mathbf{P}\{\sigma\sqrt{\delta}Z \le x\}(1-\lambda\delta). \end{aligned}$$

The problem is thus reduced to calculate the distribution of the random variable $\sigma\sqrt{\delta}Z + X$, an independent sum of a normal random variable and a power exponential random variable. In what follows we refer the problem of calculating the distribution of such a sum as *Problem NP(1)*, with "1" meaning that there is only one power exponentially random variable involved.

Let us now look at the option pricing problem. Following the idea of Merton [**M**] and/or Duffie [**D**], we shall start with the *risk-neutral* measure $\mathbf{P}^*$, under which the return process is rewritten as

$$\frac{dS_t}{S_t} = (r - \lambda\mathbf{E}(V-1))dt + \sigma dW_t + dJ_t = (\mu - \lambda\alpha)dt + \sigma dW_t + dJ_t, \tag{3.10}$$

where $\alpha = \mathbf{E}(e^X) - 1$, and $r$ is the shot rate of the riskless asset. Note that the unique solution to the SDE (3.10) is given by

$$\begin{aligned} S_t &= S_0 \exp\left\{(r - \frac{1}{2}\sigma^2 - \lambda\alpha)t + \sigma W_t\right\}\prod_{i=1}^{N_t} V_i \\ &= S_0 \exp\left\{(r - \frac{1}{2}\sigma^2 - \lambda\alpha)t + \sigma\sqrt{t}Z + \sum_{i=1}^{N_t} X_i\right\}, \end{aligned} \tag{3.11}$$

where $Z \sim N(0,1)$. For an option of the form $g(S_T)$ the hedging price at time 0 is given by

$$\begin{aligned} V_0 &= \mathbf{E}^*\left\{e^{-\mu T} g\left(S_0 \exp\left\{\left(r - \frac{\sigma^2}{2} - \lambda\alpha\right)T + \sigma\sqrt{T}Z\right\}\prod_{i=1}^{N_t} V_i\right)\right\} \\ &= \sum_{n=0}^{\infty} \mathbf{E}^*\left\{e^{-rT} g\left(S_0 e^{(r-\frac{\sigma^2}{2}-\lambda\alpha)T+\sigma\sqrt{T}Z}\prod_{i=1}^{N_t} V_i\right) \middle| N_T = n\right\}\mathbf{P}\{N_T = n\} \\ &= e^{-rT}\sum_{m=0}^{\infty} e^{-\lambda T}\frac{(\lambda T)^m}{m!}\mathbf{E}^*\left\{g\left(S_0 e^{-\lambda\alpha T} e^{\left(r-\frac{\sigma^2}{2}\right)T+\sigma\sqrt{T}Z+\sum_{i=1}^m X_i}\right)\right\}. \end{aligned}$$

Clearly, the calculation of the right hand side above would require the knowledge of the distribution of the random variable $\sigma\sqrt{T}Z + \sum_{i=1}^{m} X_i$, $m = 1, 2, \cdots$, which are essentially the independent sums of one normal random variable and $m$ independent power exponentially distributed random variables. In what follows we refer to such a problem as *Problem* $NP(m)$, $m = 2, 3, \cdots$.

We remark that the solution to the Problem NP($m$) is nothing but the $m+1$-fold convolution of a normal random variable and $k$ independent power exponential random variables. Our main purpose is to find the convolution formulae in terms of either Hermite polynomials or parabolic cylinder functions so as to numerically simulate the underlying prices and calculate option prices. We should also note such series solution may not even converge in general. For this reason, we borrow the notation of the well-known Taylor series expansion. For example, we shall denote in general a Hermite polynomial expansion as

$$f(z) \cong \sum_{n=0}^{\infty} \alpha_n(z) H_n(\beta_n z)$$

or

$$f(z) \cong \sum_{n=0}^{\infty} \alpha_n(z)[D_{\beta_n}(z) + D_{\beta_n}(-z)],$$

and change the sign "$\cong$" to "$=$" after we verify the convergence of the series. The essence here, however, is that with the explicit form of the series expansion, one can always use the partial sum to approximate the density function, even without actually proving the convergence(!).

To end this section we take a closer look at the quantity $\alpha \stackrel{\triangle}{=} \mathbf{E}\{e^X\}$ which plays an important role in the formula (3.10). The following series expansion for $\alpha$ gives an idea for what we are trying to do in the rest of the paper.

PROPOSITION 3.1. *Suppose that* $X \sim P(\beta, 0, \phi)$. *Then for all* $\beta \geq 1/2$ *and* $\phi < 1/2$ *the following formula follows*

$$\mathbf{E}(e^X) = \frac{C(\phi,\beta)}{\beta} \sum_{n=0}^{\infty} \frac{\Gamma\left(\frac{2n+1}{2\beta}\right)}{(2n)!} 2^{\frac{2n+1}{2\beta}} \phi^{2n+1}, \tag{3.12}$$

*where*

$$C(\phi,\beta) = \frac{1}{\phi\Gamma\left(1+\frac{1}{2\beta}\right)2^{1+\frac{1}{2\beta}}}. \tag{3.13}$$

PROOF. Suppose $X \sim P(\beta, 0, \phi)$. Then its density function $f(x) = f_X(x)$ is given by (2.1) with normalizing constant $C(\phi,\beta)$. Assume that $\beta \geq 1/2$, then first using the Taylor expansion for $e^x$ and then formally applying Fubini's theorem we have

$$\mathbf{E}(e^X) = \int_{\mathbb{R}} e^x f_X(x)dx \cong C(\phi,\beta)\sum_{n=0}^{\infty}\frac{1}{n!}\int_{\mathbb{R}} x^n e^{-\frac{1}{2\phi^{2\beta}}|x|^{2\beta}}dx.$$

Note that the integrals on the right hand side above are all absolutely convergent, and it holds that

$$\int_{\mathbb{R}} x^n e^{-\frac{1}{2\phi^{2\beta}}|x|^{2\beta}}dt = \begin{cases} 2\int_0^{\infty} x^n e^{-\frac{1}{2\phi^{2\beta}}x^{2\beta}}dx = \Gamma\left(\frac{n+1}{2\beta}\right)\frac{1}{\beta}\phi^{n+1}2^{\frac{n+1}{2\beta}} & \text{if } n \text{ is even,} \\ 0 & \text{if } n \text{ is odd.} \end{cases}$$

Thus we have

$$\mathbf{E}(e^X) \cong \frac{C(\phi,\beta)}{\beta}\sum_{n=0}^{\infty}\frac{\Gamma\left(\frac{2n+1}{2\beta}\right)}{(2n)!}2^{\frac{2n+1}{2\beta}}\phi^{2n+1}.$$

To show that the equality actually holds in the above (whence (3.12)), first recall an important limit (see, e.g., [**Gr-R**])

$$\lim_{|z|\to\infty}\frac{\Gamma(z+a)}{\Gamma(z)}z^{-a} = 1. \tag{3.14}$$

Let us define $b_n(\beta) \triangleq \frac{\Gamma(\frac{2n+1}{2\beta})}{(2n)!}2^{\frac{2n+1}{2\beta}}\phi^{2n+1}$. Then, applying the ratio test we have, for any $n$,

$$\left|\frac{b_{n+1}(\beta)}{b_n(\beta)}\right| = \frac{2^{\frac{1}{\beta}}\phi^2}{(2n+2)(2n+1)}\frac{\Gamma\left(\frac{2n+3}{2\beta}\right)}{\Gamma\left(\frac{2n+1}{2\beta}\right)}. \tag{3.15}$$

Thus for $\beta > 1/2$, if we let $z = \frac{n}{\beta} + \frac{1}{2\beta}$ and $a = \frac{1}{\beta}$ in (3.14), then by (3.15) we obtain that

$$\lim_{n\to\infty}\left|\frac{b_{n+1}(\beta)}{b_n(\beta)}\right| = \lim_{n\to\infty}\frac{2^{\frac{1}{\beta}}\phi^2(\frac{n}{\beta}+\frac{1}{2\beta})^{\frac{1}{\beta}}}{(2n+2)(2n+1)}\left\{\left(\frac{n}{\beta}+\frac{1}{2\beta}\right)^{-\frac{1}{\beta}}\frac{\Gamma\left(\frac{2n+3}{2\beta}\right)}{\Gamma\left(\frac{2n+1}{2\beta}\right)}\right\} = 0. \tag{3.16}$$

Hence $\sum b_n(\beta) < \infty$, and the series in (3.12) converges for all $\beta > 1/2$.

Finally, we note that when $\beta = 1/2$, the limit in (3.16) is actually equal to $(2\phi)^2$, thus the series (3.12) converges for $\phi < 1/2$. $\square$

We remark that when $\beta = 1/2$ the series (3.12) is equal to $\frac{4\phi}{(1-(2\phi)^2)}$, and it coincides with the result in [**K-W**] for the case of "double exponential". Since the double exponential distribution has the "heaviest tail" among all the power exponential distributions with $\beta \geq 1/2$, this result essentially shows that the double exponential distribution is more or less the farthest one can go in such jump diffusion models, whenever $\alpha = E[e^X] < \infty$ is satisfied.

## 4. Solution of Problem NP(1)

We now turn to the (Hermite) series solution to Problem NP(1). Let us assume that $X \sim P(\beta, \mu_1, \phi)$ and $Y \sim N(\mu_2, \sigma^2)$. That is, the density functions of $X$ and $Y$ are, respectively,

$$f_X(x) = \frac{1}{\phi\Gamma\left(1+\frac{1}{2\beta}\right)2^{1+\frac{1}{2\beta}}}e^{-\frac{1}{2}\left|\frac{x-\mu_1}{\phi}\right|^{2\beta}} = C(\phi,\beta)e^{-\frac{1}{2}\left|\frac{x-\mu_1}{\phi}\right|^{2\beta}},$$

where $C(\phi, \beta)$ is given by (3.13), and

$$f_Y(y) = \frac{1}{\sqrt{2\pi}\sigma}e^{-\frac{(y-\mu_2)^2}{2\sigma^2}}.$$

To simplify notation we assume from now on $\mu_1 = \mu_2 = 0$, the general case can be argued in the same way.

Recall that the complete (Hermite-series) solution of NP(1) consists of two parts: a closed form series representation and its convergence analysis. Since the convergence analysis is usually lengthy, and the series representation is sometimes already sufficient for practical purposes, we thus present the results separately. We first give a formal Hermite-expansion theorem without studying the convergence.

THEOREM 4.1. *Suppose that $X \sim P(\beta, 0, \phi)$ and $Y \sim N(0, \sigma^2)$, and that $X$ and $Y$ are independent. Then the density function of $X + Y$ has the following series representation:*

*(i) for $\beta > 1$,*

$$f_{X+Y}(z) \cong \frac{C(\phi,\beta)}{\sqrt{2\pi}\sigma} e^{\frac{-z^2}{2\sigma^2}} \sum_{n=0}^{\infty} \frac{\Gamma\left(\frac{2n+1}{2\beta}\right)}{(2n)!(2\sigma^2)^n} \phi^{2n+1} 2^{\frac{2n+1}{2\beta}} H_{2n}\left(\frac{z}{\sqrt{2}\sigma}\right); \tag{4.1}$$

*(ii) for $0 < \beta < 1$,*

$$\begin{aligned} f_{X+Y}(z) \quad &\cong \quad \frac{C(\phi,\beta)}{\sqrt{2\pi}} e^{-\frac{z^2}{4\sigma^2}} \sum_{n=0}^{\infty} (-1)^n \frac{\Gamma(2\beta n+1)\sigma^{2\beta n}}{n! 2^n \phi^{2\beta n}} \\ &\quad \cdot \left[D_{-(2\beta n+1)}\left(\frac{z}{\sigma}\right) + D_{-(2\beta n+1)}\left(-\frac{z}{\sigma}\right)\right], \end{aligned} \tag{4.2}$$

*where $H_n$'s and $D_\mu$'s are the Hermite polynomials and the parabolic cylinder functions, respectively.*

PROOF. First assume $\beta > 1$. From (2.3) it is easily seen that

$$e^{\frac{1}{2\sigma^2}(-t^2+2zt)} = \sum_{n=0}^{\infty} \frac{t^n}{n!(\sqrt{2}\sigma)^n} H_n\left(\frac{z}{\sqrt{2}\sigma}\right),$$

where $H_n$'s are the Hermite polynomials. Formally applying the Fubini theorem to the convolution formula we have

$$\begin{aligned} f_{X+Y}(z) &= \int_{\mathbb{R}} f_X(x) f_Y(z-x) dx \\ &\cong \frac{C(\phi,\beta)}{\sqrt{2\pi}\sigma} e^{-\frac{z^2}{2\sigma^2}} \sum_{n=0}^{\infty} \frac{H_n(\frac{z}{\sqrt{2}\sigma})}{n!(\sqrt{2}\sigma)^n} \int_{\mathbb{R}} t^n e^{-\frac{1}{2\phi^{2\beta}}|t|^{2\beta}} dt. \end{aligned}$$

We note that the integrals inside the summation on the right hand side above are obviously absolutely convergent, and they can be calculated explicitly as

$$\int_{\mathbb{R}} t^n e^{-\frac{1}{2\phi^{2\beta}}|t|^{2\beta}} dt = \begin{cases} 2\int_0^{\infty} t^n e^{-\frac{1}{2\phi^{2\beta}}t^{2\beta}} dt = \Gamma\left(\frac{n+1}{2\beta}\right) \frac{1}{\beta} \phi^{n+1} 2^{\frac{n+1}{2\beta}} & \text{if } n \text{ is even}, \\ 0 & \text{if } n \text{ is odd}. \end{cases}$$

Consequently we have

$$f_{X+Y}(z) \cong \frac{C(\phi,\beta)}{\sqrt{2\pi}\sigma} e^{-\frac{z^2}{2\sigma^2}} \sum_{n=0}^{\infty} \frac{\Gamma\left(\frac{2n+1}{2\beta}\right)}{(2n)!(2\sigma^2)^n} \phi^{2n+1} 2^{\frac{2n+1}{2\beta}} H_{2n}\left(\frac{z}{\sqrt{2}\sigma}\right),$$

proving (4.1).

Now let us assume $0 < \beta < 1$. As we will see in the next theorem, the usual technique used to prove the convergence of the series representation for $\beta > 1$ does not work for this case. Thus a different approach is in order. In this case let us first use the Taylor expansion for the exponential function

$$e^{-\frac{1}{2\phi^{2\beta}}|t|^{2\beta}} = \sum_{n=0}^{\infty} \frac{(-1)^n}{n! 2^n \phi^{2\beta n}} |t|^{2\beta n}.$$

Formally applying Fubini's theorem again in the convolution formula we obtain that

$$f_{X+Y}(z) \cong \frac{C(\phi,\beta)}{\sqrt{2\pi}\sigma} e^{\frac{-z^2}{2\sigma^2}} \sum_{n=0}^{\infty} \frac{(-1)^n}{n! 2^n \phi^{2\beta n}} \int_{\mathbb{R}} |t|^{2\beta n} e^{-\frac{1}{2\sigma^2}(t^2-2tz)} dt. \tag{4.3}$$

Note that the integrals inside the summation on the right hand side are still absolutely convergent. In fact, we have the following closed form formula (see, for example, [**Gr-R**]): for $\mu > 0$, $\nu > 0$, and $x \in \mathbb{R}$,

$$I(x;\mu,\nu) \triangleq \int_0^{\infty} t^{\nu-1} e^{-\mu t^2 - xt} dt = (2\mu)^{-\frac{\nu}{2}} \Gamma(\nu) e^{\frac{x^2}{8\mu}} D_{-\nu}\left(\frac{x}{\sqrt{2\nu}}\right). \tag{4.4}$$

Thus the integral on the right hand side of (4.3) should read

$$\begin{aligned} &\int_{\mathbb{R}} |t|^{2\beta n} e^{-\frac{1}{2\sigma^2}(t^2-2tz)} dt \\ = \; & I(-\frac{z}{\sigma^2}; \frac{1}{2\sigma^2}, 2\beta n+1) + I(\frac{z}{\sigma^2}; \frac{1}{2\sigma^2}, 2\beta n+1) \qquad (4.5) \\ = \; & \sigma^{2\beta n+1} \Gamma(2\beta n+1) e^{\frac{z^2}{4\sigma^2}} \Big[ D_{-(2\beta n+1)}\Big(-\frac{z}{\sigma}\Big) + D_{-(2\beta n+1)}\Big(\frac{z}{\sigma}\Big) \Big]. \end{aligned}$$

Plugging (4.5) into (4.3) we derive (4.2). □

We now try to replace the sign "≅" by "=" in (4.1) and (4.2). The following theorem explains why we need to use different representations for the cases $\beta > 1$ and $\beta < 1$.

THEOREM 4.2. *Suppose that $X$ and $Y$ are random variables as defined in Theorem 4.1. Then, the following convergence results hold:*

*(i) if $\beta > 1$ and $\phi \le \sigma$, then the Hermite series (4.1) converges absolutely and uniformly in $z \in \mathbb{R}$;*

*(ii) if $0 < \beta < 1$ and $\sigma \le \phi$, then the Hermite series (4.2) converges absolutely and uniformly in $z \subset \mathbb{R}$;*

*Consequently, in all cases above the "≅" signs in both (4.1) and (4.2) can be replaced by equalities.*

PROOF. (i) Assume $\beta > 1$ and $\phi \le \sigma$. Denote the right hand side of (4.1) by $I_1(z)$. Then we have

$$\begin{aligned} |I_1(z)| \; &\le \; \frac{C(\phi,\beta)}{\sqrt{2\pi}} e^{-\frac{z^2}{2\sigma^2}} \sum_{n=0}^{\infty} \frac{\Gamma\left(\frac{2n+1}{2\beta}\right)}{(2n)! 2^n} \left(\frac{\phi}{\sigma}\right)^{2n+1} 2^{\frac{2n+1}{2\beta}} \left| H_{2n}\Big(\frac{z}{\sqrt{2}\sigma}\Big)\right| \\ &\le \; \frac{C(\phi,\beta)}{\sqrt{2\pi}} e^{-\frac{z^2}{2\sigma^2}} \sum_{n=0}^{\infty} \frac{\Gamma\left(\frac{2n+1}{2\beta}\right)}{(2n)! 2^n} 2^{\frac{2n+1}{2\beta}} \left| H_{2n}\Big(\frac{z}{\sqrt{2}\sigma}\Big)\right|. \qquad (4.6) \end{aligned}$$

Thus it suffices to show the convergence of the series

$$\sum_{n=0}^{\infty} \frac{\Gamma\left(\frac{2n+1}{2\beta}\right)}{(2n)! 2^n} 2^{\frac{2n+1}{2\beta}} \left| H_{2n}\Big(\frac{z}{\sqrt{2}\sigma}\Big)\right| e^{-\frac{z^2}{2\sigma^2}}. \tag{4.7}$$

Note that the basic estimate of Hermite polynomial (2.8) implies that for all $z \in \mathbb{R}$,

$$\begin{aligned}\frac{\Gamma\left(\frac{2n+1}{2\beta}\right)}{(2n)!2^n}2^{\frac{2n+1}{2\beta}}\Big|H_{2n}\Big(\frac{z}{\sqrt{2}\sigma}\Big)\Big|e^{-\frac{z^2}{2\sigma^2}} &\le \frac{\Gamma\left(\frac{2n+1}{2\beta}\right)}{(2n)!}2^{\frac{2n+1}{2\beta}}(2n-1)!!e^{-\frac{z^2}{4\sigma^2}} \\ &\le \frac{\Gamma\left(\frac{2n+1}{2\beta}\right)}{(2n)!}2^{\frac{2n+1}{2\beta}}(2n-1)!! \triangleq a_n(\beta).\end{aligned}$$

But note that

$$\left|\frac{a_{n+1}(\beta)}{a_n(\beta)}\right| = \frac{2^{\frac{1}{\beta}}}{(2n+2)}\frac{\Gamma\left(\frac{2n+3}{2\beta}\right)}{\Gamma\left(\frac{2n+1}{2\beta}\right)}. \tag{4.8}$$

Since $\beta > 1$, if we set $z = \frac{n}{\beta} + \frac{1}{2\beta}$ and $a = \frac{1}{\beta}$ in the identity (3.14), then from (4.8) we see that

$$\lim_{n\to\infty}\left|\frac{a_{n+1}(\beta)}{a_n(\beta)}\right| = \lim_{n\to\infty}\frac{2^{\frac{1}{\beta}}(\frac{n}{\beta}+\frac{1}{2\beta})^{\frac{1}{\beta}}}{(2n+2)}\left\{\Big(\frac{n}{\beta}+\frac{1}{2\beta}\Big)^{-\frac{1}{\beta}}\frac{\Gamma\left(\frac{2n+3}{2\beta}\right)}{\Gamma\left(\frac{2n+1}{2\beta}\right)}\right\} = 0. \tag{4.9}$$

Hence $\sum a_n(\beta) < \infty$, and the series in (4.7) converges absolutely and uniformly for all $z$, proving part (i).

(ii) $0 < \beta < 1$. Again we write the right hand side of (4.2) as $I_2(z)$, using the assumption $\sigma \le \phi$ we then have

$$\begin{aligned}|I_2(z)| &\le \sum_{n=0}^{\infty}\frac{1}{2^n n!}\Gamma(2\beta n+1)\left[|D_{-(2\beta n+1)}(z)| + |D_{-(2\beta n+1)}(-z)|\right] \\ &\triangleq I_2^1(z) + I_2^2(z),\end{aligned} \tag{4.10}$$

where

$$\begin{aligned}I_2^1(z) &\triangleq \sum_{n=0}^{\infty}\frac{1}{2^n n!}\Gamma(2\beta n+1)|D_{-(2\beta n+1)}(z)| \\ I_2^2(z) &\triangleq \sum_{n=0}^{\infty}\frac{1}{2^n n!}\Gamma(2\beta n+1)|D_{-(2\beta n+1)}(-z)|.\end{aligned}$$

It then suffices to show that both $I_2^1(\cdot)$ and $I_2^2(\cdot)$ converge uniformly in $z$. It is readily seen that we need only check the uniform convergence of $I_2^1(z)$ for $z > 0$, thanks to the symmetry.

To this end, let $p = -(2\beta n+1)$. By (2.12) we have

$$D_{-(2\beta n+1)}(z) = \frac{e^{-\frac{z^2}{4}}}{\Gamma(2\beta n+1)}\int_0^\infty t^{2\beta n}e^{-tz-\frac{t^2}{2}}dt.$$

Putting this into the right side of $I_2^1$ we can easily check that, for $z > 0$,

$$\begin{aligned}I_2^1(z) &\le \sum_{n=0}^{\infty}\frac{1}{2^n n!}e^{-\frac{z^2}{4}}\int_0^\infty t^{2\beta n}e^{-tz-\frac{t^2}{2}}dt \\ &\le \sum_{n=0}^{\infty}\frac{1}{2^n n!}e^{-z^2/4}\int_0^\infty t^{2\beta n}e^{-t^2/2}dt \le \sum_{n=0}^{\infty}\frac{2^{\frac{2\beta n-1}{2}}\Gamma\left(\frac{2\beta n+1}{2}\right)}{2^n n!}.\end{aligned}$$

Again, let us define

$$a_n(\beta) \triangleq \frac{2^{\frac{2\beta n-1}{2}}\Gamma\left(\frac{2\beta n+1}{2}\right)}{2^n n!}.$$

Then, applying (3.14) again (with $z = \frac{2\beta n+1}{2}$ and $a = \beta$) we get that, for $\beta < 1$,

$$\text{(4.11)} \quad \lim_{n\to\infty}\left|\frac{a_{n+1}(\beta)}{a_n(\beta)}\right| = \lim_{n\to\infty}\frac{2^{\beta-1}}{n+1}\frac{\Gamma\left(\frac{2\beta n+1+2\beta}{2}\right)}{\Gamma\left(\frac{2\beta n+1}{2}\right)}$$
$$\leq \lim_{n\to\infty}\frac{(\beta n+\frac{1}{2})^\beta}{(n+1)}\left\{(\beta n+\frac{1}{2})^{-\beta}\frac{\Gamma\left(\frac{2\beta n+1+2\beta}{2}\right)}{\Gamma\left(\frac{2\beta n+1}{2}\right)}\right\}=0.$$

Thus, by the ratio test we see that $\sum a_n(\beta) < \infty$, and hence $I_2^1(z)$ converges absolutely and uniformly for $z > 0$. The proof is now complete. □

## 5. Two special cases ($\beta = 1$ and $\beta = \frac{1}{2}$)

Observe that in Theorems 4.1 and 4.2 we did not discuss the case $\beta = 1$. In fact in this case neither argument works, since the limits in both ratio tests equal to 1(!). We shall nevertheless prove that in this case the Hermite series (4.1) still converges, and it is in fact the unique solution to the Cauchy problem of a second order ODE. Further, since the case $\beta = 1/2$ corresponds to the double exponential distribution, for which an explicit formula is given by Kuo-Wang [**K-W**], we shall prove in this section that our Hermite series expansion (4.2) produces exactly the same thing, although starting from a seemingly different formula.

**5.1. The normal case ($\beta$= 1).** We first note that when $\beta = 1$, $P(1, 0, \phi) = N(0, \phi^2)$, thus the solution to problem NP(1) is nothing but a normal distribution. Furthermore, since both $X$ and $Y$ are normal random variables in this case, in the convolution formula the role of $\sigma$ and $\phi$ are interchangeable. Hence we might as well assume in this section that $\phi = \sigma = 1$. That is, $X \sim P(1, 0, 1) = N(0, 1)$ and $Y \sim N(0, 1)$, hence the independent sum $X + Y \sim N(0, 2)$. In other words, we have

$$\text{(5.1)} \qquad f_{X+Y}(z) = \frac{1}{2\sqrt{\pi}}e^{\frac{-z^2}{4}}.$$

On the other hand, setting $\beta = 1$, $\phi = 1$ in (4.1) we have

$$\text{(5.2)} \qquad f_{X+Y}(z) \cong \frac{C(1,1)}{\sqrt{2\pi}}e^{\frac{-z^2}{2}}\sum_{n=0}^{\infty}\frac{H_{2n}(\frac{z}{\sqrt{2}})}{(2n)!2^n}\Gamma\left(\frac{2n+1}{2}\right)2^{\frac{2n+1}{2}}.$$

Since, by (3.13), we have

$$C(1,1) = \frac{1}{\Gamma\left(1+\frac{1}{2}\right)2^{1+\frac{1}{2}}} = \frac{1}{\sqrt{2\pi}},$$

we have the following result.

THEOREM 5.1. *Suppose that $X$ and $Y$ are two independent $N(0,1)$-random variables. Then it holds that*

$$(5.3)\quad f_{X+Y}(z) = \frac{1}{2\sqrt{\pi}}e^{\frac{-z^2}{4}} = \frac{C(1,1)}{\sqrt{2\pi}}e^{\frac{-z^2}{2}}\sum_{n=0}^{\infty}\frac{H_{2n}(\frac{z}{\sqrt{2}})}{(2n)!2^n}\Gamma\left(\frac{2n+1}{2}\right)2^{\frac{2n+1}{2}}$$

$$(5.4)\quad = \frac{1}{\sqrt{2\pi}}e^{\frac{-z^2}{2}}\sum_{n=0}^{\infty}\frac{H_{2n}(\frac{z}{\sqrt{2}})}{(2n)!}\Gamma\left(\frac{2n+1}{2}\right).$$

*Furthermore, the series*

$$\psi(u) = \sum_{n=0}^{\infty}\frac{H_{2n}(u)}{(2n)!}\Gamma\left(\frac{2n+1}{2}\right) = \sqrt{\frac{\pi}{2}}e^{\frac{u^2}{2}}$$

*is the unique solution of the following second order homogenerous ODE:*

$$(5.5)\quad \begin{cases} \psi''(u) - u\psi'(u) - \psi(u) = 0, \\ \psi(0) = \sqrt{\frac{\pi}{2}}, \quad \psi'(0) = 0. \end{cases}$$

PROOF. First, letting $u = \frac{z}{\sqrt{2}}$, we see that it suffices to show the following identity:

$$(5.6)\quad \sum_{n=0}^{\infty}\frac{H_{2n}(u)}{(2n)!}\Gamma\left(\frac{2n+1}{2}\right) = \sqrt{\frac{\pi}{2}}e^{\frac{u^2}{2}}.$$

We first analyze the particular case $u = 0$. By (2.5) we have

$$H_{2n}(0) = (-1)^n 2^n (2n-1)!! \quad \text{and} \quad \Gamma\left(\frac{2n+1}{2}\right) = \frac{\sqrt{\pi}}{2^n}(2n-1)!!.$$

Therefore (5.6) is equivalent to

$$(5.7)\quad \sum_{n=0}^{\infty}\frac{(-1)^n(2n-1)!!(2n-1)!!}{(2n)!} = \frac{1}{\sqrt{2}}.$$

To prove (5.7) let us make the following elementary observations. First, consider the Taylor expansion:

$$(5.8)\quad \frac{1}{\sqrt{1-4x}} = \sum_{n=0}^{\infty}\frac{(2n)!}{n!n!}x^n, \qquad -\frac{1}{4} \le x < \frac{1}{4}.$$

Now, setting $x = \frac{-1}{4}$ in (5.8) and noting that $(2n)! = (2n-1)!!(2n)!! = (2n-1)!!2^n n!$, we obtain that

$$\frac{1}{\sqrt{2}} = \sum_{n=0}^{\infty}\frac{(-1)^n(2n)!}{(2^n n!)^2} = \sum_{n=0}^{\infty}\frac{(-1)^n(2n)!((2n-1)!!)^2}{((2n)!)^2}.$$

The identity (5.7) thus follows.

In the general case we shall make use of the properties of the Hermite polynomials given in (2.4), and the stability result of viscosity solutions (see, e.g., Fleming and Soner [**F-S**]). For any integer $N$, define

$$\psi_N(u) \triangleq \sum_{n=0}^{N}\frac{H_{2n}(u)}{(2n)!}\Gamma\left(\frac{2n+1}{2}\right) = \sqrt{\pi} + \sum_{n=1}^{N}\frac{H_{2n}(u)}{(2n)!}\Gamma\left(\frac{2n+1}{2}\right).$$

Differentiating both sides above and using the recursive relation (2.4) we obtain that

$$\psi_N'(u) = \sum_{n=1}^{N} \frac{H_{2n}'(u)}{(2n)!}\Gamma\left(\frac{2n+1}{2}\right) = \sum_{n=1}^{N} 4n\frac{H_{2n-1}(u)}{(2n)!}\Gamma\left(\frac{2n+1}{2}\right). \tag{5.9}$$

Differentiating (5.9) again and applying (2.4) we obtain

$$\begin{aligned} \psi_N''(u) &= \sum_{n=1}^{N} \frac{4n2(2n-1)H_{2n-2}(u)}{(2n)!}\Gamma\left(\frac{2n+1}{2}\right) \\ &= \sum_{n=1}^{N} \frac{4n2(2n-1)(-H_{2n}(u)+2uH_{2n-1}(u))}{(2n)!2(2n-1)}\Gamma\left(\frac{2n+1}{2}\right) \\ &= -\sum_{n=1}^{N} \frac{4nH_{2n}(u)}{(2n)!}\Gamma\left(\frac{2n+1}{2}\right) + 2u\psi_N'(u). \end{aligned} \tag{5.10}$$

On the other hand, it is easily seen that

$$\begin{aligned} \psi_N''(u) &= \sum_{n=1}^{N} \frac{4n2(2n-1)H_{2n-2}(u)}{(2n)!}\Gamma\left(\frac{2n+1}{2}\right) \\ &= \sum_{n=1}^{N} \frac{2(2n-1)H_{2n-2}(u)}{(2n-2)!}\Gamma\left(\frac{2n-1}{2}\right) \\ &= 2\sum_{n=1}^{N} \frac{H_{2n-2}(u)}{(2n-2)!}\Gamma\left(\frac{2n-1}{2}\right) + \sum_{n=1}^{N} \frac{4(n-1)H_{2n-2}(u)}{(2n-2)!}\Gamma\left(\frac{2n-1}{2}\right) \\ &= 2[\psi_N(u) - \frac{H_{2n}(u)}{(2n)!}\Gamma\left(\frac{2n+1}{2}\right)] + \sum_{n=1}^{N-1} \frac{4nH_{2n}(u)}{(2n)!}\Gamma\left(\frac{2n+1}{2}\right). \end{aligned} \tag{5.11}$$

Adding (5.10) and (5.11) and dividing both sides by 2 we obtain that

$$\psi_N''(u) = \psi_N(u) + u\psi_N'(u) - (2n+1)\frac{H_{2n-2}(u)}{(2n)!}\Gamma\left(\frac{2n+1}{2}\right). \tag{5.12}$$

This amounts to saying that $\psi_N(\cdot)$ is a (classical) solution to the (non-homogeneous) ODE (5.12) for each $N$. Since

$$\alpha_N(u) \triangleq -(2n+1)\frac{H_{2n-2}(u)}{(2n)!}\Gamma\left(\frac{2n+1}{2}\right) \to 0, \qquad \text{as } N \to \infty,$$

uniformly in $u$ on compacts, and $\psi_N(\cdot)$ converges to $\psi(u) \triangleq \sum_{n=0}^{\infty} \frac{H_{2n}(u)}{(2n)!}\Gamma\left(\frac{2n+1}{2}\right)$, uniformly near $u = 0$, we conclude from the stability of the viscosity solution that $\psi$ is a viscosity solution to the ODE (5.5), at least near $u = 0$. But the uniqueness of the viscosity solution would then imply that the $\psi$ must coincide with $\sqrt{\frac{\pi}{2}}e^{\frac{u^2}{2}}$, the unique solution of (5.5), whenever the series converges. The result then follows from some standard arguments using the extension of the solution for ODEs and analytic functions. □

**5.2. The double exponential case ($\beta$=1/2).** In the case $\beta = 1/2$ the power exponential distribution $P(\frac{1}{2}, 0, \phi)$ becomes the "*double exponential*", that is, the density function is given by

$$f_X(x) = \frac{1}{2\eta} e^{-\frac{|x|}{\eta}},$$

where $\eta = 2\phi$ . Such a case was studied by Kuo-Wang [**K-W**]. In particular, in [**K-W**] it is proved that in this case the density function of $X + Y$ is given by the following formula

$$f_{X+Y}(t) = \frac{1}{\eta} e^{\sigma^2/(2\eta^2)} \left\{ \frac{1}{2} e^{-t/\eta} \Phi\left( \frac{t\eta - \sigma^2}{\sigma\eta} \right) + \frac{1}{2} e^{t/\eta} \Phi\left( -\frac{t\eta + \sigma^2}{\sigma\eta} \right) \right\}. \tag{5.13}$$

In what follows we show that our Hermite series representation (4.2) gives exactly the same formula. To simplify numerical calculations, we shall consider only the case when $\phi = 1/4$ and $\sigma = 1$. The general cases can be obtained by a simple change of variables. We have the following result.

THEOREM 5.2. *Suppose that $\beta = \frac{1}{2}$, $\phi = \frac{1}{4}$, and $\sigma = 1$. Then the Hermite series (4.2) takes the following form:*

$$\begin{aligned} \frac{1}{\sqrt{2\pi}} e^{-z^2/4} \sum_{n=0}^{\infty} (-1)^n 2^n \left[ D_{-(n+1)}(z) + D_{-(n+1)}(-z) \right] \\ = e^2 \left( e^{-2z} \Phi(z-2) + e^{2z} \Phi(-z-2) \right), \end{aligned} \tag{5.14}$$

*where $\Phi(\cdot)$ denotes the standard normal distribution function.*

*Furthermore, if we define*

$$\varphi(z) = \frac{1}{\sqrt{2\pi}} e^{-z^2/4} \sum_{n=0}^{\infty} (-1)^n 2^n \left[ D_{-(n+1)}(z) + D_{-(n+1)}(-z) \right],$$

*then $\varphi$ is the unique solution to the following second order ODE:*

$$\begin{cases} \varphi''(z) = 4\varphi(z) - \frac{4}{\sqrt{2\pi}} e^{-z^2/2} \\ \varphi(0) = 2e^2 \Phi(-2), \qquad \varphi'(0) = 0. \end{cases} \tag{5.15}$$

PROOF. First, setting $\beta = \frac{1}{2}$, $\phi = \frac{1}{4}$, and $\sigma = 1$ in (4.2), and noting that $C(\frac{1}{4}, \frac{1}{2})) = 1$ and $\Gamma(n+1) = n!$ we have

$$\begin{aligned} f_{X+Y}(z) &= \frac{1}{\sqrt{2\pi}} e^{-z^2/4} \sum_{n=0}^{\infty} \frac{(-1)^n 2^n}{n!} \Gamma(n+1) \left[ D_{-(n+1)}(z) + D_{-(n+1)}(-z) \right] \\ &= \frac{1}{\sqrt{2\pi}} e^{-z^2/4} \sum_{n=0}^{\infty} (-1)^n 2^n \left[ D_{-(n+1)}(z) + D_{-(n+1)}(-z) \right]. \end{aligned} \tag{5.16}$$

On the other hand, setting $\beta = \frac{1}{2}$, $\phi = \frac{1}{4}$, and $\sigma = 1$ in (5.13), we have

$$f_{X+Y}(z) = e^2 \left( e^{-2z} \Phi(z-2) + e^{2z} \Phi(-z-2) \right). \tag{5.17}$$

Thus the first conclusion follows from Theorem 4.2 and the result of [**K-W**].

We now prove that the function $\varphi$, defined as the right hand side of (5.16), satisfies the ODE (5.15). Again, we first analyze the case $z = 0$. In this case we note that

$$\frac{1}{\sqrt{2\pi}} \sum_{n=0}^{\infty} (-1)^n 2^{n+1} D_{-(n+1)}(0) = 2e^2 \Phi(-2).$$

In the general case we follow the same argument as in the previous theorem to show that the function

$$\varphi(u) \stackrel{\triangle}{=} \frac{1}{\sqrt{2\pi}} e^{-u^2/4} \sum_{n=0}^{\infty} (-1)^n 2^n \left[D_{-(n+1)}(u) + D_{-(n+1)}(-u)\right]$$

is at least the unique solution to the ODE (5.15). To see this, first note that by taking partial sum if necessary, we can (formally) differentiate the function $\varphi$ to get

$$\begin{aligned} \varphi'(u) &= \frac{-u}{2}\varphi(u) + \frac{1}{\sqrt{2\pi}} e^{-u^2/4} \left( \sum_{n=0}^{\infty} (-1)^n 2^n \frac{u}{2} \left[D_{-(n+1)}(u) + D_{-(n+1)}(-u)\right] \right. \\ & \quad \left. - \sum_{n=0}^{\infty} (-1)^n 2^n \left[D_{-n}(u) - D_{-n}(-u)\right] \right) \\ \text{(5.18)} \quad &= -\frac{1}{\sqrt{2\pi}} e^{-u^2/4} \sum_{n=0}^{\infty} (-1)^n 2^n \left[D_{-n}(u) - D_{-n}(-u)\right] \\ &= -2\varphi(u) + \frac{4}{\sqrt{2\pi}} e^{-u^2/4} \sum_{n=0}^{\infty} (-1)^n 2^n D_{-(n+1)}(u), \end{aligned}$$

and by using the recursive relation (2.11) to get

$$\begin{aligned} \text{(5.19)} \quad \varphi''(u) &= -2\varphi'(u) + \frac{4}{\sqrt{2\pi}} e^{-u^2/4} \sum_{n=0}^{\infty} (-1)^n 2^n D'_{-(n+1)}(u) \\ & \quad - \frac{2u}{\sqrt{2\pi}} e^{-u^2/4} \sum_{n=0}^{\infty} (-1)^n 2^n D_{-(n+1)}(u) \\ &= -2\varphi'(u) - \frac{4}{\sqrt{2\pi}} e^{-u^2/4} \sum_{n=0}^{\infty} (-1)^n 2^n D_{-n}(u) \\ \text{(5.20)} \quad &= -2\varphi'(u) - \frac{4}{\sqrt{2\pi}} e^{-u^2/4} \left( e^{-u^2/4} - 2 \sum_{n=0}^{\infty} (-1)^n 2^n D_{-(n+1)}(u) \right) \\ &= -2\varphi'(u) - \frac{4}{\sqrt{2\pi}} e^{-u^2/2} + e^{-u^2/4} \frac{8}{\sqrt{2\pi}} \sum_{n=0}^{\infty} (-1)^n 2^n D_{-(n+1)}(u). \end{aligned}$$

Combining (5.18) and (5.19) and following the stability arguments if needed, one shows that the function $\varphi$ is at least a viscosity solution to the ODE (5.15). We can then repeat the same arguments as those in the previous theorem to conclude that $\varphi$ must coincide with the unique analytic solution of (5.15), completing the proof. □

## 6. Convolution formulae for the NP(m) cases

In the last part of this paper we shall give the series presentations of the solution to the Problem NP($m$). As we will see, such a presentation will turn out to be quite complicated in its form, although the idea is rather straightforward. Consequently, we will pay more attention to the actually computational aspect of the convolution formulae, rather than the detailed convergence analysis. We shall verify the convergence by performing the actual numerical simulation, and compare the results.

To begin with, let $X$ be a normal random variable, and let $Y_1, \cdots, Y_m$ be $m$ *i.i.d.* random variables with a same power exponential distribution. We will study the convolution formula for the random variable $Z = X + \sum_{i=1}^{m} Y_i$. We still consider the cases of $\beta > 1$ and $\beta < 1$ separately again.

**Case 1.** $\beta > 1$. First assume $m = 2$. Then applying Theorem 4.1 we have

$$
\begin{aligned}
(6.1)\qquad f_{X+Y_1+Y_2}(z) &= \int_{-\infty}^{\infty} f_{X+Y_1}(z-t) f_{Y_2}(t)dt \\
&\cong \frac{C^2(\phi,\beta)}{\sqrt{\pi}\sigma} \int_{-\infty}^{\infty} e^{-\frac{|z-t|^2}{2\sigma^2}} \sum_{n=0}^{\infty} \frac{H_{2n}(\frac{z-t}{\sqrt{2}\sigma})\Gamma(\frac{2n+1}{2\beta})\phi^{2n+1}2^{\frac{2n+1}{2\beta}} e^{-\frac{1}{2}|\frac{t}{\phi}|^{2\beta}}}{(2n)!(2\sigma^2)^n} dt \\
&= \frac{C^2(\phi,\beta)}{\sqrt{\pi}\sigma} \sum_{n=0}^{\infty} \frac{\Gamma(\frac{2n+1}{2\beta})\phi^{2n+1}2^{\frac{2n+1}{2\beta}}}{(2n)!(2\sigma^2)^n} \int_{-\infty}^{\infty} H_{2n}\Big(\frac{z-t}{\sqrt{2}\sigma}\Big) e^{-\frac{|z-t|^2}{2\sigma^2}} e^{-\frac{1}{2}|\frac{t}{\phi}|^{2\beta}} dt.
\end{aligned}
$$

Note that

$$
(6.2)\qquad H_{2n}\Big(\frac{z-t}{\sqrt{2}\sigma}\Big) = \sum_{j=0}^{n} 2^{2n-j} \Big(\frac{z-t}{\sqrt{2}\sigma}\Big)^{2n-2j} \binom{2n}{2j} (2j-1)!!
$$

and

$$
e^{-\frac{|z-t|^2}{2\sigma^2}} = \sum_{k=0}^{\infty} \frac{(-1)^k (z-t)^{2k}}{(2\sigma^2)^k},
$$

we get that

$$
\begin{aligned}
f_{X+Y_1+Y_2}(z) &= \frac{C^2(\phi,\beta)}{\sqrt{\pi}\sigma} \sum_{n=0}^{\infty} \frac{\Gamma\left(\frac{2n+1}{2\beta}\right)\phi^{2n+1}2^{\frac{2n+1}{2\beta}}}{(2n)!} \\
&\quad \times \sum_{k=0}^{\infty}\sum_{j=0}^{n} (-1)^k \frac{\binom{2n}{2j}(2j-1)!!}{2^k \sigma^{2(2n+k-j)}} \int_{-\infty}^{\infty} (z-t)^{2(k+n-j)} e^{-\frac{1}{2}|\frac{t}{\phi}|^{2\beta}} dt \\
&= \frac{C^2(\phi,\beta)}{\sqrt{\pi}\sigma} \sum_{n=0}^{\infty} \frac{\Gamma(\frac{2n+1}{2\beta})\phi^{2n+1}2^{\frac{2n+1}{2\beta}}}{(2n)!} \sum_{k=0}^{\infty}\sum_{j=0}^{n} (-1)^k \frac{\binom{2n}{2j}(2j-1)!!}{2^k \sigma^{2(2n+k-j)}} \\
&\quad \cdot \sum_{i=0}^{2(k+n-j)} (-1)^i \binom{2(k+n-j)}{i} z^{2(k+n-j)-i} \int_{-\infty}^{\infty} t^i e^{-\frac{1}{2}|\frac{t}{\phi}|^{2\beta}} dt.
\end{aligned}
$$

It is fairly easy to calculate the last integral in the above as

$$
(6.3)\qquad \int_{-\infty}^{\infty} t^i e^{-\frac{1}{2}|\frac{t}{\phi}|^{2\beta}} dt = [1+(-1)^i]\frac{1}{2\beta}\left[\frac{1}{2\phi^{2\beta}}\right]^{\frac{i+1}{2\beta}} \Gamma\left(\frac{i+1}{2\beta}\right).
$$

We conclude that

$$\begin{aligned} f_{X+Y_1+Y_2}(z) &= \frac{C^2(\phi,\beta)}{\sqrt{\pi}\sigma}\sum_{n=0}^{\infty}\frac{\Gamma(\frac{2n+1}{2\beta})\phi^{2n+1}2^{\frac{2n+1}{2\beta}}}{(2n)!}\sum_{k=0}^{\infty}\sum_{j=0}^{n}(-1)^k\frac{\binom{2n}{2j}(2j-1)!!}{2^k\sigma^{2(2n+k-j)}} \\ &\cdot\sum_{i=0}^{k+n-j}\binom{2(k+n-j)}{2i}\frac{1}{\beta}\left[\frac{1}{2\phi^{2\beta}}\right]^{\frac{2i+1}{2\beta}}\Gamma\left(\frac{2i+1}{2\beta}\right)z^{2(k+n-j-i)}. \end{aligned}$$

For the general $n$ we can repeat this procedure to obtain the following formula.

$$\begin{aligned} &f_{X+Y_1+\ldots+Y_n}(z) \\ =\ &\frac{C^{n-1}(\phi,\beta)}{\sqrt{\pi}\sigma}\sum_{n=0}^{\infty}\frac{\Gamma\left(\frac{2n+1}{2\beta}\right)\phi^{2n+1}2^{\frac{2n+1}{2\beta}}}{(2n)!}\sum_{k=0}^{\infty}\sum_{j=0}^{n}(-1)^k\frac{\binom{2n}{2j}(2j-1)!!}{2^k\sigma^{2(2n+k-j)}} \\ &\cdot\sum_{i_3=0}^{k+n-j}\binom{2(k+n-j)}{2i_3}\frac{1}{\beta}\left[\frac{1}{2\phi^{2\beta}}\right]^{\frac{2i_3+1}{2\beta}}\Gamma\left(\frac{2i_3+1}{2\beta}\right) \\ &\cdot\sum_{i_4=0}^{k+n-j-i_3}\binom{2(k+n-j-i_3)}{2i_4}\frac{1}{\beta}\left[\frac{1}{2\phi^{2\beta}}\right]^{\frac{2i_4+1}{2\beta}}\Gamma\left(\frac{2i_4+1}{2\beta}\right) \\ &\cdots\sum_{i_n=0}^{k+n-j-i_3-i_4-\ldots-i_{n-1}}\binom{2(k+n-j-i_3-i_4-\ldots-i_{n-1})}{2i_n} \\ &\times\frac{1}{\beta}\left[\frac{1}{2\phi^{2\beta}}\right]^{\frac{2i_n+1}{2\beta}}\Gamma\left(\frac{2i_n+1}{2\beta}\right)z^{2(k+n-j-i_3-i_4-\ldots-i_n)} \end{aligned}$$

**Case 2.** $\beta < 1$. Again we first consider the case $m = 2$. In this case we use the expansion in terms of PCF's to get

$$\begin{aligned} f_{X+Y_1+Y_2}(z) &= \int_{-\infty}^{\infty} f_{X+Y_1}(t)f_{Y_2}(z-t)dt \\ &= \frac{C^2(\phi,\beta)}{\sqrt{2\pi}}\int_{-\infty}^{\infty} e^{\frac{-1}{2}\left|\frac{t}{\phi}\right|^{2\beta}} e^{\frac{-(t-z)^2}{4\sigma^2}}\sum_{n=0}^{\infty}\frac{(-1)^n}{n!2^n}\left(\frac{\sigma}{\phi}\right)^{2\beta n}\Gamma(2\beta n+1) \\ &\times\left[D_{-(2\beta n+1)}\left(\frac{(t-z)}{\sigma}\right)+D_{-(2\beta n+1)}\left(\frac{-(t-z)}{\sigma}\right)\right]. \end{aligned} \tag{6.4}$$

Now recall from §2-C that

$$\begin{aligned} &\left[D_{-(2\beta n+1)}\left(\frac{t-z}{\sigma}\right)+D_{-(2\beta n+1)}\left(\frac{-(t-z)}{\sigma}\right)\right] \\ =\ &\sqrt{2\pi}2^{-(\beta n+\frac{1}{4})}\left[\frac{2^{-\frac{1}{4}}\left(y_1(2\beta n+\frac{1}{2},\frac{t-z}{\sigma})+y_1(2\beta n+\frac{1}{2},\frac{-t+z}{\sigma})\right)}{\Gamma(\beta n+1)}\right. \\ &\left.-\frac{2^{\frac{1}{4}}\left(y_2(2\beta n+\frac{1}{2},\frac{t-z}{\sigma})-y_1(2\beta n+\frac{1}{2},\frac{-t+z}{\sigma})\right)}{\Gamma(\beta n+\frac{1}{2})}\right], \end{aligned}$$

where

$$\begin{aligned} y_1(a,u)+y_1(a,-u) &= 2e_1^{u^2/4}F_1(a/2+1/4,1/2;u^2/2), \\ y_2(a,u)+y_2(a,-u) &= u2e_1^{u^2/4}F_1(-a/2+3/4,3/2;-u^2/2), \end{aligned}$$

we obtain that

$$\begin{aligned} & f_{X+Y_1+Y_2}(z) \\ = & \; C^2(\phi,\beta)\sum_{n=0}^{\infty}\frac{(-1)^n}{n!2^n2^{\beta n}}\left(\frac{\sigma}{\phi}\right)^{2\beta n}\Gamma(2\beta n+1) \\ & \cdot\sum_{k=0}^{\infty}\frac{\Gamma(\beta n+\frac{1}{2}+k)}{\Gamma(\beta n+1)\Gamma(\beta n+\frac{1}{2})(2k-1)!!k!}-\frac{4(-1)^k\Gamma(-\beta n+\frac{1}{2}+k)}{\Gamma(\beta n+\frac{1}{2})\Gamma(-\beta n+\frac{1}{2})(2k+1)!!k!} \\ & \cdot\frac{1}{\sigma^{2k}}\int_{-\infty}^{\infty}(t-z)^{2k}e^{\frac{-1}{2}|\frac{t}{\phi}|^{2\beta}}dt. \end{aligned}$$

Since $(t-z)^{2k}=\sum_{\ell=0}^{2k}\binom{2k}{\ell}(-1)^\ell z^{2k-\ell}t^\ell$, we have

$$\begin{aligned} & f_{X+Y_1+Y_2}(z) \\ = & \; C^2(\phi,\beta)\sum_{n=0}^{\infty}\frac{(-1)^n}{n!2^n2^{\beta n}}\left(\frac{\sigma}{\phi}\right)^{2\beta n}\Gamma(2\beta n+1)\cdot \\ & \cdot\sum_{k=0}^{\infty}\left\{\frac{\Gamma(\beta n+\frac{1}{2}+k)}{\Gamma(\beta n+1)(2k-1)!!}-\frac{4(-1)^k\Gamma(-\beta n+\frac{1}{2}+k)}{\Gamma(-\beta n+\frac{1}{2})(2k+1)!!}\right\}\frac{1}{\Gamma(\beta n+\frac{1}{2})k!} \\ & \cdot\frac{1}{\sigma^{2k}}\sum_{k=0}^{2k}\binom{2k}{\ell}(-1)^\ell z^{2k-\ell}\int_{-\infty}^{\infty}t^\ell e^{-\frac{1}{2}|\frac{t}{\phi}|^{2\beta}}dt. \end{aligned}$$

Now using (6.3) we finally obtain that

$$\begin{aligned} & f_{X+Y_1+Y_2}(z) \\ = & \; C^2(\phi,\beta)\sum_{n=0}^{\infty}\frac{(-1)^n}{n!2^n2^{\beta n}}\left(\frac{\sigma}{\phi}\right)^{2\beta n}\Gamma(2\beta n+1) \\ & \cdot\sum_{k=0}^{\infty}\left\{\frac{\Gamma(\beta n+1/2+k)}{(2k-1)!!}-\frac{4(-1)^k\Gamma(-\beta n+1/2+k)}{\Gamma(-\beta n+1/2)(2k+1)!!}\right\}\frac{1}{\Gamma(\beta n+1/2)k!} \\ & \cdot\frac{1}{\sigma^{2k}}\sum_{\ell=0}^{k}\binom{2k}{2\ell}z^{2(k-\ell)}\frac{2}{2\beta}\left[\frac{1}{2\phi^{2\beta}}\right]^{\frac{k+1}{2\beta}}\Gamma\left(\frac{\ell+1}{2\beta}\right). \end{aligned}$$

Finally, for general $n$ we have the following formula: denoting $s(k,\ell)=i_k+\cdots+i_\ell$,

$$\begin{aligned} & f_{X+Y_1+Y_2}(z) \\ \cong & \; C^{n-1}(\phi,\beta)\sum_{n=0}^{\infty}\frac{(-1)^n}{n!2^n2^{\beta n}}\left(\frac{\sigma}{\phi}\right)^{2\beta n}\Gamma(2\beta n+1) \\ & \cdot\sum_{k=0}^{\infty}\left\{\frac{\Gamma(\beta n+\frac{1}{2}+k)}{\Gamma(\beta n+1)(2k-1)!!}-\frac{4(-1)^k\Gamma(-\beta n+\frac{1}{2}+k)}{\Gamma(-\beta n+\frac{1}{2})(2k+1)!!}\right\}\frac{1}{\Gamma(\beta n+\frac{1}{2})k!} \end{aligned}$$

$$\cdot \frac{1}{\sigma^{2k}} \sum_{\ell=0}^{k} \binom{2k}{2\ell} z^{2(k-\ell)} \frac{2}{2\beta} \left[\frac{1}{2\phi^{2\beta}}\right]^{\frac{\ell+1}{2\beta}} \Gamma\left(\frac{\ell+1}{2\beta}\right)$$

$$\cdot \sum_{i_3=0}^{k-\ell} \binom{2(k-\ell)}{2i_3} \frac{1}{\beta} \left[\frac{1}{2\phi^{2\beta}}\right]^{\frac{2i_3+1}{2\beta}} \Gamma\left(\frac{2i_3+1}{2\beta}\right)$$

$$\cdot \sum_{i_4=0}^{k-\ell-i_3} \binom{2(k+\ell-j-i_3)}{2i_4} \frac{1}{\beta} \left[\frac{1}{2\phi^{2\beta}}\right]^{\frac{2i_4+1}{2\beta}} \Gamma\left(\frac{2i_4+1}{2\beta}\right)$$

$$\cdots \sum_{i_n=0}^{k-\ell-S(3,n-1)} \binom{2(k+n-j-s(3,n-1))}{2i_n} \frac{1}{\beta} \left[\frac{1}{2\phi^{2\beta}}\right]^{\frac{2i_n+1}{2\beta}} \Gamma\left(\frac{2i_k+1}{2\beta}\right)$$

$$\cdot z^{2(k-\ell-s(3,n))}.$$

## 7. Numerical Illustrations

In this section we illustrate our results by numerical experiments. We will be interested in the following three cases: 1) NP(1) with $\beta = 1$; 2) NP(1) with $\beta \neq 1$; and 3) NP(2) with all $\beta$.

**7.1. NP(1) with $\beta$=1.** This is a special case worth mentioning. On the one hand this is the case the convergence analysis of Theorem 4.2 does not apply. But on the other hand, in this case one actually has a convolution of two standard Normal random variables, thus the $NP(1)$ sum is simply a $N(0,2)$ random variable. We nevertheless did an experiment just to see how efficient (or inefficient) the Hermite expansion could be in this case. We should note that this numerical computation is only for theoretical purposes.

In Figure 1 the defaulting parameters are $\mu_1 = \mu_2 = 0$ and $\sigma = \phi = 1$. We see that while the Hermite series (5.2) actually does converge, it is extremely slow. For example, the result is far from satisfactory when $n = 200$; and it only becomes more acceptable when $n = 500$.

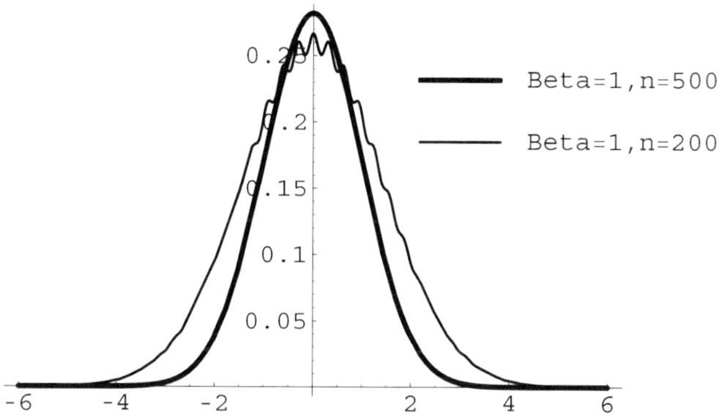


FIGURE 1. NP(1) with $\beta = 1$ ($\sim N(0,2)$)

**7.2. NP(1), $\beta \neq 1$.** In this case we have the convolution of a normal random variable and a Power exponential. We would like to see two things: the speed of convergence and the shape of "tails" for different values of $\beta$. We fix the default parameters $\mu_1 = \mu_2 = 0$ and $\sigma = \phi = 1$, but let $\beta$ vary. In Figure 2 we combine the graphs of those with $\beta = 3/10$, $1/2$, $1$, $2$, and $5/2$. In all the cases (except for $\beta = 1$) we find that $n = 100$ is already sufficient for the satisfactory results. As we can see that the smaller the $\beta$ value is, the heavier the tail becomes. The example for $\beta = 3/10$ shows that there might be cases in which a power exponential distribution can be more efficient than double exponential ($\beta = 1/2$), if the heavier tails are desired. In general, all power exponentials with $\beta < 1$ may be of independent interest in robustness studies. Among other things, they could be useful in modeling the errors in *Regression Analysis* and *Time Series*, for example.

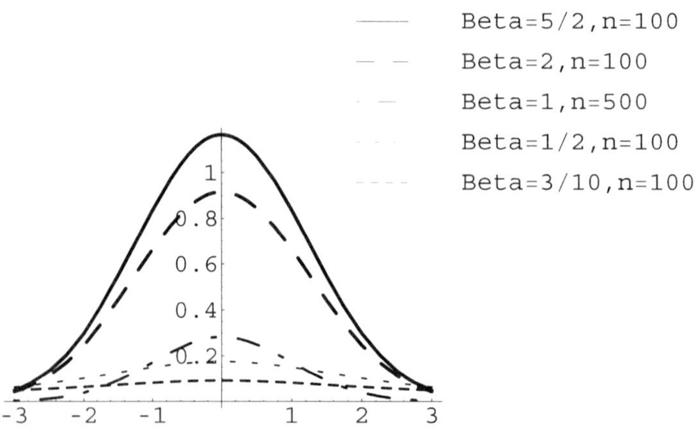


FIGURE 2. NP(1) for different values of $\beta$

**7.3. NP(2) with all $\beta$'s.** In this experiment we expect to see the same features of case (2), and we would also like to see the difference in tail shape when $m$, the number of power exponentials, increases. In Figure 3 we show the graphs of convolution of one $N(0,1)$ random variable and two Power exponentials with $\mu_1 = \mu_2 = 0$, $\sigma = \phi = 1$. It is a little surprising that although the closed form expression of the solution to NP(2) is much more complicated as we saw in the previous section, the speed of convergence is almost no worse than the case of NP(1). In fact, $n = 100$ is again sufficient for a satisfactory result. One should also note that the solutions of NP(2) have even heavier tails than those of NP(1) with the same $\beta$ values. We believe that "heaviness" of the tails increases as $m$ increases, but we did not perform further numerical experiment as it goes beyond our original purpose of this paper.

**Acknowledgements** We would like to thank the anonymous referee for the careful reading of the original manuscript and many useful suggestions. Part of this work was completed when the third author was visiting the Department of Mathematics, Purdue University, whose hospitality is greatly appreciated.

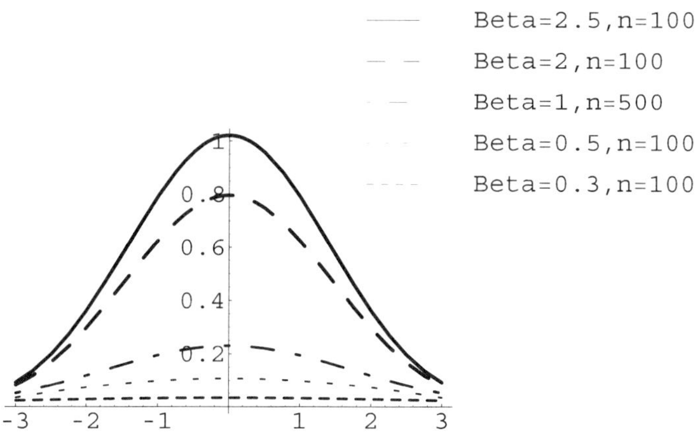


FIGURE 3. NP(2) for different values of $\beta$

## References


[A] G. Agró, *Maximum likelihood estimation for the exponential power function parameters,* Comm. Statist. Simulation Comput. **24**, no. 2 (1995), pp. 523–536.

[B-T] G. Box and G. Tiao, *Bayesian Inference in Statistical Analysis,* Addison–Wesley, Reading, (1973).

[D] D. Duffie, *Dynamic asset Pricing Theory,* Second Edition, Princeton University Press, Princeton, (1995).

[F-S] W. H. Fleming and H. M. Soner, *Controlled Markov Processes and Viscosity Solutions,* Springer–Verlag, (1993).

[Go-R] D. Gokhale and M. Rahman, *A note on combining parametric and non-parametric regression,* Comm. Statist. Simulation Comput. **26**, no. 2 (1997), pp. 519–529.

[G-G-M] E. Gómez, M. A. Gómez-Villegas and J. M. Marín, *A multivariate generalization of the power exponential family of distributions,* Comm. Statist. Theory Methods **27**, no. 3 (1998), pp. 589–600.

[Gr-R] I. M. Gradshteyn and I. M. Ryzhik, *Table of Integrals Series and Products,* Sixth Edition, Academic Press, (2000).

[Ha] E. Hansen, *A table of Series and Products,* Englewood Cliffs, Prentice-Hall, New Jersey, (1975).

[Ho] R. Hogg, *Adaptive robust procedures: a partial review and some suggestions for future applications and theory. With comments by H. Leon Harter, Joseph L. Gastwirth and Peter J. Huber, and with a rejoinder by Robert V. Hogg,* J. Amer. Statist. Assoc. **69** (1974), pp. 909–927.

[K-W] S. G. Kou and H. Wang, *Option Pricing Under a Double Exponential Jump Diffusion Model,* Preprint, (2001).

[M] R. C. Merton, *Option pricing when underlying stock returns are discontinuous,* J. of Financial Economics **3** (1976), pp. 125–144.

[P] P. Protter, *Stochastic Integration and Stochastic Differential Equations, A New Approach,* Springer–Verlag, (1990).

[R] M. Rahman, *Bayesian estimation of the scale parameter of the complete even power exponential distribution,* Egyptian Statist. J. **41**, no. 1 (1997), pp. 60–68.

[S] M. Subbotin, *On the law of frequency of errors,* Mat. Sb. **31** (1923), pp. 296–301

[T] N. M. Temme, *Numerical and asymptotic aspects of parabolic cylinder functions,* J. Comput. Appl. Math. **121** (2000), pp. 221–246.

Departamento de Estadística, Universidad de Valparaíso, Casilla 5030, Valparaíso, Chile.
*E-mail address*: manuel.galea@uv.cl.

Department of Mathematics, Purdue University, West Lafayette, IN 47907-1395.
*E-mail address*: majin@math.purdue.edu.

Departamento de Estadística, Universidad de Valparaíso, Casilla 5030, Valparaíso, Chile.
*E-mail address*: soledad.torres@uv.cl.

Contemporary Mathematics
Volume **336**, 2003

# Conditions for Nonconservativity in Quantum Dynamical Semigroups

Julio C. García and Roberto Quezada

ABSTRACT. Necessary and sufficient conditions for nonconservativity of a class of quantum dynamical semigroups are given.

## 1. Introduction

Since its introduction by A. Kossakowski ([**K**]) and E.B. Davies ([**D**]), quantum dynamical semigroups (qds) have been intensively studied and applied to describe the reduced evolution of quantum open systems. Special emphasis has received the minimal qds as well as conditions to ensure its conservativity, ([**Ch**], [**F**]) and some examples of nonconservative minimal qds have been considered in [**B-S**].
In this work we attempt to develop a general approach to study the class of nonconservative qds. We provide a necessary condition as well as a sufficient condition for nonconservativity. Both of them are imposed on the formal Lindblad generator and become generalizations of similar conditions when applied to the class of formal Lindblad generators studied by A.M Chebotarev ([**Ch**]) and F. Fagnola ([**F**]).
In Section 2 we set general definitions and recall some important results. In Section 3 we establish the main result of the paper: a necessary and a sufficient condition for nonconservativity of the minimal qds. In Section 4 we develop an example taken from pure birth stochastic processes and discuss the possibility of applying our result to an example taken from mathematical physics.

## 2. Preliminaries

Along this work $\mathcal{H}$ will denote a separable complex Hilbert space with the inner product $\langle\ ,\ \rangle$ and the norm $\|\cdot\|$; $\mathcal{B} = \mathcal{B}(\mathcal{H})$ will denote the von Neumann algebra of all bounded linear operators in $\mathcal{H}$ and $\|\cdot\|_\infty$ will denote the norm in this space; $\mathcal{B}_+$ will be the cone of positive bounded operators in $\mathcal{H}$.

DEFINITION 2.1. A quantum dynamical semigroup on $\mathcal{B}$ is a semigroup $P = (P_t)_{t\geq 0}$ of bounded operators in $\mathcal{B}$ with the following properties

2000 *Mathematics Subject Classification.* Primary: 81S25; Secondary: 47N50.
*Key words and phrases.* Quantum Dynamical Semigroups, Conservativity.
Partially supported by CONACYT, Grant 37491-E.

(1) Complete Positivity (CP). $P_t$ is completely positive for every $t \geq 0$, i.e. for every pair of finite sequences $(x_i), (y_j)$ in $\mathcal{B}$

$$\sum_{i,j} y_i^* P_t(x_i^* x_j) y_j \geq 0.$$

(2) (Normality or $\sigma$-weak continuity). For every increasing net $(x_\alpha)$ of positive elements in $\mathcal{B}$ with an upper bound we have

$$P_t(\sup_\alpha x_\alpha) = \sup_\alpha P_t(x_\alpha)$$

for every $t \geq 0$.

(3) (Ultraweak or weak* continuity in $t$). For every trace class operator and every $x \in \mathcal{B}$ we have

$$\lim_{t \to 0^+} tr(\rho P_t(x)) = tr(\rho x).$$

(4) $P_t(I) \leq I$ for all $t \geq 0$.

A qds $(P_t)_{t \geq 0}$ is *conservative* (*markovian* or *unital*) if $P_t(I) = I$, for all $t \geq 0$. When a qds is uniformly continuous, i.e., $\lim_{t \to 0^+} \sup_{\|x\|_\infty = 1} \|P_t(x) - x\|_\infty = 0$, then its infinitesimal generator is a bounded linear operator $\mathcal{L} : \mathcal{B} \to \mathcal{B}$ and there exist a CP normal bounded map $\phi : \mathcal{B} \to \mathcal{B}$ and a bounded self adjoint operator $H$ such that

$$\mathcal{L}(x) = \phi(x) - G^* x - xG \tag{2.1}$$

with $G = (1/2)\phi(I) - iH$. And conversely any linear operator $\mathcal{L}$ with the structure (2.1) is the infinitesimal generator of a uniformly continuous qds. Furthermore, this qds is conservative if and only if $\mathcal{L}(I) = 0$. This is an important result due to Lindblad and Gorini-Kossakowski-Sudarshan (see [**Par**] or [**F**]), and the references therein. In this work we shall consider unbounded formal generators $\mathcal{L}$ which associate with every $x \in \mathcal{B}$ an unbounded sesquilinear form with the structure

$$\mathcal{L}(x)[u, v] = \phi(x)[u, v] - \langle Gu, xv \rangle - \langle u, xGv \rangle, \tag{2.2}$$

$u, v \in \text{dom}\, G$, where

**i)** $-G$ is the generator of a $C_o$-semigroup of contractions in $\mathcal{H}$, $(W_t)_{t \geq 0}$.

**ii)** $\phi$ is a linear completely positive (CP) and normal map, i.e., for every $x \in \mathcal{B}$, $\phi(x)$ is a sesquilinear form defined on dom$G \times$ dom $G$ such that

  **ii.1)** For any pair of finite sequences $(u_i) \in$ dom $G$ and $(x_i) \subset \mathcal{B}$ we have that

$$\sum_{i,j} \phi(x_i^* x_j)[u_i, u_j] \geq 0.$$

  **ii.2)** For every $u \in$ dom $G$, $\phi(\cdot)[u]$ is a normal linear functional on $\mathcal{B}$, i.e., for any increasing net $(x_\alpha)$ of positive elements of $\mathcal{B}$ with an upper bound,

$$\phi\left(\sup_\alpha x_\alpha\right)[u] = \sup_\alpha \phi(x_\alpha)[u],$$

  where $\phi(\cdot)[u] = \phi(\cdot)[u, u]$ is the quadratic form associated with $\phi(\cdot)$.

**iii)** The estimate

$$0 \leq \phi(I)[u] \leq Re\langle Gu, u \rangle$$

or, equivalently, $\mathcal{L}(I)[u] \leq 0$, holds for every $u \in$ dom $G$.

Conditions **i)–iii)** are sufficient to construct (see [**F**] or [**Ch**]) a minimal qds $(P_t^{\min})_{t\geq 0}$ that satisfies the equation

$$\frac{d}{dt}\left\langle u, P_t^{\min}(x)v\right\rangle = \mathcal{L}\left(P_t^{\min}(x)\right)[u,v], \quad P_0^{\min}(x) = x, \tag{2.3}$$

$u, v \in \operatorname{dom} G,\ \ x \in \mathcal{B}$, which amounts to be equivalent with the integral equation

$$\frac{d}{dt}\left\langle u, P_t^{\min}(x)v\right\rangle = \langle u, W_t^* x W_t v\rangle + \int_0^t d\tau\phi\left(P_\tau^{\min}(x)\right)[W_{t-\tau}u, W_{t-\tau}v],$$

$u, v \in \operatorname{dom} G,\ x \in \mathcal{B}$. The minimal qds $(P_t^{\min})_{t\geq 0}$ is not necessarily conservative and the problem of finding necessary and sufficient conditions for its conservativity has received the attention of the people working in this topic. A. M. Chebotarev ([**Ch**]) and F. Fagnola ([**F**]) have found necessary and sufficient or only sufficient conditions for the conservativity of the class of minimal qds whose formal generator satisfies the additional necessary condition

**iii')** $\mathcal{L}(I)[u,v] = 0, \quad \forall\, u, v \in \operatorname{dom} G,$

which is a stronger formulation of **iii)**. Our aim in this work is to study necessary and sufficient conditions for nonconservativity of the class of minimal qds whose formal generator satisfies only the conditions **i)-iii)**.

DEFINITION 2.2. Let $\mathcal{Q}_\lambda$, $\mathcal{A}_\lambda$ and $\mathcal{R}_\lambda^{\min}$ be the maps with domain $\mathcal{B}\times\operatorname{dom} G\times\operatorname{dom} G$ and taking values on $\mathbb{C}$

$$\begin{aligned}
\mathcal{Q}_\lambda(x)[u,v] &= \int_0^\infty e^{-\lambda t}\phi(x)[W_t u, W_t v]dt \\
\mathcal{A}_\lambda(x)[u,v] &= \int_0^\infty e^{-\lambda t}\langle W_t u, x W_t v\rangle dt \\
\mathcal{R}_\lambda^{\min}(x)[u,v] &= \int_0^\infty e^{-\lambda t}\langle u, P_t^{\min}(x)v\rangle dt
\end{aligned}$$

PROPOSITION 2.3. *The maps* $\mathcal{Q}_\lambda, \mathcal{A}_\lambda, \mathcal{R}_\lambda^{\min}$ *have the following properties:*

(1) *All them three are linear, completely positive and normal in the operator variable and sesquilinear in the vector variables.*
(2) *For each* $x \in \mathcal{B}$, *all the three sesquilinear forms are bounded and hence can be extended, in a unique way, as sesquilinear forms associated to bounded operators. We still denote by* $\mathcal{Q}_\lambda(x), \mathcal{A}_\lambda(x), \mathcal{R}_\lambda^{\min}(x)$ *the corresponding operators.*
(3) *Because of (2), we can think about* $\mathcal{Q}_\lambda, \mathcal{A}_\lambda, \mathcal{R}_\lambda^{\min}$ *as maps from* $\mathcal{B}$ *into itself. These maps are linear, completely positive and normal.*
(4) *If* $\mathcal{Q}_\lambda^k = \underbrace{\mathcal{Q}_\lambda \circ \cdots \circ\ \mathcal{Q}_\lambda}_{k}$ *then for all* $x \in \mathcal{B}$,

$$\sum_{k=0}^{\infty} \mathcal{Q}_\lambda^k \circ \mathcal{A}_\lambda(x) \;=\; \mathcal{R}_\lambda^{\min}(x). \tag{2.4}$$

(5) *For any positive constant* $\lambda > 0$,

$$I = \lambda\,\mathcal{R}_\lambda^{\min}(I) + \lim_{n\to\infty}\mathcal{Q}_\lambda^n(I) + \lambda\sum_{k=0}^{\infty}\mathcal{Q}_\lambda^k \circ \ell_\lambda(I). \tag{2.5}$$

*Where*

$$\ell_\lambda(I)[u] = -\int_0^\infty dt\, e^{-\lambda t} \int_0^t ds \mathcal{L}(I)[W_{t-s}u] \ \geq 0.$$

PROOF. The proof of (1)-(4) can be found in [**F**]. The proof of (5) is in [**Q**] □

REMARK 2.4. $\mathcal{R}_\lambda^{\min}$ is called the resolvent of $(P_t^{\min})_{t\geq 0}$.

As a consequence, we have the following necessary but not sufficient condition for conservativity for the class of qds we consider in this work.

COROLLARY 2.5. *If the minimal solution is conservative then for all $x \in \mathcal{B}$,*

$$\lim_{n\to\infty} Q_\lambda^n(x) = 0,$$

*in the ultraweak topology of $\mathcal{B}$.*

PROOF. As the minimal qds is conservative then $P_t^{\min}(I) = I$ for all $t \geq 0$. This implies $\mathcal{R}_\lambda^{\min}(I) = \frac{1}{\lambda} I$, $\forall \lambda > 0$. If $x = I$, the conclusion is immediate from this observation and point (5) in Proposition 2.3. If $x \in \mathcal{B}_+$, then $x \leq \|x\|\ I$. Since $Q_\lambda$ is a linear CP map then, for all $n \in \mathbb{N}$,

$$0 \leq Q_\lambda^n(x) \leq \|x\|\ Q_\lambda^n(I).$$

By taking limit, we have the result. For general $x \in \mathcal{B}$, let us recall that it is a linear combination of four positive operators □

An example about the non sufficiency of this condition is found in [**Q**].

## 3. Criteria for explosion of the minimal qds.

Now, we can state and proof the main result.

THEOREM 3.1. *The following holds:*

a) *A necessary condition for the minimal qds, $(P_t^{\min})_{t\geq 0}$, to be nonconservative is that, for any $\lambda > 0$, there exists a bounded positive operator, $x = x_\lambda$, such that*

$$\mathcal{L}(x)[u] = \frac{1}{\lambda}\mathcal{L}(I)[u] + \lambda\langle u, xu\rangle \qquad \forall\, u \in \text{dom}\, G.$$

b) *A sufficient condition for the minimal qds, $(P_t^{\min})_{t\geq 0}$, to be nonconservative is the existence of a constant $\lambda > 0$, a bounded positive operator, $j$, and a bounded selfadjoint operator, $b$, such that $\|j\| > \|b\|$ and*

$$\mathcal{L}(j)[u] \geq \lambda\langle u, (j-b)u\rangle \qquad \forall u \in \text{dom}\, G.$$

PROOF. To prove a), let us introduce the following elements in $\mathcal{B}_+$: let $t \geq 0$, $\lambda > 0$,

$$\mathcal{E}_t(I) := I - P_t^{\min}(I), \qquad \tilde{\mathcal{E}}_\lambda(I) := \int_0^\infty e^{-\lambda t}\mathcal{E}_t(I)\,dt.$$

We call $\mathcal{E}_t(I)$ *the probability of explosion at time* $t$; $\tilde{\mathcal{E}}_\lambda(I)$ is its Laplace Transform. In [**Q**], it is proved that $\tilde{\mathcal{E}}_\lambda(I)$ is a well defined positive and bounded operator; in fact, it is equal to

$$\frac{1}{\lambda}I - \mathcal{R}_\lambda^{\min}(I).$$

Since $(P_t^{\min})_{t\geq 0}$ is nonconservative, then for any constant $\lambda > 0$ we have $\tilde{\mathcal{E}}_\lambda(I) > 0$; otherwise,$(P_t^{\min})_{t\geq 0}$ would be conservative. Let $x = \tilde{\mathcal{E}}_\lambda(I)$.
Since $\mathcal{L}$ is a normal map in the operator variable, then for any $u \in \text{dom}\, G$,

$$\begin{aligned}
\mathcal{L}(x)[u] &= \int_0^\infty e^{-\lambda t}\mathcal{L}(I - P_t^{\min}(I))[u]dt \\
&= \int_0^\infty e^{-\lambda t}\mathcal{L}(I)[u]dt - \int_0^\infty e^{-\lambda t}\mathcal{L}\left(P_t^{\min}(I)\right)[u]dt \\
&= \frac{1}{\lambda}\mathcal{L}(I)[u] - \int_0^\infty e^{-\lambda t}\frac{d}{dt}\langle u, P_t^{\min}(I)u\rangle dt \\
&= \frac{1}{\lambda}\mathcal{L}(I)[u] + \|u\|^2 - \lambda\int_0^\infty e^{-\lambda t}\langle u, P_t^{\min}(I)u\rangle dt \\
&= \frac{1}{\lambda}\mathcal{L}(I)[u] + \lambda\int_0^\infty e^{-\lambda t}\langle u, (I - P_t^{\min}(I))u\rangle dt \\
&= \frac{1}{\lambda}\mathcal{L}(I)[u] + \lambda\langle u, \tilde{\mathcal{E}}_\lambda(I)u\rangle,
\end{aligned}$$

where we have taken into account that $P_t^{\min}$ is solution of the equation (2.3) and integrated by parts.

To prove b) let us notice, by definition of the formal Lindblad generator $\mathcal{L}$ (equation (2.2)), that for all $u \in \text{dom}\, G$,

$$\phi(j)[u] \geq \langle Gu, ju\rangle + \langle u, jGu\rangle + \lambda\langle u, ju\rangle + \lambda\langle u, (-b)u\rangle$$

Since $W_t u \in \text{dom}\, G$ for any $u \in \text{dom}\, G$, we have

$$\begin{aligned}
\mathcal{Q}_\lambda(j)[u] &= \int_0^\infty e^{-\lambda t}\phi(j)[W_t u]dt \\
&\geq \int_0^\infty e^{-\lambda t}\left\{\langle GW_t u, jW_t u\rangle + \langle W_t u, jGW_t u\rangle + \lambda\langle W_t u, jW_t u\rangle\right\}dt \\
&\quad + \lambda\int_0^\infty e^{-\lambda t}\langle W_t u, (-b)W_t u\rangle dt \\
&= -\int_0^\infty \frac{d}{dt}\left\{e^{-\lambda t}\langle W_t u, jW_t u\rangle\right\}dt + \lambda\langle u, \mathcal{A}_\lambda(-b)u\rangle \\
&= \langle u, ju\rangle + \lambda\langle u, \mathcal{A}_\lambda(-b)u\rangle.
\end{aligned}$$

Since dom $G$ is dense, we can extend the above conclusion to the whole of $\mathcal{H}$ and obtain the following operator inequality,

$$\mathcal{Q}_\lambda(j) \geq j + \lambda\mathcal{A}_\lambda(-b). \tag{3.1a}$$

If we propose as inductive hypothesis,

$$\mathcal{Q}_\lambda^{n+1}(j) \geq j + \lambda \sum_{k=0}^{n} \mathcal{Q}_\lambda^k \circ \mathcal{A}_\lambda(-b), \tag{3.1b}$$

for some $n$ then, by the linearity and positivity of the map $\mathcal{Q}_\lambda$ and (3.1$a$),

$$\begin{aligned} \mathcal{Q}_\lambda^{n+2}(j) &\geq \mathcal{Q}_\lambda(j) + \lambda \sum_{k=0}^{n} \mathcal{Q}_\lambda^{k+1} \circ \mathcal{A}_\lambda(-b) \\ &\geq j + \lambda \mathcal{A}_\lambda(-b) + \lambda \sum_{k=1}^{n+1} \mathcal{Q}_\lambda^k \circ \mathcal{A}_\lambda(-b) \\ &= j + \lambda \sum_{k=0}^{n+1} \mathcal{Q}_\lambda^k \circ \mathcal{A}_\lambda(-b). \end{aligned}$$

So (3.1b) is valid for all $n$. From here we have the following, by taking limit as $n \to \infty$ and using equality (2.4),

$$\lim_{n\to\infty} \mathcal{Q}_\lambda^n(j) \geq j + \lambda \sum_{k=0}^{\infty} \mathcal{Q}_\lambda^k \circ \mathcal{A}_\lambda(-b) = j + \lambda \mathcal{R}_\lambda^{\min}(-b). \tag{3.2}$$

Now, let us apply the linear and positive map $\lambda\mathcal{R}_\lambda^{\min}(\cdot)$ to the inequality $-b \geq -\|b\|\, I$ furthermore, by (2.5), we have

$$\lambda\mathcal{R}_\lambda^{\min}(-b) \geq -\|b\|\, \lambda\mathcal{R}_\lambda^{\min}(I) \geq -\|b\|\, I.$$

In view of (3.2),

$$\lim_{n\to\infty} \mathcal{Q}_\lambda^n(j) \geq j + \lambda\mathcal{R}_\lambda^{\min}(-b) \geq j - \|b\|\, I.$$

In other words,

$$j \leq \|b\|\, I + \lim_{n\to\infty} \mathcal{Q}_\lambda^n(j).$$

Since $\|j\| > \|b\|$, we conclude that $\lim_{n\to\infty} \mathcal{Q}_\lambda^n(j) \neq 0$; otherwise, we would have $j \leq \|b\|\, I$, a contradiction. Thus, the necessary condition for conservativity (Corollary 2.4) is violated □

## 4. An Example and Discussion.

EXAMPLE 4.1. Let $\mathcal{H} = \ell_2(\mathbb{C})$ and let us consider a pure birth stochastic process with birth intensities given by a sequence of positive numbers

$$\{c_n\}_n.$$

It is well known that a quantum extension of its associated semigroup (see [**Par**]) acts on $\mathcal{B}$ leaving invariant the abelian algebra $\ell_\infty$ and the formal Lindblad generator acts on $\ell_\infty$ as

$$\mathcal{L}(\{a_n\}_n) = \{d_n\}_n \qquad \text{where} \qquad d_n = c_n(a_{n+1} - a_n). \tag{4.1}$$

The known necessary and sufficient condition for non conservativity (see [**Ch**]) is

$$\sum_{k=0}^{\infty} \frac{1}{c_k} < \infty. \tag{4.2}$$

We can give a proof of this fact by using Theorem 3.1.

Indeed, if condition (4.2) holds, then let us define the operator $j$ as multiplication by the following element of $\ell_\infty$, $\{j_n\}_n$, where

$$j_0 = 0 \qquad j_{n+1} = \sum_{k=0}^{n} \frac{1}{c_k},$$

Then

$$c_n(j_{n+1} - j_n) = 1, \tag{4.3}$$

Now take $0 < \lambda \leq (\sum_{k=0}^{\infty} \frac{1}{c_k})^{-1}$. Hence we obtain $0 \leq \lambda j \leq I$ and

$$\mathcal{L}(j)[u] = \|u\|^2 \geq \lambda \langle u, (j - b)u \rangle$$

holds with

$$b := 0 < j$$

and, by part b) of Theorem 3.1, the minimal qds is nonconservative.
Conversely assume the minimal qds is nonconservative. Let us notice that, in equation (4.1), $\mathcal{L}(I)[u] = 0$. Then for any $\lambda > 0$ there is a bounded, positive operator $x$ such that statement a) in Theorem 3.1 holds. Since the abelian algebra $\ell_\infty$ is left invariant by the minimal qds then, we can assume that $x$ is multiplication by some positive and bounded sequence, $\{x_n\}_n$. Since equation in part a) of Theorem is satisfied by $x$, and $\mathcal{L}(I) = 0$, then this equation can be written as

$$\lambda x_n = (\mathcal{L}(x))_n = c_n(x_{n+1} - x_n)$$

Furthermore, by the linearity in the operator variable, we can assume that $0 < x_n < 1$. Hence, this sequence satisfies the condition

$$c_n(x_{n+1} - x_n) \geq \lambda (x_n)^\alpha, \quad \text{for any} \quad \alpha > 1,$$

which can be restated as

$$\frac{x_{k+1} - x_k}{(x_k)^\alpha} \geq \frac{\lambda}{c_k}.$$

From here, we conclude that $\{x_n\}_n$ is a bounded and monotonic increasing sequence, hence $\lim_{n\to\infty} x_n < \infty$. On the other hand, let us observe that for all $n$,

$$\sum_{k=1}^{n} \frac{x_{k+1} - x_k}{(x_{k+1})^\alpha} \leq \int_{x_1}^{x_n} \frac{dz}{z^\alpha} \leq \sum_{k=1}^{n} \frac{x_{k+1} - x_k}{(x_k)^\alpha}$$

But since,

$$\int_{x_1}^{\infty} \frac{dz}{z^\alpha} < \infty \quad \text{and} \quad \lim_{k\to\infty} (\frac{x_{k+1}}{x_k})^\alpha = 1,$$

we conclude

$$\lambda \sum_{k=1}^{\infty} \frac{1}{c_k} \leq \sum_{k=1}^{\infty} \frac{x_{k+1} - x_k}{(x_k)^\alpha} < \infty.$$

So condition (4.2) holds.

Next, we shall make a discussion on the possibility to apply our result to an example coming from mathematical physics.

EXAMPLE 4.2. On $\mathcal{H} = L_2(0,\infty)$ let us consider operators induced by the differential form

$$\tau_f u = \frac{1}{2i}\left((fu)' + fu'\right),$$

where $f \in C^\infty(0,\infty)$, $f > 0$, $f'$ is bounded and $\int_0^\infty dx f(x)^{-1} = \infty$. We denote by $H_{1,0}$ the minimal operator induced by $\tau_f$, it is defined by

$$\mathrm{dom} H_{1,0} = C_0^\infty(0,\infty) \quad \text{and} \quad H_{1,0}u = \tau_f u, \qquad u \in \mathrm{dom} H_{1,0}.$$

The maximal operator $H_1$ induced by $\tau_f$ is defined by

$$\mathrm{dom} H_1 = \{u \in L_2(0,\infty) : u \text{ is absolutely continuous and } \tau_f u \in L_2(0,\infty)\}$$

and

$$H_1 u = \tau_f u, \quad u \in \mathrm{dom} H_1.$$

One can show that $H_{1,0}$ is a symmetric operator; hence it is closable and its closure $\bar{H}_{1,0}$ is symmetric, moreover $\bar{H}_{1,0}^* = H_{1,0}^* = H_1$. Notice that

$$\mathrm{dom}\bar{H}_{1,0} = \{u \in \mathrm{dom} H_1 : \; u(0) = 0\}.$$

Now let us consider the equations

$$H_{1,0}^* u = \pm i u, \quad u \in \mathrm{dom} H_{1,0}^*.$$

The solutions of these equations are respectively

$$u_+(x) = c_1 f(x)^{-1/2} e^{-\int_0^x \frac{d\tau}{f(\tau)}}$$

and

$$u_-(x) = c_2 f(x)^{-1/2} e^{+\int_0^x \frac{d\tau}{f(\tau)}}, \qquad c_1, c_2 \text{ nonzero}.$$

Notice that

$$\|u_+\|^2 = c_1^2 \int_0^\infty dx f(x)^{-1} e^{-2\int_0^x \frac{d\tau}{f(\tau)}} < \infty,$$

since $\int_0^\infty \frac{d\tau}{f(\tau)} = \infty$, therefore $u_+ \in L_2(0,\infty)$. Similarly one can show that $u_- \notin L_2(0,\infty)$. This proves that the defect indices of the symmetric operator $\bar{H}_{1,0}$ are $n_+(\bar{H}_{1,0}) = 1$ and $n_-(\bar{H}_{1,0}) = 0$.

Being $\bar{H}_{1,0}$ maximal symmetric, one can show ([**Q**]) that $i\bar{H}_{1,0}^* = iH_1$ satisfies the hypotheses of the Lumer Phillips Theorem ([**Paz**]), so it generates a strongly continuous semigroup of contractions.
Now consider a linear, CP and normal map $\phi : \mathcal{B} \mapsto \mathcal{B}$. Notice that $-G = \frac{1}{2}\phi(I) + iH_1$ with dom $G =$ dom $H_1$, generates a strongly continuous semigroup of contractions, $(W_t)_{t\geq 0}$. Take $\mathcal{L}$ as the formal generator defined, for every $x \in \mathcal{B}$, as the sesquilinear form

$$\mathcal{L}(x)[u,v] = \langle u, \phi(x)v\rangle - \langle Gu, x\, v\rangle - \langle u, x\, Gv\rangle, \qquad u, v \in \mathrm{dom} G.$$

Clearly, $\mathcal{L}$ satisfies conditions **i)** and **ii)** of Section 2 and

$$\begin{aligned}\mathcal{L}(I)[u] &= \langle u, \phi(I)u\rangle - \langle \frac{1}{2}u, \phi(I)u\rangle + \langle iH_1 u, u\rangle - \langle u, \frac{1}{2}\phi(I)u\rangle + \langle u, iH_1 u\rangle \\ &= 2Re\langle iH_1 u, u\rangle \leq 0, \qquad \forall u \in \mathrm{dom}\, G,\end{aligned}$$

since $iH_1$ is dissipative. So $\mathcal{L}$ satisfies also condition iii) of Section 2.
Let $(P_t^{\min})_{t\geq 0}$ be the minimal qds constructed from the formal generator $\mathcal{L}$.

Take $u_+$ as above and $j = |u_+ >< u_+|$ the projector on the subspace generated by $u_+$. Then we have for $v \in \mathrm{dom}\bar{H}_{1,0}$,

$$\begin{aligned}
\mathcal{L}(j)[v] &= \langle v, \phi(j)v\rangle - \langle v, \phi(I)v\rangle + \langle i\bar{H}_{1,0}v, j\, v\rangle + \langle v, j\, i\bar{H}_{1,0}v\rangle \\
&= \langle v, \phi(j-I)v\rangle + \langle i\bar{H}_{1,0}v, |u_+ >< u_+|\, v\rangle + \langle v, |u_+ >< u_+|\, i\bar{H}_{1,0}v\rangle \\
&= \langle v, \phi(j-I)\, v\rangle - i\langle v, H_1\, u_+\rangle\langle u_+, v\rangle + i\langle v, u_+\rangle\langle H_1\, u_+, v\rangle \\
&= \langle v, \phi(j-I)\, v\rangle - i\langle v, iu_+\rangle\langle u_+, v\rangle + i\langle v, u_+\rangle\langle iu_+, v\rangle \\
&= \langle v, \phi(j-I)v\rangle + 2\langle v, |u_+ >< u_+|v\rangle = 2\langle v, jv\rangle + \langle v, \phi(j-I)v\rangle \\
&= 2\langle v, (j-b)v\rangle,
\end{aligned}$$

with $b := \frac{1}{2}\phi(I-j) \geq 0$.

If we take $\phi$ such that $\|\phi(I)\| < 2$, we will have $\|b\| \leq \frac{1}{2}\|\phi(I)\| < 1 = \|j\|$

Notice that this formal generator $\mathcal{L}$ satisfies conditions of point b) in Theorem 3.1 for every $v \in$ dom $\bar{H}_{1,0}$, which is properly contained in dom $G$ but may not be a core of $G$. However, the above computations suggest that the corresponding minimal qds is not conservative. We will continue this research in a forthcoming paper.

## References


[B-S] B.V. R. Bhat and K.B. Sinha, *Examples of unbounded generators leading to nonconservative minimal semigroups,* Infin. Dimens. Anal. Quantum Probab. Relat. Top. **IX** (1994), pp. 89–103.

[B-R] O. Bratteli and D.W. Robinson, *Operator Algebras and Quantum Statistical Mechanics,* vol. **I**, Springer–Verlag, New York, (1987).

[Ch] A.M. Chebotarev, *Lectures on Quantum Probability,* Aportaciones Mat. Textos vol. **14**, Soc. Mat. Mexicana, México, (2000).

[D] E.B. Davies, *Quantum Theory of Open Systems,* Academic Press, London, (1976).

[F] F. Fagnola, *Quantum Markov semigroups and quantum flows,* Proyecciones **18**, no. 3 (1999), pp. 1–144.

[K] A. Kossakowski, *On quantum statistical mechanics of non-hamiltonian systems,* Rep. Math. Phys. **3** (1972), pp. 247–274.

[Par] K.R. Parthasarathy, *An Introduction to Quantum Stochastic Calculus,* Monogr. Math., vol. **85**, Birkhauser–Verlag, Basel–Boston–Berlin, (1992).

[Paz] A. Pazy, *Semigroups of Linear Operators and Applicactions to Partial Differential Equations,* Springer–Verlag, New York, Berlin Heidelberg–Tokyo, (1983).

[Q] R.Quezada, *Non-conservative minimal quantum dynamical semigroups,* http://arXiv.org/abs/math-ph/0112036.

[W] J. Weidmann, *Linear Operators in Hilbert Spaces,* Grad. Texts in Math. vol. **68**, Springer–Verlag, New York–Heidelberg–Berlin, (1980).



Departamento de Matemáticas, Universidad Autónoma Metropolitana-Iztapalapa, Av. San Rafael Atlixco 186, 09340 Col. Vicentina, México D. F.

*E-mail address*: jcgc@xanum.uam.mx, roqb@xanum.uam.mx

Contemporary Mathematics
Volume **336**, 2003

# Some Notes on a Dependency Measure

José M. González-Barrios

ABSTRACT. The problem of dependency between two random variables has been studied thoroughly in the literature. In this paper we study the behavior of a multidimensional extension of two dependency measures, finding some of their basic properties.

We also provide a sample version of one of these measures and find some interesting statistical properties such as invariance under monotone transformations, invariance of distributions under continuity assumptions and robustness.

## 1. Introduction

Clearly many random phenomena consist of dependent random variables; as a consequence the concept of dependency has been largely studied in Probability and Statistics. Even in the case of two random variables dependency can have a large variety of forms, and many dependency measures have been provided according to some specifications, for example in terms of the concepts of concordance and discordance which are scale-invariant measures of association such as the Kendall's tau and the Spearman's rho. Other measures have been proposed in terms of ranks, quadrant dependency, etc. see for example [N].
More recently the study of dependence has been largely based on the concept of copulas. Recall that a copula is a function $C : I^2 \to I$ , where $I = [0, 1]$ such that
a) For every $u, v \in I$ $C(u, 0) = 0 = C(0, v)$
and $C(u, 1) = u$ and $C(1, v) = v$.
b) For every $u_1, u_2, v_1, v_2 \in I$ such that $u_1 \leq u_2$ and $v_1 \leq v_2$,

$$C(u_2, v_2) - C(u_2, v_1) - C(u_1, v_2) + C(u_1, v_1) \geq 0.$$

There is of course a multidimensional version of copulas. One of the most important theorems in the theory of copulas is:

THEOREM 1.1 (Sklar's Theorem in $n$ dimensions). *Let $H$ be an $n$-dimensional distribution function with margins $F_1, F_2, \ldots, F_n$. Then there exists an $n$-copula $C$*

---

2000 *Mathematics Subject Classification.* Primary 60E05, 62F35, 62G30; Secondary 62E15, 62G10.

*Key words and phrases.* Copulas, Multivariate Dependency Measures, Robustness.

The author was supported in part by Conacyt grant 32705-E.

*such that for all* $(x_1, x_2, \ldots, x_n) \in I\!R^n$,

$$H(x_1, x_2, \ldots, x_n) = C(F_1(x_1), F_2(x_2), \ldots, F_n(x_n)). \tag{1.1}$$

*If* $F_1, F_2, \ldots, F_n$ *are all continuous, then* $C$ *is unique; otherwise,* $C$ *is uniquely determined on* $RanF_1 \times RanF_2 \times \cdots \times RanF_n$. *Conversely, If* $C$ *is an* $n$*-copula and* $F_1, F_2, \ldots, F_n$ *are distribution functions, then* $H$ *defined by (1.1) is an* $n$*-dimensional distribution function with margins* $F_1, F_2, \ldots, F_n$.

REMARK 1.2. For a proof of this result we refer the reader to [S]. Thanks to Sklar's Theorem, a copula $C$ can be thought of as the joint distribution of a continuous random vector $\underline{X} = (X_1, X_2, \ldots, X_n)$ with $U(0,1)$ marginals.

In this paper we propose two multivariate dependency measures and analize their basic properties. We then provide a sample version of one of them and we find some of its statistical properties.

## 2. Multidimensional Dependency Measures

Let $(\Omega, \mathcal{F}, P)$ be an arbitrary probability space and let

$$X_i : (\Omega, \mathcal{F}) \longrightarrow (\Omega_i, \mathcal{F}_i) \qquad \text{for} \qquad i \in I_n = \{1, 2, \ldots, n\}$$

be $n$ random elements. We define a dependency measure among the $X_i$'s as follows. Let

$$\eta_{X_1, X_2, \ldots, X_n} := \sup_{A_i \in \mathcal{F}_i, i \in I_n} \left| P\left(\cap_{i=1}^n \{X_i \in A_i\}\right) - \Pi_{i=1}^n P\left(\{X_i \in A_i\}\right) \right|. \tag{2.1}$$

Its general properties are:

THEOREM 2.1. *Let* $(\Omega, \mathcal{F}, P)$, $X_i$ *and* $\eta_{X_1, X_2, \ldots, X_n}$ *as above. Then* $\eta_{X_1, X_2, \ldots, X_n}$ *satisfies*

(1) $\eta_{X_1, X_2, \ldots, X_n} = \eta_{X_{\sigma(1)}, X_{\sigma(2)}, \ldots, X_{\sigma(n)}}$ *for every* $\sigma$ *permutation of* $I_n := \{1, 2, \ldots, n\}$.
(2) $\eta_{X_1, X_2, \ldots, X_n} = 0$ *if and only if* $X_1, X_2, \ldots, X_n$ *are independent.*
(3) *For every* $A_i \in \mathcal{F}_i$, $i \in I_n$

$$-\left(\frac{n-1}{n}\right)^n \leq P\left(\cap_{i=1}^n \{X_i \in A_i\}\right) - \Pi_{i=1}^n P\left(X_i \in A_i\right) \leq \left(\frac{1}{n}\right)^{\frac{1}{n-1}} \left(1 - \frac{1}{n}\right) < 1.$$

*Hence* $0 \leq \eta_{X_1, X_2, \ldots, X_n} \leq 1$. *Besides the bounds above can be attained.*
(4) $0 \leq \eta_{X_1, X_2} \leq \eta_{X_1, X_2, X_3} \leq \cdots \leq \eta_{X_1, X_2, \ldots, X_{n-1}} \leq \eta_{X_1, X_2, \ldots, X_n}$

For a proof of the last theorem and the results that follow see [F-G].
Recall that a sequence $\{X_n \,|\, n \in \mathbf{Z}\}$ of random variables is $m$-dependent if and only if for each $J \in \mathbf{Z}$ the sets $\{X_n \,|\, n \leq J\}$ and $\{X_n \,|\, n \geq J + m\}$ are independent. Now we bound our measure for sequences of $m$-dependent random variables.

PROPOSITION 2.2. *Let* $\{X_n \,|\, n \in \mathbf{Z}\}$ *be a sequence of* $m$ *dependent random variables. Then*

$$\eta_{X_J, X_{J+1}, \ldots, X_{J+n}} \leq \left(\frac{1}{m}\right)^{\frac{1}{m-1}} \left(1 - \frac{1}{m}\right),$$

*for all* $J \in \mathbf{Z}$ *and* $n \geq m$.

For computations an easier definition of a dependency measure among real random variables can be given by

$$\delta_{X_1,X_2,\ldots,X_n} = \sup_{(x_1,x_2,\ldots,x_n)\in \mathbb{R}^n} |F_{X_1,\ldots,X_n}(x_1,\ldots,x_n) - \Pi_{i=1}^n F_{X_i}(x_i)|, \tag{2.2}$$

where $F_{X_1,\ldots,X_n}$ is the joint distribution function of the $X_i's$ and $F_{X_i}$ is the distribution function of $X_i$. Comparing both definitions, we can see that $\delta_{X_1,X_2,\ldots,X_n}$ satisfies

THEOREM 2.3. *Let $(\Omega, \mathcal{F}, P), X_i$ real random variables and $\delta_{X_1,X_2,\ldots,X_n}$ as above. Then $\delta_{X_1,X_2,\ldots,X_n}$ satisfies*

(1) $\delta_{X_1,X_2,\ldots,X_n} = \delta_{X_{\sigma(1)},X_{\sigma(2)},\ldots,X_{\sigma(n)}}$ *for every $\sigma$ permutation of $I_n$.*
(2) $\delta_{X_1,X_2,\ldots,X_n} = 0$ *if and only if $X_1, X_2, \ldots, X_n$ are independent.*
(3) *For every $x_i \in \mathbb{R}$, $i \in I_n$*

$$-\left(\frac{n-1}{n}\right)^n \leq F_{X_1,X_2,\ldots,X_n}(x_1,x_2,\ldots,x_n) - \Pi_{i=1}^n F_{X_i}(x_i) \leq \left(\frac{1}{n}\right)^{\frac{1}{n-1}}\left(1-\frac{1}{n}\right).$$

*Hence $0 \leq \delta_{X_1,X_2,\ldots,X_n} \leq 1$. Besides the bounds above can be attained.*
(4) $0 \leq \delta_{X_1,X_2} \leq \delta_{X_1,X_2,X_3} \leq \cdots \leq \delta_{X_1,X_2,\ldots,X_{n-1}} \leq \delta_{X_1,X_2,\ldots,X_n}$
(5) $\delta_{X_1,X_2,\ldots,X_n} \leq \eta_{X_1,X_2,\ldots,X_n}$ *for all $X_1, X_2, \ldots, X_n$ real random variables.*

We compute the value of $\delta_{X_1,X_2,\ldots,X_n}$ in several examples in [F-G].

LEMMA 2.4. *Let $X_1, X_2, \ldots, X_n, X_{n+1}$ be $n+1$ random elements defined on $(\Omega, \mathcal{F}, P)$. If $X_{n+1}$ is independent of $(X_1, \ldots, X_n)$, then*

$$\eta_{X_1,X_2,\ldots,X_n} = \eta_{X_1,X_2,\ldots,X_n,X_{n+1}}.$$

*If all the $X_i's$ are real random variables then*

$$\delta_{X_1,X_2,\ldots,X_n} = \delta_{X_1,X_2,\ldots,X_n,X_{n+1}}.$$

PROPOSITION 2.5. *Let $\mathbf{X} = (X_1, X_2, \ldots, X_n)$ be an $n$-dimensional random vector with distribution function $F_{X_1,X_2,\ldots,X_n}$ and respective continuous margins $F_{X_i}$ for $i = 1, 2, \ldots, n$ and corresponding copula $C$. Then*

$$\delta_{X_1,X_2,\ldots,X_n} = \sup_{(u_1,u_2,\ldots,u_n)\in[0,1]^n} |C(u_1,u_2,\ldots,u_n) - \Pi_{i=1}^n u_i|.$$

Now for the case where $n = 2$ we analize results about the robustness of $\delta_{Y,X}$. We first recall some basic definitions, see for example [A] or [H-R-R-S]. Let $F$ be a joint distribution function of two random variables $Y$ and $X$. Let $0 < \epsilon < 1$ and define

$$\mathcal{F}_\epsilon^F := \{ (1-\epsilon)F + \epsilon G \mid \text{where } G \text{ is a distribution function in } \mathbb{R}^2 \}.$$

Then $\mathcal{F}_\epsilon^F$ is a set of distribution functions which contains an $\epsilon$ percentage of contamination coming from an arbitrary distribution $G$. Let us denote by $\delta_H = \delta_{Y,X}$ the value of the statistic $\delta$ when $H$ is the joint distribution of the vector $(Y, X)$. Then the breakdown point of the statistic at the distribution function $F$ is defined by

$$\epsilon^*(F, \delta) = \sup\{\epsilon > 0 \mid \sup_{H\in\mathcal{F}_\epsilon^F} |\delta_F - \delta_H| = 1/4\}.$$

Since $\delta_H \leq 1/4$ for any joint distribution function $H$ of $Y$ and $X$, and $\delta_H$ measures dependence between $Y$ and $X$, we want to evaluate $\epsilon^*(F, \delta)$ in $F(y, x)$ corresponding to $Y$ and $X$ independent random variables. We have the following

LEMMA 2.6. *Let $(Y, X)$ be a continuous random vector with independent coordinates and joint distribution $F$. Then*

$$\epsilon^*(F,\delta) = \frac{1}{2}.$$

PROOF. Without losing generality we can assume that $Y$ and $X$ are uniform $(0,1)$ independent random variables. Then

$$F(y,x) = \begin{cases} 0 & \text{if } \ y < 0 \text{ or } x < 0 \\ xy & \text{if } \ 0 \le y < 1 \text{ and } 0 \le x < 1 \\ x & \text{if } \ y \ge 1 \text{ and } 0 \le x < 1 \\ y & \text{if } \ 0 \le y < 1 \text{ and } x \ge 1 \\ 1 & \text{if } \ y \ge 1 \text{ and } x \ge 1. \end{cases}$$

Let us consider the contamination $G$ to be the joint distribution of $Y$ and $X$ independent uniform $(1,2)$ random variables. Then

$$G(y,x) = \begin{cases} 0 & \text{if } \ y < 1 \text{ or } x < 1 \\ (x-1)(y-1) & \text{if } \ 1 \le y < 2 \text{ and } 1 \le x < 2 \\ (x-1) & \text{if } \ y \ge 2 \text{ and } 1 \le x < 2 \\ (y-1) & \text{if } \ 1 \le y < 2 \text{ and } x \ge 2 \\ 1 & \text{if } \ y \ge 2 \text{ and } x \ge 2. \end{cases}$$

Take $0 < \epsilon < 1$ and define $H = (1-\epsilon)F + \epsilon G$. Then

$$H(y,x) = \begin{cases} 0 & \text{if } \ y < 0 \text{ or } x < 0 \\ (1-\epsilon)xy & \text{if } \ 0 \le y < 1 \text{ and } 0 \le x < 1 \\ (1-\epsilon)x & \text{if } \ y \ge 1 \text{ and } 0 \le x < 1 \\ (1-\epsilon)y & \text{if } \ 0 \le y < 1 \text{ and } x \ge 1 \\ (1-\epsilon) + \epsilon(x-1)(y-1) & \text{if } \ 1 \le y < 2 \text{ and } 1 \le x < 2 \\ (1-\epsilon) + \epsilon(y-1) & \text{if } \ 1 \le y < 2 \text{ and } x \ge 2 \\ (1-\epsilon) + \epsilon(x-1) & \text{if } \ y \ge 2 \text{ and } 1 \le x < 2 \\ 1 & \text{if } \ y \ge 2 \text{ and } x \ge 2. \end{cases}$$

To obtain $\delta_H$ we define the following regions on the plane

$$A_1 = \{(y,x) \in I\!R^2 \,|\, y < 0 \,\text{or}\, x < 0\}$$
$$A_2 = \{(y,x) \in I\!R^2 \,|\, 0 \le y < 1 \,\text{and}\, 0 \le x < 1\}$$
$$A_3 = \{(y,x) \in I\!R^2 \,|\, 0 \le y < 1 \,\text{and}\, 1 \le x < 2\}$$
$$A_4 = \{(y,x) \in I\!R^2 \,|\, 0 \le y < 1 \,\text{and}\, x \ge 2\}$$
$$A_5 = \{(y,x) \in I\!R^2 \,|\, 1 \le y < 2 \,\text{and}\, 0 \le x < 1\}$$
$$A_6 = \{(y,x) \in I\!R^2 \,|\, 1 \le y < 2 \,\text{and}\, x \ge 2\}$$
$$A_7 = \{(y,x) \in I\!R^2 \,|\, 1 \le y < 2 \,\text{and}\, 1 \le x < 2\}$$
$$A_8 = \{(y,x) \in I\!R^2 \,|\, 1 \le y < 2 \,\text{and}\, x \ge 2\}$$
$$A_9 = \{(y,x) \in I\!R^2 \,|\, y \ge 2 \,\text{and}\, 1 \le x < 2\}$$

and

$$A_{10} = \{(y,x) \in I\!R^2 \,|\, y \ge 2 \,\text{and}\, x \ge 2\}.$$

Define $\delta_i = \sup_{(y,x)\in A_i} |H(y,x) - H_1(y)H_2(x)|$ where $H_1$ and $H_2$ are the marginals of $Y$ and $X$ respectively. Then

$$\delta_1 = |0 - 0 \cdot 0| = 0$$

$$\delta_2 = |(1-\epsilon)yx - (1-\epsilon)y(1-\epsilon)x| = \epsilon(1-\epsilon)yx \le \epsilon(1-\epsilon)$$
$$\delta_3 = |(1-\epsilon)y - ((1-\epsilon)y((1-\epsilon)+\epsilon(x-1)))| = \epsilon(1-\epsilon)y(2-x) \le \epsilon(1-\epsilon)$$
$$\delta_4 = |(1-\epsilon)y(1-\epsilon)y \cdot 1| = 0$$
$$\delta_5 = |(1-\epsilon)x - (((1-\epsilon)+\epsilon(y-1))(1-\epsilon)x)| = (1-\epsilon)\epsilon x(2-y) \le \epsilon(1-\epsilon)$$
$$\delta_6 = |(1-\epsilon)x - 1 \cdot (1-\epsilon)x| = 0$$
$$\delta_7 = |(1-\epsilon)+\epsilon(y-1)(x-1) - (((1-\epsilon)+\epsilon(y-1))((1-\epsilon)+\epsilon(x-1)))| = \epsilon(1-\epsilon)(2-y)(2-x) \le \epsilon(1-\epsilon)$$
$$\delta_8 = |(1-\epsilon)+\epsilon(y-1) - ((1-\epsilon)+\epsilon(y-1))| = 0$$
$$\delta_9 = |(1-\epsilon)+\epsilon(x-1) - ((1-\epsilon)+\epsilon(x-1))| = 0$$

and

$$\delta_{10} = |1 - 1 \cdot 1| = 0.$$

Therefore

$$\delta_H = \max_{i=1}^{10} \delta_i = \epsilon(1-\epsilon),$$

and since for $0 < \epsilon < 1$ the function $f(\epsilon) = \epsilon(1-\epsilon)$ attains a maximum at $\epsilon = 1/2$, and $\delta_F = 0$ we get that $\epsilon^*(F, \delta) = \frac{1}{2}$. □

From the last lemma we conclude that we need 50 % of contamination in order to ruin the behavior of $\delta_F$. This is compatible with the results for the median, which is known to be one of the most robust location statistics.

## 3. A Sample Multidimensional Dependency Measure

Assume we have a $n$-dimensional sample of size $m$, $\mathbf{X}_i = (X_{i_1}, X_{i_2}, \ldots, X_{i_n})$ for $i = 1, 2, \ldots, m$, coming from a joint distribution $F_{\mathbf{X}}(x_1, x_2, \ldots, x_n)$ with margins given by $F_{X_j}(x_j)$, $j = 1, 2, \ldots, n$. Denote by $F_m(x_1, \ldots, x_n)$ the joint empirical distribution function and by $F_{m,j}(x_j)$ the empirical distributions of $X_j$ for $j = 1, 2, \ldots, n$. We propose as a sample multidimensional dependency measure the following

$$\delta^m_{X_1, X_2, \ldots, X_n} := \sup_{(x_1, x_2, \ldots, x_n) \in \mathbb{R}^n} |F_m(x_1, x_2, \ldots, x_n) - \Pi^n_{j=1} F_{m,j}(x_j)|. \tag{3.1}$$

Then this sample version mimics the populational version of the dependency measure defined above. In fact this proposal makes sense since when $m \to \infty$, $F_m$ approaches the population joint distribution, and $F_{m,j}$ approaches the population margin distribution.
We have a strong law of large numbers for this statistic under the hypothesis of independence among samples. For the proof of the following results see [F-G].

THEOREM 3.1. *Let $(X_{i_1}, X_{i_2}, \ldots X_{i_n})$ for $i = 1, 2, \ldots, m$ be an independent sample of size $m$, coming from a common $n$-dimensional joint distribution $F_{\mathbf{X}}$ with corresponding marginal distribution functions given by $F_{X_j}$ for $j = 1, 2, \ldots, n$, let us use the same notation as above for the empirical distribution functions. Then*

$$\delta^m_{X_1, X_2, \ldots, X_n} = \sup_{(x_1, x_2, \ldots, x_n) \in \mathbb{R}^n} |F_m(x_1, x_2, \ldots, x_n) - \Pi^n_{j=1} F_{m,j}(x_j)| \to \delta_{X_1, X_2, \ldots, X_n},$$

*almost surely when $m \to \infty$.*

We can also find an upper bound for the population version of the statistic.

LEMMA 3.2. *For every $n \geq 2$ and any $m > n$,*

$$\delta^m_{X_1,X_2,\ldots,X_n} \leq \max_{0\leq k\leq m} \left( \frac{k}{m} - \left(\frac{k}{m}\right)^n \right) =: \mathcal{K}_{m,n}.$$

Using the last Lemma we can provide a sample dependency coefficient which does not depend on the existence of any moments.

DEFINITION 3.3. Assume we have a $n$-dimensional sample of size $m$, $\mathbf{X}_i = (X_{i_1}, X_{i_2}, \ldots, X_{i_n})$ for $i = 1, 2, \ldots, m$, coming from a joint distribution

$F_{\mathbf{X}}(x_1, x_2, \ldots, x_n)$ with margins given by $F_{X_j}(x_j)$, $j = 1, 2, \ldots, n$. We define the Sample Dependency Coefficient by

$$\mathcal{C}_{\mathbf{X}} := \frac{\delta^m_{X_1,X_2,\ldots,X_n}}{\mathcal{K}_{m,n}}.$$

Of course this coefficient lies between zero and one, and it is an indicator of the dependency among the $n$ coordinates, given a sample of size $m$. It is more general than the correlation coefficient, because we do not need existence of any moments to define it.

Another important property of the statistic $\delta^m_{X_1,X_2,\ldots,X_n}$ is that it is unvariant under increasing transformations.

PROPOSITION 3.4. *For $i = 1, 2, \ldots, m$, let $(X_{i_1}, X_{i_2}, \ldots, X_{i_n})$ be a random sample as in Theorem 3.4, and let $f_j : I\!R \to I\!R$ for $j = 1, 2, \ldots, n$ be strictly increasing functions. Then*

(1) *For continuous random variables $X_1, X_2, \ldots, X_n$*
$\delta^m_{X_1,X_2,\ldots,X_n} = \delta^m_{f_1(X_1),f_2(X_2),\ldots f_n(X_n)}$.
(2) *If $f_1$ and $f_2$ are both strictly increasing or decreasing functions, and $X_1, X_2$ are continuous, then*
$$\delta^m_{X_1,X_2} = \delta^m_{f_1(X_1),f_2(X_2)}.$$
(3) *For every $n \geq 2$ and any $m > n$ and any different subindices $i_1, i_2, \ldots, i_k \in \{1, 2, \ldots, n\}$ with $k \leq n$,*
$$\delta^m_{X_{i_1},X_{i_2}} \leq \cdots \leq \delta^m_{X_{i_1},X_{i_2},\ldots,X_{i_k}} \leq \cdots \leq \delta^m_{X_1,X_2,\ldots,X_n}.$$

For a proof of this proposition see [G-R]. This result is very useful in applications since we can apply transformations such as standarizations, or Box-Cox transformations to each of the the components of the vector $\underline{X} = (X_1, X_2, \ldots, X_n)$ and the statistic $\delta^m_{X_1,X_2,\ldots,X_n}$ remains unchanged. Another important feature of the statistic is that its value increases when we add coordinates to a vector in any order.

Now we state a lemma giving results about the extreme values of $\delta^m_{X_1,X_2,\ldots,X_n}$.

LEMMA 3.5. *Let $\mathbf{X}_i = (X_{i_1}, X_{i_2}, \ldots, X_{i_n})$ be an $n$-dimensional sample. Then*

(1) $\delta^m_{X_1,X_2,\ldots,X_n} = 0$ *if and only if the sample lies in a perfect grid.*
(2) *Denote by $[a]$ the greatest integer less than or equal to $a$. If $n = 2$ then $\delta^m_{X_1,X_2} = 1/4$ if and only if, for $j = [m/2]$, there exists $(x_0, y_0) \in I\!R^2$ such that $F_m(x_0, y_0) = j/m$, $F_{m,1}(x_0) = j/m$ and $F_{m,2}(y_0) = j/m$.*

PROOF. 1) Let us assume that $n = 2$ and let $(x_1, y_1), (x_2, y_2), \ldots, (x_m, y_m)$ be the sample. Assume the sample points form a perfect grid. Then there exist $k, l$ positive integers such that $m = k \cdot l$, and using Proposition 3.4 we can assume that the sample is given by $\mathbf{X} = \{(i,j) \mid 1 \le i \le k, \text{ and } i \le j \le l\}$. Then $F_m(i,j) = ij/m$, $F_{m,1}(i) = li/m$ and $F_{m,2} = jk/m$ and

$$F_m(i,j) - F_{m,1}(i)F_{m,2}(j) = \frac{ij}{m} - \frac{li}{m}\frac{jk}{m} = \frac{ij}{kl} - \frac{lijk}{k^2l^2} = 0,$$

for every $(i,j) \in \mathbf{X}$. Therefore $\delta^m_{X,Y} = 0$.
Now assume that $\delta^m_{X,Y} = 0$, using Proposition 3.4 again we can assume that the sample values of the $X$ coordinates are $1, 2, \ldots, k$ where $k \le m$, and the values of the $Y$ coordinates are $1, 2, \ldots, l$ where $l \le m$. Since $\delta^m_{X,Y} = 0$ then we have that for any $(i,j)$ with $1 \le i \le k$ and $1 \le j \le l$, there exist integers $c, C_1, C_2$ such that

$$F_m(i,j) = \frac{c}{m} = \frac{c_1}{m}\frac{c_2}{m} = F_{m,1}(i)F_{m,2}(j) \tag{3.2}$$

Now if we assume that $F_m(1,1) = 0$, since $F_{m,1}(1) > 0$ and $F_{m,2}(1) > 0$ then (3.2) does not hold. Therefore $F_m(1,1) = 1/m$ and from (3.2) we get that $m = c_1c_2$ and that the point $(1,1)$ belongs to the sample. Now assume that $(1,2)$ does not belong to the sample then $F_m(1,2) = 1/m$, since $F_{m,2}(2) > F_{m,2}(1)$ from (3.2) we get

$$F_m(1,2) = \frac{1}{m} = F_{m,1}(1)F_{m,2}(1) = F_{m,1}(1)F_{m,2}(2),$$

which is a contradiction. Therefore $F_m(1,2) = 2/m$ and $(1,2)$ belongs to the sample. We then proceed inductively to see that $(i,j)$ belongs to the sample where $1 \le i \le k$ and $1 \le j \le l$, and therefore the sample lies in a perfect grid.
The proof for the multivariate case is analogous.
2) Let us assume that there exists $(x_0, y_0) \in I\!R^2$ such that $F_m(x_0, y_0) = j/m$, $F_{m,1}(x_0) = j/m$ and $F_{m,2}(y_0) = j/m$ for $j = [m/2]$. In other words if $S$ denotes the sample there exists $(x_0, y_0) \in I\!R^2$ such that if $A = \{(x,y) \mid x \le x_0, y \le y_0\}$, $B = \{(x,y) \mid x > x_0, y > y_0\}$, $C = \{(x,y) \mid x \le x_0, y > y_0\}$ and $D = \{(x,y) \mid x > x_0, y \le y_0\}$. Then $S \subset A \cup B$, where each of $A$ and $B$ includes roughly half of the sample, or $S \subset C \cup D$, where again each of $C$ and $D$ includes roughly half of the sample. Then $\delta^m_{X,Y} = 1/4$. □

From the last lemma we see that $\delta^m_{X_1,X_2}$ measures quadrant dependency, see [N]. It is also important to notice that the statistic $\delta^m_{X_1,X_2}$ is not invariant under rotations, see Lemma 3.5 i). However, it is well known that dependence itself is not invariant under rotations, as can be easily seen by rotating a joint uniform $(0,1) \times (0,1)$ distribution about the point $(1/2, 1/2)$. Of course the invariance under rotations of the independent multivariate normal distribution can not be generalized to other distributions. We now state an important result.

THEOREM 3.6. *$(X_{i_1}, X_{i_2}, \ldots, X_{i_n})$ for $i = 1, 2, \ldots, m$ be a random sample of size $m$ coming from a common joint distribution $F_{\underline{X}}(x_1, x_2, \ldots, x_n)$, where $\underline{X} = (X_1, X_2, \ldots, X_n)$. Then*

(1) *If $X_1, X_2, \ldots, X_n$ are continuous random variables then $\delta^m_{X_1,X_2,\ldots,X_n} \neq 0$ a.s. for every $m \ge n$.*
(2) *If $n = 2$ and $F_m(x_i, y_j) = k/m$, for $0 \le k \le m$, $F_{m,1}(x_i) = k_1/m$ and $F_{m,2}(y_j) = k_2/m$. Then $\min\{k_1, k_2\} \ge k$ and $m + k - k_1 - k_2 \ge 0$.*
(3) *Assume that $F_{\underline{X}}(x_1, x_2, \ldots, x_n) = x_1x_2\cdots x_n$, that is $X_1, X_2, \ldots, X_n$ are independent uniform $(0,1)$ random variables. Then it is always possible to find the distribution of $\delta^m_{X_1,X_2,\ldots,X_n}$ for every $m \ge n$.*

(4) *Assume that* $F_{\underline{X}}(x_1, x_2, \ldots, x_n) = F_{X_1}(x_1)F_{X_2}(x_2)\cdots F_{X_n}(x_n)$, *that is* $X_1, X_2, \ldots, X_n$ *are independent random variables with corresponding distribution functions* $F_{X_i}(\cdot)$ *for* $i = 1, 2, \ldots, n$, *which we will assume to be continuous. Then*

$$\delta^m_{X_1,X_2,\ldots,X_n} \stackrel{\text{dist}}{=} \delta^m_{F_{X_1}(X_1),F_{X_2}(X_2),\ldots,F_{X_n}(X_n)}.$$

Therefore the last theorem states that assuming continuity the distribution of the estimator $\delta^m_{X_1,X_2,\ldots,X_n}$, under the hypothesis of independence, relies only on the sample size and the dimension, but not on the distribution.

## 4. Final Remarks

The study of dependence for more than two random variables has been neglected compared to the the study of dependence of two random variables. This work presents two proposals which have been known for some time. However they are analized here in detail, finding some new properties which proved to be very useful in applications when transfered to a sample version.

The sample version $\delta^m_{X_1,X_2,\ldots,X_n}$ has very nice properties which allow us to apply this statistic in many areas such as principal component analysis, see [G-R], time series analysis, see [C-G], and regression analysis [G-R1]. The main result of this paper was to present an alternative to measure sample dependence in an easy way, with a statistic whose distribution depends only on the sample size and the dimension, but not on the distribution, at least in the continuous case. We also find that the statistic is quite robust and analyze for what kind of random samples we attain the bounds.

A Fortran program that evaluates $\delta^m_{X_1,X_2,\ldots,X_n}$ for any $n$ and $m$ for any given sample, even if the sample comes from a discrete distribution, can be obtained from the author.

## References

[A] J. Adrover, *Notas de Curso de Robustez* Preprint, (2002).

[C-G] A. Contreras and J. González-Barrios, *The Determination of the Appropriate Order in a General Class of Time Series Models Comm. in Statist. (Theory and Methods)* **32**, no.4, (2003), pp. 875-891.

[F-G] B. Fernández-Fernández and J. González-Barrios, *Multivariate Dependency Measures*, To appear in J. of Multiv. Anal. (2003).

[G-R] J. González-Barrios and S. Ruiz-Velasco, *The Application of a New Dependency Measure to Principal Component Analysis*, To appear in Comm. in Statist. (Comp. and Sim.) **32**, No. 3, (2003).

[G-R1] J. González-Barrios and S. Ruiz-Velasco, *Regression Analysis and Dependence*, Submitted to Metrika, (2003).

[H-R-R-S] R.P. Hampel, E.M. Ronchetti, P.J. Rousseuw, and W.A. Stahel, W.A., *Robust Statistics - The Approach Based on Influence Functions*, Ed. John Wiley & Sons, Ser. in Probab. and Math. Statist., New York, 1986.

[N] R.B. Nelsen, *An Introduction to Copulas*, Ed. Springer-Verlag, New York, 1999

[S] A. Sklar, *Random Variables, Distribution Functions and Copulas-a Personal Look Backward and Forward*, In Distributions With Fixed Marginals and Related Topics. L. Rüschendorf, B. Schweizer and M.D. Taylor, editors. Institute of Mathematical Statistics, Hayward, CA, (1996), pp. 1-14.

Department of Probability and Statistics, Instituto de Investigaciones en Matemáticas Aplicadas y en Sistemas, Universidad Nacional Autónoma de México, Apdo. Postal 20-726, Admon. No. 20, Deleg. Alvaro Obregón, México, D.F. 01000, México
*E-mail address*: gonzaba@sigma.iimas.unam.mx

Contemporary Mathematics
Volume **336**, 2003

# An Example of an Averaged Markov Decision Process without Stable Policies

Juan González-Hernández

ABSTRACT. We give an example of a Markov decision process with expected average cost where the optimum value is not reached by a stable policy and a variation of this where there are not stables policies.

## 1. Introduction

In Markov decision processes with average cost is very useful to work with policies such that instead of taking limsup (or liminf) in the calculus of the performance index of a policy (Definition 2.2 below) we just take limit. The Individual Ergodic Theorem (see for instance Theorem E.8 in [**H-L**] or pp. 388 in [**Y**]) implies that stable policies (Definition 2.3 below) have this property. That is why they are used in: ( [**B**], [**H-G-L**], Chapter 5 of [**H-L**]) and the relatives references therein. However stable policies do not always exist as we show with Example 2.4 or if they exist it is possible that the optimal value is not reached by any stable measure as we show in Example 2.1.

## 2. Example of a Markov decision process

For a deeper exposition of the Markov decision concepts see [**A**], [**H-L**], [**H**], [**P**] or [**R**].

EXAMPLE 2.1. *of a Markov decision model:*

Let $\mathbf{X} = \mathbf{Q} \cap (-1, 1)$ be the denumerable state space, let $\mathbf{A} = A(x) = \{-1, 1\}$ be the finite action space and the set of avalaible action in the state space $x$ for every $x \in \mathbf{X}$. The dynamic of the system is given by the conditional probability $Q(\frac{x+a}{2} \mid x, a) = 1$ for every $x \in \mathbf{X}$. The nonnegative continuous measurable cost function to be minimized is $c(x, a) := 1 + x$.

We will need some notation. The sets of histories are defined by $\mathbb{H}_0 := \mathbf{X}$ and $\mathbb{H}_m := (\mathbf{X} \times \mathbf{A})^m \times \mathbf{X}$ for $m = 1, 2, \ldots$ .

DEFINITION 2.2. *Policies, stationary and deterministic policies.*

2000 *Mathematics Subject Classification.* Primary: 90C40, Secondary: 93E20.

*Key words and phrases.* Stable Measures, Counterexample, Markov Decision Process, Average Cost.

(a) A *policy* $\pi = (\pi_m)_{m=0,1,2,...}$ is a sequence of conditional probabilities $\pi_m$ on $\mathbf{A}$ given $\mathbb{H}_m$ for $m = 1, 2, ...$ . We will denote by $\mathbf{\Pi}$ as the set of all the policies.

(b) A policy it is said to be *randomized* (*stationary*) if there is a conditional probability $\varphi$ in $\mathbf{X}$ given $\mathbf{A}$ such that $\pi_m(\cdot \mid h_m) = \varphi(\cdot \mid x_m)$ for $m = 0, 1, ...$ The set of all of such conditional probabilities is denoted by $\Phi$. We will identify a stationary policy $\pi = (\varphi, \varphi, ...)$ with the conditional probabilityl $\varphi$, and the sets all the randomized policies with $\Phi$.

(c) A policy it is said to be *deterministic* (*stationary*) if there is a function $f$ from $\mathbf{X}$ to $\mathbf{A}$ such that $\varphi(\cdot \mid x_m) = \delta_{(\cdot)}(f(x))$ for $m = 1, 2, ...$ . The set of all functions is denoted by $\mathbf{F}$, and an element in $\mathbf{F}$ is called selector. We will identify a deterministic policy $\pi = (f, f, ...)$ with the selector $f$, and the set of all deterministic policies with $\mathbf{F}$.

In this way we have that:

$$\mathbf{F} \subset \Phi \subset \Pi. \tag{2.1}$$

*Canonical construccion.*

Le us consider $\mathbf{S} := (\mathbf{X} \times \mathbf{A})^{\infty}$ with the $\sigma - a\lg ebra$ product $\mathcal{A} = \mathcal{B}(\mathbf{X} \times \mathbf{A})^{\infty}$. The Kolmogorov Consistency Theorem (pp. 94 of [**L**] or pp. 27-33 of [**K**]) or The Ionescou Tulcea Theorem (Proposition C.10 of [**H-L**] or pp. 138 of [**L**]) imply that for each $\pi = (\pi_t) \in \Pi$, $\exists P_\pi^\nu \in \mathcal{P}(\mathbf{S})$ and its marginals satisfy that

$P_\pi^\nu [x_0 \in B] = \nu(B)$ for every $B \subset \mathbf{X}$,

$P_\pi^\nu [a_t \in C \mid h_t] = \pi_t(C \mid h_t)$ for every $C \subset \mathbf{A}$, history $h_t \in \mathbb{H}_t$ and $t = 0, 1, 2, ...$ , and

$P_\pi^\nu [x_{t+1} \in B \mid h_t, a_t] = Q(B \mid h_t, a_t)$ for every $B \subset \mathbf{X}$, history $h_t \in \mathbb{H}_t$ and $t = 0, 1, 2, ...$ .

In this way we have for each policy $\pi \in \Pi$ a stochastic process:

$$(S, \mathcal{A}, P_\pi^\nu, (x_t)) \tag{2.2}$$

And by the expected operator $E_\pi^\nu$ we mean the expected value with respect to the probability $P_\pi^\nu$. If $\nu$ is concentrated in $x$ we will denote by $P_\pi^x$ and $E_\pi^x$ instead.

*Performance index.*

The performance index of average expected cost $c$ is defined as follows:

$$J(\pi, \nu) := \limsup \frac{1}{n} E_\pi^\nu \left[ \sum_{t=0}^{N-1} \alpha^t c(x_t, a_t) \right] \quad \text{for } \pi \in \Pi \text{ and } \nu \in \mathcal{P}(\mathbf{X}). \tag{2.3}$$

The Control Problem (**C. P.**) we are interested in is

$$\textbf{C. P.} \quad \textbf{minimize} \quad J(\pi, \nu) \quad \text{for } \pi \in \Pi \text{ and } \nu \in \mathcal{P}(\mathbf{X}). \tag{2.4}$$

A policy that minimize the right side of equation (2.4) is said to be an optimal policy.

Any measure $\mu$ on $\mathbf{X} \times \mathbf{A}$ can be disintegrated as $\mu(x, a) = \varphi(a \mid x)\hat{\mu}(x)$, where $\hat{\mu}(x) := \mu(\{x\} \times \mathbf{A})$ is the marginal or projection of $\mu$ on $\mathbf{X}$. And $\varphi$ is a conditional probability on $\mathbf{X}$ given $\mathbf{A}$.

DEFINITION 2.3. stable measures and stable policies (see Defn. 3.4 in [**H-G-L**] or Defn. 5.7.7 in [**H-L**]).

Let $\mu(x,a) = \varphi(a \mid x)\hat{\mu}(x)$ a measure on $\mathbf{X} \times \mathbf{A}$. Then the probability measure $\mu$ and the stationary policy $\varphi$ are said to be *stable* if:

$\hat{\mu}(x) = \sum_{y\in\mathbf{X}} \sum_{a\in\{-1,1\}} Q(x \mid y, a)\varphi(a \mid y)\hat{\mu}(y)$ for every $x \in \mathbf{X}$, and

$J(\varphi, \hat{\mu}) = \sum_{x\in\mathbf{X}} \sum_{a\in\{-1,1\}} c(x,a)\varphi(a \mid x)\hat{\mu}(x)$

If we take two points with the same image $x = \frac{x_0+a_0}{2} = \frac{x_1+a_1}{2}$, then the only solution in $\mathbf{X}$ is $a_0 = a_1$ and $x_0 = x_1$. Hence the stable policies are deterministic policies. And we have that for each point $x_1$ with positive measure $\hat{\mu}(x)$ that Definition 2.2 (a) becomes

$\hat{\mu}(x_1) = Q(\frac{x_0+f(x_0)}{2} \mid x_0, f(x_0))\hat{\mu}(x_0) = \hat{\mu}(x_0)$.

Now we apply the same argument to $x_1$ to obtain $x_2$ and so on. In this way we obtain a sequence $(x_i)_{i=0,1,2...}$ such that $\hat{\mu}(x_0) = \hat{\mu}(x_1) = \hat{\mu}(x_2) = ...$ . We can not have an infinite number of equiprobable points, so there is $n$ such that $x_0 = x_n$. Since

$$x_n = \frac{x_{n-1}}{2} + \frac{f(x_{n-1})}{2} = \frac{x_{n-2}}{4} + \frac{f(x_{n-2})}{4} + \frac{f(x_{n-1})}{2} = \cdots,$$

we have:

$$a_0\frac{1}{2^n} + a_1\frac{1}{2^{n-1}} + ... + a_{n-1}\frac{1}{2} = x(1 - \frac{1}{2^n}), \tag{2.5}$$

where $a_i = \pm 1 = f(x_i)$ for $i = 0, 1, ..., n-1$. This equation give us all the stable measures. Then the points in a "cycle" are given by

$$x_i = \frac{x_0}{2^i} + a_0\frac{1}{2^i} + a_1\frac{1}{2^{i-1}} + ... + a_{i-1}\frac{1}{2} \quad \text{for } i = 1, 2, ..., n. \tag{2.6}$$

where $a_i = \pm 1 = f(x_i)$ for $i = 0, 1, ..., n-1$.

The average expected cost for one cycle of (2.5) is

$J(\pi, \nu) = \frac{1}{n}((1+x_0) + (1 + \frac{x_0}{2} + a_0\frac{1}{2}) + ... + (1 + \frac{x_0}{2^{n-1}} + a_0\frac{1}{2^{n-1}} + .... + a_{n-1}\frac{1}{2}))$,

which is positive, and where the initial distribution is supported on the points $x_i$ defined in (2.6). On the other hand let us define the next measurable selector: $f(x) \equiv -1$. Let $x_0$ be any element in $\mathbf{X}$, and let us define $x_i$ again as in (2.6), but now not in a cycle. Then

$x_i = \frac{x_0}{2^i} - 1 + (\frac{1}{2})^i$ and $c(x_i, -1) > c(x_{i+1}, -1)$ for $i = 1, 2, ...$ .

It is immediate that $J(\pi, \nu) = 0$, where $\pi^* = (f, f, ...)$ and $\nu$ is any initial such that is supported on the points $x_i$. Hence the optimal policy is not a stable policy.

EXAMPLE 2.4. Now let us consider the same model as in Example 2.1 but with the space $\mathbf{Y} = (\mathbf{Q} + \sqrt{2}) \cap (-1, 1)$, then the same formula (2.5) holds for the stable policies. But its solutions are rational numbers. Hence there are not solutions in $\mathbf{Y}$. There are not stable policies. On the other hand the deterministic stationary policy $\pi^*$ given by $f(x) \equiv -1$ still is an optimal policy.

For additional counterexamples see [**R**], pp. 141-144 and Appendix 2.

## References


[A] E. Altman, *Constrained Markov Decision Processes*, Chapman & Hall/CRC, Boca Raton, Florida, (1999).

[B] V.S. Borkar, *Ergodic control of Markov chains with constraints-the general case*, SIAM J. Control Optim. **32** (1994), pp. 176-186.

[H-G-L] O. Hernández-Lerma, J. González-Hernández and R.R. López Martínez, *Constrained average cost Markov control processes in Borel spaces*, to appear in SIAM J. of Control Optim. (2003).

[H-L] O. Hernández-Lerma and J.B. Lasserre, *Discrete-Time Markov Control Processes: Basic Optimality Criteria*, Springer-Verlag, New York, (1996).

[H] K. Hinderer, *Foundations of Non-stationary Dynamic Programming with Discrete-Time Parameter*, Lectures Notes Oper. Res. Math. Syst. vol. 33, Springer-Verlag, Berlin, (1970).

[K] A.N. Kolmogorov, *Foundations of the Theory of Probability*, Chelsea Publishing Company, New York, (1956).

[L] M. Loève, *Probability Theory* , vol. I, 4th edition, Springer Verlag, New York, (1977).

[P] A.B. Piunovskiy, *Optimal Control of Random Sequences in Problems with Constraints*, Kluwer Academic Publishers, Dordrecht, Netherlands, (1997).

[R] S.M. Ross, *Applied Probability Models with Optimization Applications*, Dover Publications, New York, (1992).

[Y] K.Yosida, *Functional Analysis*, 5th edition, Springer-Verlag, Berlin, (1978).



IIMAS, UNAM, Apdo. Postal 20-726, Admon. No. 20, Deleg. Alvaro Obregón, 01000 México, D.F., México

*E-mail address*: `juan@sigma.iimas.unam.mx`

Contemporary Mathematics
Volume **336**, 2003

# Closeness Estimates for Sums of Independent Random Variables

Evgueni Gordienko, Mario Mendieta, and Juan Ruiz de Chávez

ABSTRACT. Let $(X_k,\ k \geq 1)$, $(Y_k,\ k \geq 1)$ be two sequences of i.i.d. random variables and $S_n = X_1 + \cdots + X_n$, $\tilde{S}_n = Y_1 + \cdots + Y_n$. Under the condition: $EX_1^i = EY_1^i$, $i = 1, 2, \ldots, r-1$ $(r \geq 3)$ and a certain "smoothness" assumption about the distribution of $Y_1$ we prove that

$$\rho(S_n, \tilde{S}_n) \equiv \sup_{x \in \mathbf{R}} \left| F_{S_n}(x) - F_{\tilde{S}_n}(x) \right| \leq \frac{c}{n^{\frac{r}{2}-1}} \mu_r(X_1, Y_1), \quad n = 1, 2, \ldots,$$

where $\mu_r$ is an appropriate probability metric.

## 1. Motivation

Let $X_1, X_2, \ldots\ ; Y_1, Y_2, \ldots$ be two sequences of i.i.d.r.v.'s with respective distribution function $F_X$ and $F_Y$, and let $\nu$ be a nonnegative integer r.v., independent of $(X_k,\ k \geq 1)$ and $(Y_k,\ k \geq 1)$. We will use the notation:

$$S_\nu = \sum_{k=1}^{\nu} X_k, \quad \tilde{S}_\nu = \sum_{k=1}^{\nu} Y_k.$$

Let

$$\rho(X, Y) \equiv \rho(F_X, F_Y) := \sup_{x \in \mathbb{R}} \left| F_X(x) - F_Y(x) \right| \tag{1.1}$$

the uniform or Kolmogorov's metric.

The problem of finding upper bounds for $\rho(S_\nu, \tilde{S}_\nu)$ has received a considerable attention at least in two areas of probability. The first relates to estimation of the rate of convergence in the central limit theorem. In this case $\nu = n$; $Y_1, Y_2, \ldots$ are *normally distributed* (and so, $\tilde{S}_n$ is a normal r.v.). It is known (see, for instance [**S**] [**S1**]), that if $EX = EY$; $EX^2 = EY^2$; $E|X|^3 < \infty$ then,

$$\rho(S_n, \tilde{S}_n) \leq cn^{-1/2} \max \left\{ \rho(X, Y), \mathbf{k}_1(X, Y),\ \frac{1}{6}\, \mathbf{k}_3(X, Y) \right\}, n = 1, 2, \ldots \tag{1.2}$$

---

2000 *Mathematics Subject Classification.* Primary 60E15, 60G50, 60K10.

*Key words and phrases.* Sums of Independent Random Variables, Estimates of Proximity, Uniform and Zolotarev's Metrics, Pseudomoments, Asymptotic Expansion.

where

$$\mathbf{k}_r(X,Y) := r\int_{-\infty}^{\infty} |x|^{r-1}\big|F_X(x) - F_Y(x)\big|dx \tag{1.3}$$

is the difference pseudomoment of order $r$.

Second, the above mentioned problem arises in the study of, so-called, stability (continuity) of applied stochastic models involving geometric sums (queues, risk models, reliability etc., see for example, [**A**],[**Be**],[**Bo**],[**Ge**],[**Go**],[**G-R**],[**Gr**],[**K**], [**M**],[**Z**]). In this context $P(\nu = k) = q(1-q)^{k-1}$, $k \geq 1$ and $S_\nu$, $\tilde{S}_\nu$ are called geometric sums. Examples of published results are the below inequalities. (See also [**Be**],[**Bo**],[**Go**],[**G-R**],[**R**] for related results.) From [**K**], Ch. 5 we learn that for all sufficiently small distances $\rho(X,Y)$

$$\sup_{q\in(0,1)} \rho(S_\nu, \tilde{S}_\nu) \leq c\sqrt{\rho(X,Y)},$$

provided that

$$EX = EY;\ EX^2, EY^2 < \infty. \tag{1.4}$$

Imposing (together with (1.4)) some additional conditions on the distributions of $X$ and $Y$ we obtained in [**G-R1**] that

$$\sup_{q\in(0,1)} \rho(S_\nu, \tilde{S}_\nu) \leq c\max\{\rho(X,Y), \frac{1}{2}\,\mathbf{k}_2(X,Y)\}.$$

The aim of this paper is to extend the bounds in (1.2) along the following two lines. Instead of normally distributed r.v.'s $Y_1, Y_2, \ldots$ we consider r.v.'s with an arbitrary distribution satisfying Assumption 2.1 in Section 2. Note that in the stability (continuity) problem a distribution function $F_Y$ is treated as a known (available) approximation to an unknown distribution function $F_X$. For this it is desirable to have a constant $c$ in inequalities as (1.6) below to be independent of the distribution function $F_X$. Along with (1.4) we pay attention to the following more restrictive moment conditions.

ASSUMPTION 1.1. For a given integer $r \geq 3$

$$EX^i = EY^i;\ i = 1,2,\ldots,r-1; \quad E|X|^r,\ E|Y|^r < \infty. \tag{1.5}$$

If $r > 3$ we get the better rate of the proximity bound as $n \to \infty$. To be more explicit, we prove the following inequalities:

$$\rho(S_n, \tilde{S}_n) \leq cn^{-\frac{r}{2}+1}\mu_r(X,Y),\ \ n = 1,2,\ldots, \tag{1.6}$$

where

$$\mu_r(X,Y) := \max\left\{\rho(X,Y),\ \frac{1}{r!}\,\mathbf{k}_r(X,Y)\right\} \tag{1.7}$$

and the constant $c$ in (1.6) is determined by the "known" distribution function $F_Y$. The latter allows to apply (1.6) (and its versions with a random number $\nu$ of summands) for stability estimation of several stochastic models. We examine a couple of such models in Section 2 (compound Poisson's and renewal processes). Finally, we notice that (1.6) was proved only under condition (2.2) requiring a prior nearness of the distribution of $Y$ and $X$.

## 2. Results and Examples of Applications.

Let $r \geq 3$ from Assumption 1.1 be fixed in what follows. We shall use the metric $\mu_r$ defined in (1.1), (1.3), (1.7) and the total variation metric $\mathbf{V}$:

$$\mathbf{V}(X,Y) = 2\sup\{|P(X \in B) - P(Y \in B)| : \ B \text{ is a Borel set } \}.$$

We denote: $\sigma^2 = \mathrm{Var(X)} = \mathrm{Var(Y)}$, considering the nontrivial case $\sigma > 0$.

ASSUMPTION 2.1. There is an integer $s \geq 1$ such that the r.v. $Y_1 + Y_2 + \cdots + Y_s$ has a density $p \in \mathbb{C}_b^{r-1}$ (the space of bounded functions with $(r-1)$th continuous derivative). Moreover the derivative $p^{(r)}$ exists almost everywhere on $\mathbb{R}$, $p^{(r)} \in L_1(\mathbb{R})$ and for some $\alpha > 0$

$$\int_{|x|>\alpha n^{\frac{r-1}{2}}} |p^{(r)}(x)|dx = \mathcal{O}(n^{-\frac{r}{2}}) \quad \text{as} \quad n \to \infty. \tag{2.1}$$

THEOREM 2.2. *Suppose that Assumptions 1.1 and 2.1 hold. Then there are finite constants* $\overline{c}_r$ *and* $\overline{\overline{c}}_r$ *such that the condition*

$$\overline{c}_r \mathbf{V}(X,Y) + \overline{\overline{c}}_r \mu_r(X,Y) \leq \frac{1}{3}, \tag{2.2}$$

*implies the following inequalities*

$$\rho(S_n, \tilde{S}_n) \leq c_r \mu_r(X,Y) n^{-\frac{r}{2}+1}, \quad n = 1, 2, \ldots, \tag{2.3}$$

*where* $c_r$ *is the constant defined in (2.4).*

REMARK 2.3. As it is seen from the proof of the Theorem 2.2, the constants $c_r, \overline{c}_r$ and $\overline{\overline{c}}_r$ in (2.2), (2.3) are completely specified by the integers $r, s$ from Assumptions 1.1 and 2.1, the value of $\sigma$ and certain characteristics of the "known" distribution function $F_Y$. (See formulas (2.4)-(2.10) below.)

REMARK 2.4. The equality of the first $r-1$ moments of $X$ and $Y$ (see (1.5)) is essential to obtain the order $-\frac{r}{2}+1$ in the right hand side of (2.3). Indeed, it is known (see [**I**]) that if $Y_1, Y_2, \ldots$ are normally distributed and

$$\rho(S_n, \tilde{S}_n) = \mathcal{O}(n^{-\frac{r}{2}+1}) \text{ as } n \to \infty,$$

then (1.5) holds.

REMARK 2.5. By the total probability formula one can extend (2.3) to get upper bounds for $\rho(S_\nu, \tilde{S}_\nu)$. For instance, if $P(\nu = k) = q(1-q)^{k-1}$, $k \geq 1$ then, applying inequalities (2.3) and a Tauberian theorem we can find constants $c'$, $c''$ such that

$$\begin{aligned}
\rho(S_\nu, \tilde{S}_\nu) &\leq c' q^{\frac{1}{2}} \mu_3(X,Y), \ (r=3); \\
\rho(S_\nu, \tilde{S}_\nu) &\leq c'' q \log(1/q) \mu_4(X,Y); \ (q \in (0, \tfrac{1}{2}], \ r = 4).
\end{aligned}$$

These bounds are useful for small values of $q$. The latter is of special interest in a number of models in queues, risk theory and reliability (see [**A**], [**Ge**], [**Gr**], [**K**]).

Returning to the constants $c_r, \overline{c}_r, \overline{\overline{c}}_r$ in (2.2), (2.3) we can see from the proof of Theorem 2.2 that:

$$c_r = \max\left\{(2s-1)^{r/2}, \frac{3}{2}\frac{b_r}{\sigma^r}\left(2^{\frac{r}{2}-1}+c_{1,r}(s)\right)\right\}; \tag{2.4}$$

$$\overline{c}_r = \sup_{n\geq 2s}\sum_{j=0}^{s-1}\left(\frac{n}{n-j-1}\right)^{\frac{r}{2}-1}; \tag{2.5}$$

$$\overline{\overline{c}}_r = \frac{d_r}{\sigma^r}c_{2,r}(s); \tag{2.6}$$

where the constants $c_{1,r}(s)$ and $c_{2,r}(s)$ depend only on $r, s$ and they are calculated as follows

$$c_{1,r}(s) = \sup_{n\geq 2s}\frac{\left[\frac{n}{2}\right]+1}{(n-1)^{r/2}}n^{\frac{r}{2}-1}, \tag{2.7}$$

$$c_{2,r}(s) = \begin{cases} \sup\limits_{n\geq 2s}\sum\limits_{j=s}^{[n/2]}\left(\dfrac{n}{n-j-1}\right)^{\frac{r}{2}-1} j^{-\frac{r}{2}}, & s>1, \\ \sup\limits_{n\geq 3}\sum\limits_{j=s}^{[n/2]}\left(\dfrac{n}{n-j-1}\right)^{\frac{r}{2}-1} j^{-\frac{r}{2}}, & s=1. \end{cases} \tag{2.8}$$

($[x]$ stands for the greatest integer less or equal $x$.) The constant $c_{2,r}(s)$ is finite since the terms in parentheses are bounded, while the series $\sum_{j=1}^{\infty} j^{-\frac{r}{2}}$ converges. On the other hand, the constants $b_r$ and $d_r$ in (2.4), (2.6) are defined as follows. In view of Assumption 2.1, the density $f_k$ of the r.v. $\dfrac{Y_1+\cdots+Y_k}{\sigma\sqrt{k}}$ exists for every $k\geq s$. Let

$$b_r := \sup_{k\geq s}\sup_{x\in\mathbb{R}}\left|f_k^{(r-1)}(x)\right|, \tag{2.9}$$

$$d_r := \sup_{k\geq s}\int_{-\infty}^{\infty}\left|f_k^{(r)}(x)\right|dx. \tag{2.10}$$

Lemma 3.1 in the next section ensures the finiteness of $b_r$ and $d_r$, but it leaves the problem of calculation of these constants open. In a number of cases (Gaussian, uniform, Gamma distributions, etc.) the densities $f_k$, $k\geq 1$ are given by explicit expressions. Thus, in these cases $b_r$ and $d_r$ can be calculated in a computer, using the fact that the terms in (2.9) and (2.10) converge to the corresponding terms obtained from the standard normal density. For example, let $r=3, \sigma=1$. Then $c_3 = 18.53$, $\overline{c}_3 = 4.91$, $\overline{\overline{c}}_3 = 2.47$ for the uniformly distributed r.v. $Y$, and $c_3 = 5.2$, $\overline{c}_3 = 2.57$, $\overline{\overline{c}}_3 = 3.75$ for the r.v. $Y$ having the Gamma density with $\alpha = 4$.

EXAMPLE 2.6 (Stability estimate for a renewal process.). Let $N(t)$ denote the total number of replacements of some unreliable unit up to time $t$, and $X_1, X_2, \ldots$ represent lifetimes of successively replaced units. Suppose that one needs to estimate $P(N(t)\geq n)$ for a given $n$ and various t in the situation when instead of an unknown distribution function $F_X$ some approximation $F_Y$ is used. Since

$P(N(t) \geq n) = P(S_n \leq t)$ and $P(\tilde{N}(t) \geq n) = P(\tilde{S}_n \leq t)$, where $\tilde{N}(t)$ is a renewal process generated by $Y_1, Y_2, \ldots$, we get that under hypotheses of Theorem 2.2

$$\sup_{t \geq 0} \left| P(N(t) \geq n) - P(\tilde{N}(t) \geq n) \right| \leq c_r n^{-\frac{r}{2}+1} \mu_r(X, Y).$$

EXAMPLE 2.7 ( Stability estimate for a compound Poisson process.). The above mentioned stability problem arises, for instance, in the study of the classical risk process (see [**A**], [**Gr**], [**K**], [**R**]). Let a r.v. $\nu$ have the Poisson distribution with parameter $\lambda t$ ($\lambda > 0,\ t \in [0, \infty)$). Admitting the convention: $\sum_{k=1}^{0} := 0$ and assuming the conditions of Theorem 2.2 hold we get from (2.3):

$$\rho(S_\nu, \tilde{S}_\nu) \leq c_r \mu_r(X, Y) \sum_{n=1}^{\infty} n^{-\frac{r}{2}+1} \frac{(\lambda t)^n}{n!} e^{-\lambda t}.$$

Straightforward calculations (given in [**M**]) show that for some constant $\tilde{c}$

$$\sum_{n=1}^{\infty} n^{-\frac{r}{2}+1} \frac{(\lambda t)^n}{n!} e^{-\lambda t} = \lambda t\, E(\nu + 1)^{-\frac{r}{2}} \leq \tilde{c}(\lambda t)^{-\frac{r}{2}+1}, \quad t \geq 0.$$

Thus

$$\rho(S_\nu, \tilde{S}_\nu) \leq c_r \tilde{c} \mu_r(X, Y)(\lambda t)^{-\frac{r}{2}+1}.$$

## 3. Proof of Main Theorem

We shall need three auxiliary results.

LEMMA 3.1. *Under Assumptions 1.1 and 2.1 the quantities $b_r$ and $d_r$ defined in (2.9), (2.10) are finite.*

The proof of Lemma 3.1 is based on the following considerations. Without loss of generality we can assume that $EX = EY = 0$. Assumptions 1.1 and 2.1 yield the fulfilment of the conditions of Theorem 7 in [**P**], Ch. VI, which in our particular case of identically distributed r.v.'s $Y_1, Y_2, \ldots$ states that for the derivatives of the density $f_k$ $\left(\text{of } \dfrac{Y_1 + \cdots + Y_k}{\sigma\sqrt{k}},\ k \geq s\right)$ the following asymptotic expansion holds ($\ell = 1, 2, \ldots, r$):

$$f_k^{(\ell)}(x) = \frac{1}{\sqrt{2\pi}} e^{-\frac{x^2}{2}} \left[ 1 + \sum_{i=1}^{r-2} \frac{Q_{i,\ell}(x)}{n^{1/2}} \right] + o\left(n^{-\frac{r-2}{2}}\right), \tag{3.1}$$

uniformly on $x \in \mathbb{R}$. In (3.1) $Q_{i,\ell}$ are certain polynomials. The finiteness of $b_r$ follows from (3.1). (See (2.9).) To prove that $d_r < \infty$, we use (3.1) together with condition (2.1). For $r = 3$ we refer the reader to [**G-R**]. The general case is examined along the same lines as for $r = 3$. The corresponding proof is given in [**M**].

In the Lemmas 3.2, 3.3 and in the proof of Theorem 2.2 we shall use Zolotarev's metric $\zeta_r$ of order $r$. By definition $\zeta_r := \sup\{|E\varphi(X) - E\varphi(Y)| : \varphi \in \mathcal{D}_r\}$. where $\mathcal{D}_r$ denotes the set of all functions $\varphi : \mathbb{R} \to \mathbb{R}$ having (almost everywhere) $r$th derivatives bounded by 1. (see [**S**], [**Z**]). The distance $\zeta_r(X, Y)$ is finite provided that $EX^i = EY^i$, $i = 1, 2, \ldots, r-1$; $E|X|^r$, $E|Y|^r < \infty$. Under these conditions

$$\zeta_r(X, Y) \leq \frac{1}{r!} \mathbf{k}_r(X, Y), \tag{3.2}$$

where the distance $\mathbf{k}_r(X,Y)$ was defined in (1.3). (See [**R**] for the proof.) The metric $\zeta_r$ possesses the so-called, "ideal property", i.e. for $(X_k,\ k\geq 1)$, $(Y_k,\ k\geq 1)$ being independent and $a>0$ one gets:

$$\zeta_r(a\sum_{k=1}^{n}X_k, a\sum_{k=1}^{n}Y_k)\leq a^r\sum_{k=1}^{n}\zeta_r(X_k,Y_k). \tag{3.3}$$

(See, for instance, [**S**], [**Z**].)

LEMMA 3.2. *Let $EX^i=EY^i$, $i=1,2,\dots,r-1$, $E|X|^r, E|Y|^r<\infty$ and $\xi$ be a r.v. independent of $(X,Y)$ possessing a density $f_\xi$. Suppose that a derivative $f_\xi^{(r)}$ exists almost everywhere on $\mathbb{R}$ and $f_\xi^{(r)}\in L_1(\mathbb{R})$. Then*

$$\mathbf{V}(X+\xi,\ Y+\xi)\leq\int_{-\infty}^{\infty}|f_\xi^{(r)}(x)|dx\zeta_r(X,Y). \tag{3.4}$$

For $r=3$ inequality (3.4) is proven in [**R**], Ch. 17. The general case is treated in [**M**].

LEMMA 3.3. *Under the hypotheses of Lemma 3.2 (omitting $f_\xi^{(r)}\in L_1(\mathbb{R})$)*

$$\rho(X+\xi,\ Y+\xi)\leq\sup_{x\in\mathbb{R}}|f_\xi^{(r-1)}(x)|\zeta_r(X,Y). \tag{3.5}$$

The proof of (3.5) is rather similar to that of (3.4). The particular case of (3.5) with $r=3$ and a normally distributed $\xi$ is examined in [**S**], Ch. 4.

Now we turn our attention to the proof of inequalities (2.3). By the regularity of the metric $\rho$ (see [**S**]) we have for $n<2s$

$$\rho(S_n,\tilde{S}_n)\leq\sum_{k=1}^{n}\rho(X_k,Y_k)\leq(2s-1)\rho(X,Y)\leq c_r n^{-\frac{r}{2}+1}\mu_r(X,Y),$$

provided that $c_r\geq(2s-1)^{r/2}$. For $n\geq 2s$, $0\leq j\leq n$ being arbitrary, but fixed, let:

- $F_j$ denote the distribution function of $\dfrac{X_0+\cdots+X_j}{\sqrt{n}}$;
- $G_j$ denote the distribution function of $\dfrac{Y_0+\cdots+Y_j}{\sqrt{n}}$;

$(X_0\equiv Y_0\equiv 0)$; $m=[n/2]\geq 2$.

In [**S**], §4.1, §4.3 it is proved that

$$\begin{aligned}\rho(F_n,G_n)\ &\leq\ \sum_{j=0}^{m}\rho(F_1*G_j*F_{n-j-1},G_1*G_j*F_{n-j-1})\\ &+\ \rho(G_1*G_m*F_{n-m-1},G_1*G_m*G_{n-m-1}).\end{aligned} \tag{3.6}$$

Here $*$ stands for the operation of convolution.

Denoting the last summand in (3.6) by $I_n$ we get

$$\begin{aligned}I_n\ =\ \rho\Bigg(&\frac{Y_1+\cdots+Y_{m+1}}{\sqrt{n}}+\frac{X_1+\cdots+X_{n-m-1}}{\sqrt{n}},\\ &\frac{Y_1+\cdots+Y_{m+1}}{\sqrt{n}}+\frac{Y_1+\cdots+Y_{n-m-1}}{\sqrt{n}}\Bigg).\end{aligned}$$

In view of Assumption 2.1 the r.v. $\xi := \dfrac{Y_1 + \cdots + Y_{m+1}}{\sqrt{n}}$ has r times differentiable density $f_\xi$ and

$$f_\xi(x) = \frac{1}{\sigma}\left(\frac{n}{m+1}\right)^{1/2} f_{m+1}\left(\frac{1}{\sigma}\left(\frac{n}{m+1}\right)^{1/2} x\right),$$

where $f_k$ is the density of $\dfrac{Y_1 + \cdots + Y_k}{\sigma\sqrt{k}}$ $(k \geq s)$. By Lemma 3.3, Lemma 3.1, (3.3) and the definition of $b_r$ in (2.9) we have

$$\begin{aligned} I_n \;&\leq\; \frac{1}{\sigma^r}\left(\frac{n}{[n/2]+1}\right)^{r/2} \sup_x \left|f_{m+1}^{(r-1)}(x)\right| \\ &\quad \times \zeta_r\left(\frac{X_1 + \cdots + X_{n-m-1}}{\sqrt{n}}, \frac{Y_1 + \cdots + Y_{n-m-1}}{\sqrt{n}}\right) \\ &\leq\; \frac{1}{\sigma^r}\left(\frac{n}{[n/2]+1}\right)^{r/2} \frac{b_r}{n^{r/2}}(n - [n/2] - 1)\, \zeta_r(X_1, Y_1) \\ &\leq\; \frac{2^{r/2-1}}{\sigma^r} \frac{b_r}{n^{\frac{r}{2}-1}}\, \zeta_r(X, Y). \end{aligned} \tag{3.7}$$

It is known (see [**S**]) that for any distribution functions $P, Q, U, H$

$$\rho(P * U,\ Q * U) \leq \rho(U, H)\mathbf{V}(P, Q) + \rho(P * H,\ Q * H). \tag{3.8}$$

Denoting by $\mathcal{J}_j$ the $j^{th}$ summand for $j = 0, 1, \ldots, m$ in (3.6) and applying (3.8) with $P = F_1 * G_j$, $Q = G_1 * G_j$, $U = F_{n-j-1}$, $H = G_{n-j-1}$ we find that

$$\begin{aligned} \mathcal{J}_j \leq \rho(F_{n-j-1}, G_{n-j-1})&\mathbf{V}(F_1 * G_j, G_1 * G_j) \\ &+\rho(F_1 * G_{n-1}, G_1 * G_{n-1}). \end{aligned} \tag{3.9}$$

Similarly to the above inequality (3.7) we get:

$$\rho(F_1 * G_{n-1}, G_1 * G_{n-1}) \leq \frac{b_r}{\sigma^r}(n-1)^{-r/2}\zeta_r(X, Y). \tag{3.10}$$

By virtue of the regularity of the metric $\mathbf{V}$ (see [**S**]), we have for $0 \leq j \leq s-1$:

$$\mathbf{V}(F_1 * G_j, G_1 * G_j) \leq \mathbf{V}(F_1, G_1) = \mathbf{V}(X, Y). \tag{3.11}$$

On the other hand, by Lemma 3.2, Lemma 3.1 and the definition of $d_r$ in (2.10) we get for $s \leq j \leq m$:

$$\begin{aligned} &\mathbf{V}(F_1 * G_j, G_1 * G_j) \equiv \\ &\quad \mathbf{V}\left(\frac{X_1}{\sqrt{n}} + \frac{Y_1 + \cdots + Y_j}{\sqrt{n}}, \frac{Y_1}{\sqrt{n}} + \frac{Y_1 + \cdots + Y_j}{\sqrt{n}}\right) \\ &\quad \leq \int_{-\infty}^{\infty} \left|f_\xi^{(r)}(x)\right| dx\, \zeta_r\left(\frac{X_1}{\sqrt{n}}, \frac{Y_1}{\sqrt{n}}\right) \\ &\quad \leq \frac{d_r}{\sigma^r} j^{-r/2}\zeta_r(X, Y), \end{aligned} \tag{3.12}$$

since the density $f_\xi$ of $\xi := \dfrac{Y_1 + \cdots Y_j}{\sqrt{n}}$ is equal to $\left(\dfrac{n}{j}\right)^{1/2} \dfrac{1}{\sigma} f_j(x)$.

Gathering inequalities (3.6), (3.7), (3.9)–(3.12) we get the following inequality

$$\begin{aligned}\rho(S_n, \tilde{S}_n) &= \rho(F_n, G_n)\\ &\le \frac{2^{r/2-1}}{\sigma^r}\frac{b_r}{n^{r/2-1}}\zeta_r(X,Y) + \frac{b_r}{\sigma^r}\frac{[n/2]+1}{(n-1)^{r/2}}\zeta_r(X,Y)\\ &+\sum_{j=0}^{s-1}\rho(F_{n-j-1}, G_{n-j-1})\mathbf{V}(X,Y)\\ &+\sum_{j=s}^{m}\rho(F_{n-j-1}, G_{n-j-1})\frac{d_r}{\sigma^r}j^{-r/2}\zeta_r(X,Y).\end{aligned}\tag{3.13}$$

We let $\tilde{\mu}_r := \max\{\rho, \zeta_r\}$ and assume that for $s \le k < n$

$$\rho(S_k, \tilde{S}_k) = \rho(F_k, G_k) \le c_r k^{-\frac{r}{2}+1}\tilde{\mu}_r(X,Y),\tag{3.14}$$

where the constant $c_r$ will be specified later. Using the last inequalities and definitions (2.5), (2.7), (2.8) of the constants $\bar{c}_r$, $c_{1,r}(s)$ and $c_{2,r}(s)$ we can rewrite (3.13) as follows

$$\begin{aligned}\rho(S_n, \tilde{S}_n) \le \Bigg\{&\frac{2^{\frac{r}{2}-1}b_r}{\sigma^r} + \frac{b_r c_{1,r}(s)}{\sigma^r} + c_r\bar{c}_r\mathbf{V}(X,Y)\\ &+c_r\frac{d_r c_{2,r}(s)}{\sigma^r}\tilde{\mu}_r(X,Y)\Bigg\} n^{-\frac{r}{2}+1}\tilde{\mu}_r(X,Y).\end{aligned}\tag{3.15}$$

If we choose

$$c_r = \max\left\{(2s-1)^{r/2}, \frac{3}{2}\frac{b_r}{\sigma^r}\left(2^{\frac{r}{2}+1} + c_{1,r}(s)\right)\right\}$$

and assume that

$$\bar{c}_r\mathbf{V}(X,Y) + \frac{d}{\sigma^r}c_{2,r}(s)\tilde{\mu}_r(X,Y) \le \frac{1}{3},$$

then (3.15) yields inequality (3.14) with $k = n$.

To complete the proof it is enough to apply the bound (3.2) and to recall definition (1.7).

**Acknowledgements** The authors would like to thank the referee for a number of valuable suggestions to improve the paper.

## References

[A] S. Asmussen, *Ruin Probabilities*, World Scientific Publishing Co. Inc., River Edge, NJ, (2000).

[Be] J. Beirlant and S.T. Rachev, *The problem of stability in insurance mathematics,* Insurance: Math. Econ. **6** (1987), pp. 179-188.

[Bo] J.L. Bon and V. Kalashnikov, *Bounds for geometric sums used for evaluation of reliability of regenerative models*, J. Math. Sci. (New York) **93** (1999), pp. 501–510.

[Ge] I.B. Gertsbakh, *Asymptotic methods in reliability theory: A review*, Adv. Appl. Prob. **16** (1984), pp. 147-175.

[Go] E.I. Gordienko, *Estimates of stability of geometric convolutions*, Appl. Math. Letters **12**, no. 5, (1999), pp. 103-106.

[G-R] E.I. Gordienko and J. Ruiz de Chávez, *New estimates of continuity in $M|GI|1|\infty$ queues*, Queueing Systems **29** (1998), pp. 175-188.

[G-R1] E.I. Gordienko and J. Ruiz de Chávez, *New continuity estimates of geometric sums*, Journ. Appl. Math. and Stoch. Analysis **15** (2002), pp. 235–249.

[Gr] J. Grandell, *Aspects of Risk Theory*, Springer-Verlag, Heidelberg, (1991).

[I] I.A. Ibragimov, *On the accuracy of approximation to the distribution functions of the sums of independent variables by normal distribution*, Theory Probab. Appl. **11** (1966), pp. 632–655.

[K] V.V. Kalashnikov, *Geometric Sums: Bounds for Rare Events with Applications*, Kluwer Academic Publishers, Dordrecht, (1997).

[M] M. Mendieta, *Nuevas estimaciones de la estabilidad de procesos de Poisson compuestos y sumas geométricas*, Master's degree dissertation. Universidad Autónoma Metropolitana-I, México, D.F., (2002).

[P] V.V. Petrov, *Sums of Independent Random Variables*, Springer-Verlag, Berlin, (1975).

[R] S.T. Rachev, *Probability Metrics and the Stability of Stochastic Models*, John Wiley & Sons, Chichester, (1991).

[S] V.V. Senatov, *Normal Approximation: New Results*, Methods and Problems, VSP, Utrecht, (1998).

[S1] V.V. Senatov, *Uniform estimates of the rate of convergence in the multi-dimensional central limit theorem*, Theory Probab. Appl. **25**, no. 4 (1980), pp. 745-759.

[Z] V. Zolotarev, *Ideal metrics in the problems of probability theory*, Austral. J. Statist. **21** (1979), pp. 193-208.

Universidad Autonoma Metropolitana Unidad Iztapalapa. Av. San Rafael Atlixco, no. 186 Col. Vicentina, Mexico D.F. c.p. 09340, MEXICO
*E-mail address*: gord@xanum.uam.mx

Universidad Anahuac. Lomas Anahuac Huixquilucan, Edo. de Mexico c.p. 52786
*E-mail address*: mmendieta@anahuac.mx

Universidad Autonoma Metropolitana Unidad Iztapalapa. Av. San Rafael Atlixco, no. 186 Col. Vicentina, Mexico D.F. c.p. 09340, MEXICO
*E-mail address*: jrch@xanum.uam.mx

Contemporary Mathematics
Volume **336**, 2003

# An Example of Infinite Dimensional Quasi–Helix

Christian Houdré and José Villa

ABSTRACT. A generalization of fractional Brownian motion, which happens to be a quasi-helix in the sense of Kahane, is presented. A brief study of its path properties is also initiated.

## 1. Introduction

Since its introduction by Kolmogorov and Lévy, the coining of its name and its study by Mandelbrot and Van Ness, fractional Brownian Motion (fBm) has enjoyed many successes as a modeling tool in various areas of applications such as turbulence, finance, telecommunications ([**M-V**], [**S-T**]). These successes are mainly due to the self-similar nature of fBm and to a lesser extend to the stationarity of its increments. For small increments, in application such as turbulence, fBm seems to be a valuable model but also appears inadequate for large increments. It is thus very natural to explore the existence of processes which keep some of the properties of fBm (self-similarity, stationarity of small increments, gaussianity) but also enlarge our modelling tool kit. (As well known, fBm is the only self-similar, stationary increments Gaussian process which is (a.s.) zero at the origin.) It is the purpose of these notes to propose such a process (although we do not directly address the modelling aspect) and to briefly study some of its basic properties.

## 2. Definition of bifractional Brownian motion

We start with the following result which is certainly well known, but whose proof we were unable to find in the literature.

PROPOSITION 2.1. *For any,* $0 < H \leq 1$, $0 < K \leq 1$, *the function* $R : \mathbb{R} \times \mathbb{R} \to \mathbb{R}$, *given by*

$$R(t,s) = (|t|^{2H} + |s|^{2H})^K - |t-s|^{2HK}, \tag{2.1}$$

*is positive definite.*

---

2000 *Mathematics Subject Classification.* Primary: 60G15; Secondary: 60G18, 60G12.
*Key words and phrases.* Gaussian processes, fractional Brownian motion, quasi-helices.
Research supported in part by the NSF Grant 9803239.
We would like to thank CIMAT for its facilities and its hospitality.

PROOF. First recall the following (easily verified) identity

$$\lambda^K = \frac{K}{\Gamma(1-K)} \int_0^\infty (1-e^{-\lambda x}) x^{-1-K} dx, \tag{2.2}$$

valid for $\lambda \geq 0$, $0 < K < 1$, where as usual $\Gamma$ is the gamma function, i.e., $\Gamma(x) = \int_0^\infty e^{-u} u^{x-1} du$, $x > 0$. To prove the proposition, note first that for $K = 1$, (2.1) is just the covariance function of fBm and so is positive definite. Let us thus assume that $0 < K < 1$. Then, for any $c_1, \dots, c_n \in \mathbb{R}$,

$$\begin{aligned}
&\sum_{i=1}^n \sum_{j=1}^n c_i c_j R(t_i, t_j) \qquad (2.3)\\
&= \frac{K}{\Gamma(1-K)} \int_0^\infty \sum_{i=1}^n \sum_{j=1}^n c_i c_j \left( -e^{-x(|t_i|^{2H}+|t_j|^{2H})} + e^{-x|t_i-t_j|^{2H}} \right) x^{-1-K} dx \\
&= \frac{K}{\Gamma(1-K)} \int_0^\infty \sum_{i=1}^n \sum_{j=1}^n c_i e^{-x|t_i|^{2H}} c_j e^{-x|t_j|^{2H}} \\
&\quad \cdot \left( e^{x(|t_i|^{2H}+|t_j|^{2H}-|t_i-t_j|^{2H})} - 1 \right) x^{-1-K} dx.
\end{aligned}$$

Since the function $|t|^{2H} + |s|^{2H} - |t-s|^{2H}$ is positive definite, so is

$$\left( e^{x(|t_i|^{2H}+|t_j|^{2H}-|t_i-t_j|^{2H})} - 1 \right),$$

for all $x \geq 0$. Hence, the integrand in (2.3) is non-negative and so $(|t|^{2H}+|s|^{2H})^K - |t-s|^{2HK}$ is itself positive definite. □

In the above, we made use of the fact that for $K = 1$ the function in (2.1) is positive definite. However, this can be proved by the same method as above (outside of the trivial case $H = 1$), making the proof of the proposition more self contained.

Since $R$ is positive definite, Kolmogorov's consistency theorem asserts the existence of a centered Gaussian process $X$ having

$$R_{H,K}(t,s) = \frac{1}{2^K} \left\{ \left(|t|^{2H} + |s|^{2H}\right)^K - |t-s|^{2HK} \right\} \operatorname{Var} X(1), \tag{2.4}$$

as its covariance function. This process $X$ will be called bifractional Brownian motion (bBm) and will be denoted by $B_{H,K}$ (a similar definition holds if the index set $\mathbb{R}$ is replaced, for example, by $[0, +\infty)$).

## 3. Some basic facts about bBm

For $K = 1$, bBm is just fBm (and Bm for $K = 1$, $H = \frac{1}{2}$). It is also clear that bBm is self-similar of index $I = HK$, thus $B_{H,K}(0) = 0$ (a.s.) and moreover Var $B_{H,K}(t) = |t|^{2HK}$ Var $B_{H,K}(1)$. From now on, since Var $B_{H,K}(1)$ will not play rôle in our study, we only consider the "standard case," i.e., Var $B_{H,K}(1) = 1$.

The following is (in our opinion) more surprising.

PROPOSITION 3.1. *Let $0 < H \leq 1$, $0 < K \leq 1$, then*

$$2^{-K}|t-s|^{2HK} \leq E(B_{H,K}(t) - B_{H,K}(s))^2 \leq 2^{1-K}|t-s|^{2HK}, \tag{3.1}$$

*for all $t, s \in \mathbb{R}$.*

PROOF. The right inequality in (3.1) is trivial since,

$$(3.2)\quad E(B_{H,K}(t)-B_{H,K}(s))^2 = \frac{2}{2^K}|t-s|^{2HK} + \left(|t|^{2HK}+|s|^{2HK}-\frac{2}{2^K}(|t|^{2H}+|s|^{2H})^K\right),$$

and the last term in parentheses in (3.2) is non-positive by concavity. The left inequality in (3.1) is a calculus exercise. Indeed, the left inequality in (3.1) is equivalent to

$$(3.3)\qquad |t-s|^{2HK}+2^K|t|^{2HK}+2^K|s|^{2HK}-2(|t|^{2H}+|s|^{2H})^K \geq 0,$$

which is clearly true if $t=0$ or $s=0$. So let us assume that $t \neq 0$ and $s \neq 0$. Now for $u \in \mathbb{R}$, let

$$(3.4)\qquad f(u) = |u-1|^{2HK}+2^K+2^K|u|^{2HK}-2(1+|u|^{2H})^K.$$

To prove (3.3), it is enough to show that $f(u) \geq 0$ for $|u| \geq 1$. Assuming $t \geq s$ (the case $t \leq s$ being done similarly by symmetry), we see that
(i) For $t \geq s > 0$, (hence $\frac{t}{s} \geq 1$), (3.3) becomes

$$(3.5)\qquad |s|^{2HK} f\left(\frac{t}{s}\right) \geq 0.$$

(ii) For $s \leq t < 0$, (hence $\frac{s}{t} \geq 1$), (3.3) becomes

$$(3.6)\qquad |t|^{2HK} f\left(\frac{s}{t}\right) \geq 0.$$

(iii) For $s < 0 < t$ and $t \geq -s$, (3.3) becomes (3.5) with $\frac{t}{s} \leq -1$, while for $s < 0 < t$ and $t \leq -s$, (3.3) becomes (3.6) with $\frac{s}{t} \leq -1$. We thus just need to prove that

$$(3.7)\qquad f(u) \geq 0 \qquad \text{for } |u| \geq 1.$$

Let us first treat the case $u \geq 1$. Then

$$(3.8)\qquad f(u) = (u-1)^{2HK}+2^K+2^Ku^{2HK}-2(1+u^{2H})^K,$$

and

$$(3.9)\qquad f'(u) = 2HKu^{2HK-1}\left(\left(1-\frac{1}{u}\right)^{2HK-1}+2^K-2\left(1+\frac{1}{u^{2H}}\right)^{K-1}\right).$$

Now, if $2HK \leq 1$, and since $u \geq 1$, the term in parentheses in (3.9) is greater or equal to $1+2^K-2 \geq 0$. Hence $f'(u) \geq 0$ for all $u \geq 1$ and since $f(1)=0$, (3.7) follows in that case. Now, if $2HK \geq 1$, then also $2H \geq 1$ and the term in parentheses in (3.9) is greater or equal to

$$\left(1-\frac{1}{u}\right)+2^K-2\left(1+\frac{1}{u^2}\right)^{K-1},$$

since $2HK \leq 2$. As easily seen by differentiation, (and since also $K \geq \frac{1}{2}$) this last expression is non-negative for all $u \geq 1$, and again (3.7) follows. Let us now tackle the case $u \leq -1$ where there is little to do. Indeed (3.4) becomes

$$(1-u)^{2HK}+2^K+2^K(-u)^{2HK}-2(1+(-u)^{2H})^K, \qquad u \leq -1,$$

i.e.,

$$(3.10)\qquad (1+u)^{2HK}+2^K+2^Ku^{2HK}-2(1+u^{2H})^K, \qquad u \geq 1.$$

But (3.10) dominates (3.8) which is nonnegative as shown above. □

Proposition 3.1 tells us that bBm is an infinite dimensional quasi-helix (in the sense of Kahane [**K1**]-[**K2**]) of index $I = HK$. Various properties of quasi-helices are known and again we refer to [**K1**]-[**K2**] for further information. In particular, bBm has an a.s. continuous version as seen from either Kolmogorov continuity theorem or the Marcus and Shepp criterion or Dudley's entropy theorem, e.g., [**K2**].

It is clear that bBm has stationary (resp. self-similar) increments if and only if it is fBm. Now for $t \in \mathbb{R}$, let

$$\begin{aligned}\sigma_n^2(t) &= E\left(B_{H,K}\left(t+\frac{1}{n}\right) - B_{H,K}(t)\right)^2 \\ &= \frac{2^{1-K}}{n^{2HK}} + \left|t+\frac{1}{n}\right|^{2HK} + |t|^{2HK} - 2^{1-K}\left(\left|t+\frac{1}{n}\right|^{2H} + |t|^{2H}\right)^K.\end{aligned}$$

Hence $\lim_{n\to+\infty} n^{2HK}\sigma_n^2(t) = 2^{1-K}$ which we will reinterpret as saying that for $t \approx s$ ($t$ close to $s$), $E(B_{H,K}(t) - B_{H,K}(s))^2 \approx 2^{1-K}|t-s|^{2HK}$, i.e., the increments of $B_{H,K}$ are approximately stationary for small increments.

With probability one, the trajectories of bBm are not differentiable and in fact of unbounded variation in any finite interval (except, of course, in the degenerate case $H = K = 1$). It is clear that more refined path properties of bBm can be expected. However, the very elementary nature of the proof below entices us to present it (the method actually shows the lack of Hölder continuity of index striclty greater than 1/2).

PROPOSITION 3.2. *For* $HK \neq 1$,

$$\lim_{\epsilon\downarrow 0}\sup_{t\in[t_0-\epsilon,t_0+\epsilon]}\left|\frac{B_{H,K}(t) - B_{H,K}(t_0)}{t-t_0}\right| = +\infty,$$

*with probability one for each* $t_0 \in \mathbb{R}$.

PROOF. For $m, n \in \mathbb{N}$, let

$$A_n^{(m)} = \left\{\omega \in \Omega : \sup_{t\in[t_0-\frac{1}{n},t_0+\frac{1}{n}]}\left|\frac{B_{H,K}(t,\omega) - B_{H,K}(t_0,\omega)}{t-t_0}\right| > m\right\}.$$

Clearly, $A_n^{(m)} \supseteq A_{n+1}^{(m)}$ and letting $A^{(m)} = \bigcap_{n=1}^{\infty} A_n^{(m)}$, also $A^{(m)} \supseteq A^{(m+1)}$. So to prove the result, it is enough to show that

$$\begin{aligned}P\left(\bigcap_{m=1}^{\infty} A^{(m)}\right) &= \lim_{m\to+\infty} P(A^{(m)}) \\ &= \lim_{m\to+\infty}\lim_{n\to+\infty} P(A_n^{(m)}) = 1.\end{aligned}$$

But,

$$P(A_n^{(m)}) \geq P\left(\left|B_{H,K}\left(t_0+\frac{1}{n}\right) - B_{H,K}(t_0)\right| > \frac{m}{n}\right),$$

and we will show that

$$\lim_{n\to+\infty} P\left(\left|B_{H,K}\left(t_0+\frac{1}{n}\right) - B_{H,K}(t_0)\right| \leq \frac{m}{n}\right) = 0.$$

$B_{H,K}\left(t_0+\frac{1}{n}\right)-B_{H,K}(t_0)$ is a centered Gaussian random variable with variance $\sigma_n^2(t_0)$ such that

$$\lim_{n\to+\infty} n^2\sigma_n^2(t_0)=\lim_{n\to+\infty} n^{2(1-HK)}=+\infty.$$

Thus,

$$\begin{aligned} & P\left(\left|B_{H,K}\left(t_0+\frac{1}{n}\right)-B_{H,K}(t_0)\right|\leq\frac{m}{n}\right) \\ = & \frac{1}{\sigma_n(t_0)\sqrt{2\pi}}\int_{-m/n}^{m/n}\exp\left(-\frac{x^2}{2\sigma_n^2(t_0)}\right)dx \\ \leq & m\sqrt{\frac{2}{\pi}}\frac{1}{n\sigma_n(t_0)}, \end{aligned}$$

the result follows. □

Let us mention that the quadratic variation of bBm is zero if $\frac{1}{2}<HK<1$. Moreover, $B_{H,K}$ is not a semi-martingale for $\frac{1}{2}<HK<1$. These last facts can be proved using techniques as in Proposition 3.2 of [**L**]. Neverheless, the stochastic integral of a deterministic function with respect to bBm can be defined as shown below. The fBm case ($K=1$) is already considered in [**N-V-V**], so let $0<K<1$. For $1/2\leq H<1$, define the kernel

$$\begin{aligned} \Re_1(s,t) &= \frac{2HK}{2^K}((2H-1)|t-s|^{2HK-2} \\ & -2H(1-K)((st)^{2H-1}(s^{2H}+t^{2H})^{K-2}+|t-s|^{2HK-2})), \end{aligned}$$

and for $0<H<1/2$,

$$\begin{aligned} \Re_2(s,t) &= \frac{2HK}{2^K}(|t-s|^{2H-1}\text{sgn}(t-s)|t-s|^{2H(K-1)} \\ & -t^{2H-1}(s^{2H}+t^{2H})^{K-1}). \end{aligned}$$

The operators $J_i$, $i=1,2$, are defined by

$$(J_1f)(t)=\int_0^\infty f(s)\Re_1(s,t)ds,$$

and

$$(J_2f)(t)=\int_0^\infty \Re_2(s,t)df(s),$$

where $df$ is defined as in [**N-V-V**]. By $\langle\cdot,\cdot\rangle_{J_i}$ we denote the inner product

$$\langle f,g\rangle_{J_i}=\langle f,J_ig\rangle=\int_0^\infty f(t)(J_ig)(t)dt,$$

and $\langle\cdot,\cdot\rangle$ is the usual inner product of $L^2([0,\infty),dx)$. Let $L^2_{J_i}$ be the space of measurable functions $f$ such that $\langle f,f\rangle_{J_i}<\infty$ and let $G$ be the Gaussian space generated by $\{B_{H,K}(t),\, t\geq 0\}$.

PROPOSITION 3.3. *Let $0<K<1$. The mapping*

$$B_{H,K}(t)\mapsto 1_{[0,t)}$$

*from $\{B_{H,K}(t),\, t\geq 0\}$ to $L^2_{J_i}$ can be extended to an isometry between $G$ and $L^2_{J_i}$, $i=1,2$.*

PROOF. It is sufficient to prove that

$$E(B_{H,K}(u)B_{H,K}(v)) = \langle 1_{[0,u)}, 1_{[0,v)}\rangle_{J_i},$$

for $i=1,2$. We only give the details for $i=1$, the case $i=2$ can be worked out in an analogous way. Notice that,

$$\begin{aligned}
\langle 1_{[0,u)}, 1_{[0,v)}\rangle_{J_1} &= \int_0^u\int_0^v \Re_1(s,t)dsdt \\
&= \frac{2HK}{2^K\Gamma(1-K)}\int_0^u\int_0^v \Big((2H-1)|t-s|^{2H-2}\frac{\Gamma(1-K)}{(|t-s|^{2H})^{1-K}} \\
&\quad -2H(st)^{2H-1}\frac{(1-K)\Gamma(1-K)}{(s^{2H}+t^{2H})^{2-K}} \\
&\quad -2H|t-s|^{4H-2}\frac{(1-K)\Gamma(1-K)}{(|t-s|^{2H})^{2-K}})\Big) dsdt.
\end{aligned}$$

Using that $\Gamma(2-K)=(1-K)\Gamma(1-K)$ (remember that $K<1$) and the identity

$$\int_0^\infty x^{v-1}e^{-\mu x}dx = \frac{\Gamma(v)}{\mu^v}, \tag{3.11}$$

where $v>0$, $\mu>0$, we have by Fubini Theorem,

$$\begin{aligned}
\langle 1_{[0,u)}, 1_{[0,v)}\rangle_{J_1} &= \frac{2HK}{2^K\Gamma(1-K)}\int_0^u\int_0^v \Big((2H-1)|t-s|^{2H-2} \\
&\quad \cdot\int_0^\infty x^{(1-K)-1}e^{-x|t-s|^{2H}}dx \\
&\quad -2H(st)^{2H-1}\int_0^\infty x^{(2-K)-1}e^{-x(s^{2H}+t^{2H})}dx \\
&\quad -2H|t-s|^{4H-2}\int_0^\infty x^{(2-K)-1}e^{-x|t-s|^{2H}}dx\Big) dsdt \\
&= \frac{2HK}{2^K\Gamma(1-K)}\int_0^\infty\int_0^u\int_0^v \Big((2H-1)|t-s|^{2H-2}xe^{-x|t-s|^{2H}} \\
&\quad -2H(st)^{2H-1}x^2e^{-x(s^{2H}+t^{2H})} \\
&\quad -2H|t-s|^{4H-2}x^2e^{-x|t-s|^{2H}}\Big) dsdt\, x^{-1-K}dx.
\end{aligned}$$

Set

$$\begin{aligned}
I_1(x) &= 2H\int_0^u\int_0^v \Big( (2H-1)x|t-s|^{2H-2}e^{-x|t-s|^{2H}} \\
&\quad -2H|t-s|^{4H-2}x^2e^{-x|t-s|^{2H}}\Big) dsdt \\
&= 1+e^{-x|u-v|^{2H}} - e^{-xu^{2H}} - e^{-xv^{2H}},
\end{aligned}$$

(it is in this integral that we need $H\ge 1/2$) and

$$\begin{aligned}
I_2(x) &= 2H\int_0^u\int_0^v (-2H(st)^{2H-1}x^2e^{-x(s^{2H}+t^{2H})})dsdt \\
&= e^{-xu^{2H}} + e^{-xv^{2H}} - e^{-x(u^{2H}+v^{2H})} - 1.
\end{aligned}$$

Therefore

$$\begin{aligned}
\langle 1_{[0,u)}, 1_{[0,v)}\rangle_{J_1} &= \frac{K}{2^K\Gamma(1-K)}\int_0^\infty (I_1+I_2)(x)x^{-1-K}dx \\
&= \frac{K}{2^K\Gamma(1-K)}\int_0^\infty (e^{-x|u-v|^{2H}} - e^{-x(u^{2H}+v^{2H})})x^{-1-K}dx \\
&= \frac{1}{2^K}\left(\frac{K}{\Gamma(1-K)}\int_0^\infty (1-e^{-x(u^{2H}+v^{2H})})x^{-1-K}dx\right. \\
&\quad \left. -\frac{K}{\Gamma(1-K)}\int_0^\infty (1-e^{-x|u-v|^{2H}})x^{-1-K}dx\right) \\
&= \frac{1}{2^K}\left((u^{2H}+v^{2H})^K - |u-v|^{2HK}\right),
\end{aligned}$$

where the last equality follows from (2.3). □

For $f \in L^2_{J_i}$ the integral $\int_0^\infty f(t)dB_{H,K}(t)$ can then be defined as the inverse image of $f$ in this isometry. Hence, we think that some form of stochastic calculus, as in [**N**], could be developed. In closing, we remark that in Proposition 2.1, replacing the absolute value by the Euclidean norm, we could similarly have obtained a Gaussian random field which is a quasi-helix.

## References

[K1] J. P. Kahane, *Hélices et quasi-hélices,* Volume dédié à Laurent Schwartz. Adv. Math. **7B** (1981), pp. 417–433.

[K2] J. P. Kahane, *Some Random Series of Functions,* Cambridge University Press, (1985).

[L] S. J. Lin, *Stochastic Analysis of Brownian Motions,* Stoch. Stoch. Rep. **55** (1995), pp. 121–140.

[M-V] B. B. Mandelbrot and J. W. Van Ness, *Fractional Brownian Motion, Fractional Noises and Applications,* SIAM Rev. **10**, no. 4 (1968), pp. 422–437.

[N-V-V] I. Norros, E. Valkeila and J. Virtamo, *An elementary approach to a Girsanov formula and other analytical results on fBm,* Bernoulli **5**, no. 4 (1999), pp. 571–587.

[N] D. Nualart, *Stochastic Integration with Respect to Fractional Brownian Motion and Applications,* in this issue (2003).

[S-T] G. Samorodnitsky and M. Taqqu, *Stable Non-Gaussian Random Processes*, Chapman and Hall, New York, (1994).

Laboratoire d' Analyse et de Mathématiques Appliquées, CNRS UMR 8050, Université Paris XII, 94010, Créteil France, and School of Mathematics, Georgia Institute of Technology, Atlanta, GA 30332 USA

*E-mail address*: houdre@math.gatech.edu

Department of Mathematics and Physics, UAA, Mayapan 123, Colinas del Sol, Jesús María, Aguascalientes 20900 México

*E-mail address*: villa@cimat.mx

Contemporary Mathematics
Volume **336**, 2003

# A Non–homogeneous Wave Equation Driven by a Poisson Process

Jorge A. León and Mònica Sarrà

ABSTRACT. The objective of this article is to show the existence and uniqueness for a class of stochastic wave equations with spatial parameter in $\{1, 2, 3\}$ driven by a Poisson process.

## 1. Introduction

The purpose of this paper is to study the existence of a unique solution to a class of stochastic wave equations of the form

$$\begin{aligned} LX(t,x) &= \alpha(X(t,x),t,x,h)\lambda^+(t,x,h) \\ &\quad +\beta(X(t,x),t,x) \qquad \text{in } (0,\infty)\times\mathbb{R}^d, \\ X &= 0, \qquad \partial_t X = 0 \qquad \text{on } \{t=0\}\times\mathbb{R}^d. \end{aligned} \tag{1.1}$$

Here $L$ is the d'Alembert operator $\partial_t^2 - \Delta$, $d \in \{1,2,3\}$, $h \in U$ where $(U, \mathcal{U}, \mu)$ is a measure space and $\lambda^+$ is a Poisson process defined on $\mathbb{R}_+ \times \mathbb{R}^d \times U$.

Equation (1.1) is a symbolic expression and it may be interpreted in various different ways. The solution of this equation is in general given by the stochastic evolution equation

$$\begin{aligned} X(t,x) &= \int_0^t \int_{\mathbb{R}^d} \int_U S^d(t-s, x-y)\alpha(X(s,y),s,y,h)d\lambda^+(s,y,h) \\ &\quad + \int_0^t \int_{\mathbb{R}^d} S^d(t-s,x-y)\beta(X(s,y),s,y)dyds, \end{aligned} \tag{1.2}$$

where $S^d$ is the fundamental solution of the wave equation $LX = 0$.

The existence of solutions of stochastic partial differential equations has been considered by several authors. For example, Walsh [**Wa**] introduces a stochastic integral with respect to worthy martingales to deal with the stochastic heat and wave equations driven by a space–time white noise when $d = 1$. In this case, the

---

2000 *Mathematics Subject Classification.* Primary: 60H15; Secondary: 35R60.

*Key words and phrases.* Wave equation, d'Alembert operator, stochastic and pathwise integration, Poisson process.

The first author was partially supported by CONACyT grant 37130–E.

The second author was partially supported by Subdirección General de Formación y Promoción del Conocimiento grant BFM2000–0607.

solutions are random fields. He also deals with linear stochastic equations for $d > 1$, but now the solutions only exist as random distributions.

Recently, Dalang [**D**] gives an extension of the stochastic integral of Walsh [**Wa**]. This extension allows him to prove the existence of the solutions of the stochastic wave and heat equations driven by a Gaussian process, white in time and correlated in space, for $d \in \{1, 2, 3\}$ and $d \geq 1$, respectively. (See also [**D-F**], [**K-Z**], [**M-S**], [**P-Z1**] and [**P-Z2**]).

In Section 3, using the ideas of Saint Loubert Bié [**S**], we analyze equation (1.2) when $d \in \{1, 2\}$, $\alpha : \mathbb{R} \times \mathbb{R}_+ \times \mathbb{R}^d \times U \to \mathbb{R}$ and $\beta : \mathbb{R} \times \mathbb{R}_+ \times \mathbb{R}^d \to \mathbb{R}$ are two measurable functions, and the integral with respect to $\lambda^+$ is either a path–wise integral, or a stochastic integral. In this setting the solution of equation (1.2) is a random field adapted to the underlying filtration.

In the case that $d = 3$, the fundamental solution $S^d$ is *not* a function but a distribution. Therefore the integrals appearing in (1.2) may have no meaning.

Now we point out that Saint Loubert Bié [**S**] can analyze the stochastic heat equation driven by a Poisson process for any space–dimension because the kernel related to this equation is a quite smooth function.

In Section 4, we follow the ideas of Wilcox [**Wi**] to study equation (1.1) using the theory of distributions when $d = 3$, the stochastic integral is considered as a generalized random variable, which is possible because of the regularization theorem ([**Wa**], Corollary 4.1), and $\alpha$ and $\beta$ are suitable functions (see equation (4.1) below). Now the solution is a distribution that may not be a function. We observe that under the assumptions of Section 3 (i.e., $d \in \{1, 2\}$), Wilcox [**Wi**] shows that this approach gives us the solution to a family of equations of the form (1.2) whenever $\alpha \equiv 0$.

## 2. Preliminaries

In this section we give the framework that we use in this paper.

Let $(U, \mathcal{U}, \mu)$ be a measure space and $M : \mathbb{R}^d \times U \to \mathbb{R}$ a non–negative and measurable function such that the measure

$$d\lambda^-(s, y, h) = M(y, h) ds dy d\mu(h) \tag{2.1}$$

is $\sigma$–finite on $(\mathbb{R}_+ \times \mathbb{R}^d \times U, \mathbb{B}(\mathbb{R}_+) \otimes \mathbb{B}(\mathbb{R}^d) \otimes \mathcal{U})$, where $\mathbb{B}(\mathbb{R}^d)$ denotes the Borel $\sigma$–algebra on $\mathbb{R}^d$.

In the remainder of this paper, $(\Omega, \mathcal{F}, P)$ is a probability space and $\lambda^+$ is a Poisson process (i.e., a measurable point measure–valued random variable) on $\mathbb{R}_+ \times \mathbb{R}^d \times U$ with intensity $\lambda^-$ given by (2.1) (see Neveu [**N**] for details).

For $t \geq 0$, $\mathcal{F}_t$ denotes the $\sigma$–algebra generated by the family $\{\lambda^+([0, s] \times B) : 0 \leq s \leq t \text{ and } B \in \mathbb{B}(\mathbb{R}^d) \otimes \mathcal{U}\}$.

We say that a random field $Y = \{Y(s, y, h) : (s, y, h) \in \mathbb{R}_+ \times \mathbb{R}^d \times U\}$ is predictable (resp. progressively measurable) if it is $\mathcal{P}(\mathcal{F}_t) \otimes \mathbb{B}(\mathbb{R}^d) \otimes \mathcal{U}$ (resp. $Prog(\mathcal{F}_t) \otimes \mathbb{B}(\mathbb{R}^d) \otimes \mathcal{U}$)–measurable, where $\mathcal{P}(\mathcal{F}_t)$ (resp. $Prog(\mathcal{F}_t)$) represents the predictable (resp. the progressive) $\sigma$–field.

As it was pointed out in [**S**], for every predictable random field $Y$ in $L^1(\lambda^-)$, we have

$$E\Big(\int_{\mathbb{R}_+}\int_{\mathbb{R}^d}\int_U Y(s, y, h) d\lambda^+(s, y, h)\Big) = \int_{\mathbb{R}_+}\int_{\mathbb{R}^d}\int_U E(Y(s, y, h)) d\lambda^-(s, y, h). \tag{2.2}$$

The stochastic integral is path–wise defined. This isometry relation is also true for progressively measurable random fields $Y$ such that $Y(\cdot, y, h)$ is locally integrable for a.a. $(y,h) \in \mathbb{R}^d \times U$. Indeed, in this case there is a predictable process $\overline{Y}$ such that $Y = \overline{Y}$, $\lambda^- \otimes P$–a.s. Hence it is natural to define

$$\int_{\mathbb{R}_+} \int_{\mathbb{R}^d} \int_U Y(s,y,h) d\lambda^+(s,y,h) := \int_{\mathbb{R}_+} \int_{\mathbb{R}^d} \int_U \overline{Y}(s,y,h) d\lambda^+(s,y,h). \tag{2.3}$$

Also it was pointed out in **[S]** that the Burkholder–Davis–Gundy inequality allows us to introduce a stochastic integral for progressively measurable random fields $Y$ belonging to $L^p(d\lambda^- \otimes dP)$, $1 \le p \le 2$, such that

$$\begin{aligned} E\Big( \Big| \int_{\mathbb{R}_+} \int_{\mathbb{R}^d} \int_U Y(s,y,h) d\lambda(s,y,h) \Big|^p \Big) \\ \le C_p \int_{\mathbb{R}_+} \int_{\mathbb{R}^d} \int_U E\Big( |Y(s,y,h)|^p \Big) d\lambda^-(s,y,h), \end{aligned} \tag{2.4}$$

where $\lambda = \lambda^+ - \lambda^-$ is the compensated Poisson process.

## 3. The stochastic wave equation with spatial parameter $\mathbf{d} \in \{1, 2\}$

In this section we show the existence of a unique progressively measurable random field $X$ that satisfies equation (1.2) with $d \in \{1,2\}$, first when the stochastic integral is given by (2.3), and then when we write $\lambda$ instead of $\lambda^+$ and the stochastic integral satisfies (2.4).

We recall that the fundamental solution $S^d$ of the d'Alembert operator $\partial_t^2 - \Delta$ is:

$$S^1(t,x) = \frac{1}{2} 1_{\{|x| \le t\}}, \qquad d = 1$$

and

$$S^2(t,x) = \left( 2\pi \sqrt{t^2 - |x|^2} \right)^{-1} 1_{\{|x| \le t\}}, \qquad d = 2.$$

### 3.1. Stochastic wave equation driven by $\lambda^+$.

Now we analyze equation (1.2) when the stochastic integral is defined using (2.2) and (2.3). Here we show that equation (1.2) has a unique solution in the Banach space $\mathbb{P}$ of all the progressively measurable random fields $X$ such that

$$||X||_{\mathbb{P}} := \sup_{\mathbb{R}_+ \times \mathbb{R}^d} e^{-\eta t} E(|X(t,x)|) < \infty,$$

where $\eta$ is a suitable fixed positive constant.

The coefficients $\alpha : \mathbb{R} \times \mathbb{R}_+ \times \mathbb{R}^d \times U \to \mathbb{R}$ and $\beta : \mathbb{R} \times \mathbb{R}_+ \times \mathbb{R}^d \to \mathbb{R}$ are measurable functions that satisfy the following assumptions:

(H1) There are measurable functions $f_\alpha, g_\alpha : \mathbb{R}^d \times U \to \mathbb{R}_+$ and $f_\beta, g_\beta : \mathbb{R}^d \to \mathbb{R}_+$ such that:

i) (Linear growth) For all $(z,t,y,h) \in \mathbb{R} \times \mathbb{R}_+ \times \mathbb{R}^d \times U$, we have

$$|\alpha(z,t,y,h)| \le f_\alpha(y,h)(1 + |z|)$$

and

$$|\beta(z,t,y)| \le f_\beta(y)(1 + |z|).$$

ii) (Lipschitz condition) For all $(z_1, z_2, t, y, h) \in \mathbb{R} \times \mathbb{R} \times \mathbb{R}_+ \times \mathbb{R}^d \times U$,

$$|\alpha(z_1, t, y, h) - \alpha(z_2, t, y, h)| \leq g_\alpha(y, h)|z_1 - z_2|$$

and

$$|\beta(z_1, t, y) - \beta(z_2, t, y)| \leq g_\beta(y)|z_1 - z_2|.$$

iii) The functions

$$(t, x) \mapsto \int_0^t e^{\eta s} \int_{\mathbb{R}^d} \int_U (f_\alpha(y, h) + g_\alpha(y, h)) S^d(t - s, x - y) d\lambda^-(s, y, h)$$

and

$$(t, x) \mapsto \int_0^t e^{\eta s} \int_{\mathbb{R}^d} (f_\beta(y) + g_\beta(y)) S^d(t - s, x - y) dy ds$$

belongs to the space $\mathbb{P}$ for some $\eta > 0$.

iv) For $\eta$ satisfying Condition iii), we have

$$\begin{aligned} || \int_0^{\cdot} e^{\eta s} \int_{\mathbb{R}^d} \int_U & g_\alpha(y, h) S^d(\cdot - s, \circ - y) d\lambda^-(s, y, h) \\ & + \int_0^{\cdot} e^{\eta s} \int_{\mathbb{R}^d} g_\beta(y) S^d(\cdot - s, \circ - y) dy ds ||_{\mathbb{P}} < 1. \end{aligned}$$

Observe that it is easy to see that the following conditions on $f \in \{f_\alpha, g_\alpha\}$ and $g \in \{f_\beta, g_\beta\}$ guarantee Hypotheses (H1). iii) and (H1). iv) hold:

a) In the case $d = 1$,
  a1) $fM$ belongs to either $L^1(U)$, or $L^1(\mathbb{R} \times U)$ (see (2.1)).
  a2) $g$ is either in $L^1(\mathbb{R})$, or bounded by a positive constant.

b) When $d = 2$,
  b1) $f$ satisfies either $fM \in L^1(U)$, or $(y \mapsto \int_U f(y, h) M(y, h) d\mu(h)) \in L^p(\mathbb{R}^2)$, for some $p > 2$.
  b2) $g$ is either in $L^p(\mathbb{R}^2)$ for some $p > 2$, or bounded by a positive constant.

The following is the main result of this subsection.

THEOREM 3.1. *Let us assume that $\alpha$ and $\beta$ satisfy Hypothesis (H1). Then, equation (1.2) has a unique solution in the Banach space $\mathbb{P}$.*

PROOF. Let $\phi : \mathbb{P} \to \mathbb{P}$ be given by

$$\begin{aligned} \phi(V)(t, x) &= \int_0^t \int_{\mathbb{R}^d} \int_U S^d(t - s, x - y) \alpha(V(s, y), s, y, h) d\lambda^+(s, y, h) \\ &\quad + \int_0^t \int_{\mathbb{R}^d} S^d(t - s, x - y) \beta(V(s, y), s, y) dy ds. \end{aligned}$$

Now from (2.2), (2.3) and Hypothesis (H1), we have that $\phi(V)$ is well–defined (i.e., $\phi(V) \in \mathbb{P}$ whenever $V \in \mathbb{P}$) and that $\phi$ is a contraction. Thus the result holds. □

**3.2. Stochastic wave equation driven by $\lambda$.** Here we fix $p \in [1,2]$ and $T < \infty$. Now we see that the stochastic wave equation

$$
\begin{aligned}
(3.1) \qquad Y(t,x) &= \int_0^t \int_{\mathbb{R}^d} \int_U S^d(t-s, x-y)\alpha(Y(s,y), s, y, h) d\lambda(s,y,h) \\
&\quad + \int_0^t \int_{\mathbb{R}^d} S^d(t-s, x-y)\beta(Y(s,y), s, y) dy ds
\end{aligned}
$$

has a unique solution in the space $\mathbb{P}_p$ of all the progressively measurable random fields $X$ such that

$$||X||_{\mathbb{P}_p} := \sup_{[0,T]\times\mathbb{R}^d} ||X(t,x)||_{L^p(\Omega)} < \infty.$$

Henceforth we assume that the coefficients $\alpha$ and $\beta$ satisfy (H1).i) and (H1).ii), and the following condition:

$(H2)_p$ There is a non–negative and integrable function $g : [0,T] \to \mathbb{R}$ such that for every $0 \le s \le t \le T$,

$$
\begin{aligned}
\sup_{x\in\mathbb{R}^d} \Bigg\{ \int_{\mathbb{R}^d} \int_U \Big( f_\alpha^p(y,h) &+ g_\alpha^p(y,h) \Big) M(y,h) (S^d(t-s, x-y))^p d\mu(h) dy \\
&+ \int_{\mathbb{R}^d} \Big( f_\beta^p(y) + g_\beta^p(y) \Big) S^d(t-s, x-y) dy \Bigg\} \\
&\le g(t-s).
\end{aligned}
$$

The following assumptions provide us some examples of functions $f \in \{f_\alpha, g_\alpha\}$ and $\widetilde{g} \in \{f_\beta, g_\beta\}$ that satisfy Hypothesis $(H2)_p$:

a) For $d = 1$,
   a1) $f^p M$ belongs to either $L^1(U)$, or $L^1(\mathbb{R} \times U)$.
   a2) $\widetilde{g}$ is either in $L^p(\mathbb{R})$, or bounded by a constant.

b) For $d = 2$,
   b1) $f^p M \in L^1(U)$, or $(y \mapsto \int_U f^p(y,h) M(y,h) d\mu(h)) \in L^q(\mathbb{R}^2)$, for some $q > 1$ such that $pq(q-1)^{-1} < 2$.
   b2) $\widetilde{g}^p$ is either in $L^q(\mathbb{R}^2)$ for some $q > 2$, or bounded by a constant.

Now we are ready to state the main result of this subsection.

THEOREM 3.2. *Suppose that Hypotheses (H.1).i), (H1).ii) and $(H2)_p$ are satisfied for $\alpha$ and $\beta$ for some $p \in [1,2]$. Then equation (3.1) has a unique solution in $\mathbb{P}_p$.*

PROOF. Define $\psi : \mathbb{P}_p \to \mathbb{P}_p$ by

$$
\begin{aligned}
\psi(V)(t,x) &= \int_0^t \int_{\mathbb{R}^d} \int_U S^d(t-s, x-y)\alpha(V(s,y), s, y, h) d\lambda(s,y,h) \\
&\quad + \int_0^t \int_{\mathbb{R}^d} S^d(t-s, x-y)\beta(V(s,y), s, y) dy ds.
\end{aligned}
$$

As in the proof of Theorem 3.1, it is not difficult to see that our assumptions imply that $\psi$ is well–defined. Now we consider the Picard iteration scheme defined by

$$V^{(0)} \equiv 0$$

and

$$V^{(n)} = \psi(V^{(n-1)}).$$

Thus using our assumptions again and (2.4), we obtain

$$H_n(t) \leq C_{p,T} \int_0^t g(t-s) H_{n-1}(s) ds, \quad t \in [0,T]. \tag{3.2}$$

Here $H_n(t) := \sup_{[0,t]\times\mathbb{R}^d} E\Big(|V^{(n+1)}(s,x) - V^{(n)}(s,x)|^p\Big)$.

Finally the result is an immediate consequence of Lemma 15 in [**D**]. □

## 4. The three–dimensional space case

Before introducing the equation that we study in this section, we give some notation.

For $g \in L^1_{\text{loc}}(\mathbb{R}_+ \times \mathbb{R}^3 \times U; \lambda^-)$, we define $K_g : \mathcal{D}(\mathbb{R}^4) \to L^1(\Omega)$ by

$$K_g(\phi) = \int_{\mathbb{R}_+} \int_{\mathbb{R}^3} \int_U \phi(s,y) g(s,y,h) d\lambda^+(s,y,h).$$

Here $\mathcal{D}(\mathbb{R}^4)$ denotes the set of test functions on $\mathbb{R}^4$. As a consequence of the regularization theorem (see, for example, Corollary 4.1 in [**Wa**]), $K_g$ has a $\mathcal{D}'(\mathbb{R}^4)$–valued version. Henceforth we always deal with such a version.

For $\Lambda \in \mathcal{D}'(\mathbb{R}^4)$ and $\phi \in \mathcal{D}(\mathbb{R}^4)$, $\Lambda_\phi$ denotes the convolution of $\Lambda$ and the test function $x \mapsto \phi(-x)$.

In the sequel we are interested in the stochastic wave equation of the form

$$\begin{aligned} (\partial_t^2 - \Delta) u &= T_1(u)\Lambda + T_2(u) K_g \quad \text{in} \quad \mathcal{D}'(\mathbb{R}^4), \\ &\quad \text{supp}\, u \subset \mathbb{R}_+ \times \mathbb{R}^3. \end{aligned} \tag{4.1}$$

Here $\Lambda \in \mathcal{D}'(\mathbb{R}^4)$, supp $\Lambda \subset \mathbb{R}_+ \times \mathbb{R}^3$, $T_1 : \mathcal{D}'(\mathbb{R}^4) \to C^\infty(\mathbb{R}^4)$ and $T_2(u)$ is a progressively measurable random field bounded on compact sets if $u$ is a point–wise measurable $\mathcal{D}'(\mathbb{R}^4)$–valued random variable.

By a solution of (4.1), we mean a point–wise measurable random variable $u : \Omega \to \mathcal{D}'(\mathbb{R}^4)$ such that for every $\phi \in \mathcal{D}(\mathbb{R}^4)$,

$$u(\phi) = (4\pi)^{-1} \int_{-\infty}^0 \frac{1}{-r} \int_{\partial B(0,-r)} (T_1(u)\Lambda + T_2(u)K_g)_\phi(r,\xi) dS_\xi dr, \tag{4.2}$$

where $\int_{\partial B(0,-r)} \cdot dS_\xi$ is the surface integral over $\{y \in \mathbb{R}^3 : |y| = -r\}$. We observe that this definition of solution is reasonable taking into account Wilcox [**Wi**] (proof of Theorem 4) and the fact that the function

$$v(t,x) = \int_{t_0}^t (4\pi(t-r))^{-1} \int_{\partial B(x,t-r)} f(r,y) dS_y dr,$$

$f \in C^2([t_0,\infty) \times \mathbb{R}^3)$, is the solution of the wave equation

$$(\partial_t^2 - \Delta) u = f \qquad \text{in} \qquad (t_0,\infty)\times\mathbb{R}^3,$$

$$u = 0, \quad \partial_t u = 0 \qquad \text{on} \qquad \{t = t_0\} \times \mathbb{R}^3$$

(see for instance [**E**]). Also we note that the right–hand side of (4.2) is well–defined because the map

$$\phi \mapsto \int_{-\infty}^0 \frac{1}{-r} \int_{\partial B(0-r)} \Lambda_\phi(r,\xi) dS_\xi dr$$

belongs to $\mathcal{D}'(\mathbb{R}^4)$ for any $\Lambda \in \mathcal{D}'(\mathbb{R}^4)$ by the proof of Theorem 4 in [**Wi**].

Now let us introduce the conditions that the coefficients of equation (4.1) will satisfy in the remainder of this section.

(H3) For all $\phi \in \mathcal{D}(\mathbb{R}^4)$ and $v_1, v_2 : \Omega \to \mathcal{D}'(\mathbb{R}^4)$ point–wise integrable, we have:

i) the maps

$$\phi \mapsto \int_{-\infty}^{0} (-r)^{-1} \int_{\partial B(0,-r)} (T_1(v_1)\Lambda)_\phi(r,\xi) dS_\xi dr$$

and

$$\phi \mapsto \int_{-\infty}^{0} (-r)^{-1} \int_{\partial B(0,-r)} (T_2(v_1)K_g)_\phi(r,\xi) dS_\xi dr$$

are $\mathcal{D}'(\mathbb{R}^4)$–valued point–measurable random variables.

ii) There exist positive constants $M_\phi^{(1)}$ and $M_\phi^{(2)}$, and $\phi_0 \in \mathcal{D}(\mathbb{R}^4)$ such that: $M_{\phi_0}^{(1)} + M_{\phi_0}^{(2)} < 1$,

$$(4\pi)^{-1} E\Big| \int_{-\infty}^{0} (-r)^{-1} \int_{\partial B(0,-r)} \Big( (T_1(v_1) - T_1(v_2))\Lambda \Big)_\phi (r,\xi) dS_\xi dr \Big| \leq M_\phi^{(1)} E|v_1(\phi_0) - v_2(\phi_0)|,$$

and

$$(4\pi)^{-1} E\Big| \int_{-\infty}^{0} (-r)^{-1} \int_{\partial B(0,-r)} \Big( (T_2(v_1) - T_2(v_2))K_g \Big)_\phi (r,\xi) dS_\xi dr \Big| \leq M_\phi^{(2)} E|v_1(\phi_0) - v_2(\phi_0)|.$$

iii) There is $v_0 : \Omega \to \mathcal{D}'(\mathbb{R}^4)$ point–wise integrable such that

$$E\Big| \int_{-\infty}^{0} (-r)^{-1} \int_{\partial B(0,-r)} \Big( T_1(v_0)\Lambda + T_2(v_0)K_g \Big)_{\phi_0} (r,\xi) dS_\xi dr \Big| < \infty.$$

iv) For any sequence $(\phi_n)_{n\geq 1} \subseteq \mathcal{D}(\mathbb{R}^4)$ converging to $\phi$, we have,

$$\sup_{n\geq 1} \Big( M_{\phi_n}^{(1)} + M_{\phi_n}^{(2)} \Big) < \infty.$$

It is not difficult to prove that the following family of coefficients satisfy above–mentioned hypothesis.

- $T_1(\cdot) = \overline{T}_1(\cdot) f_1$, where $f_1 \in C^\infty(\mathbb{R}^4)$ and $\overline{T}_1$ is such that for any $v_1, v_2 : \Omega \to \mathcal{D}'(\mathbb{R}^4)$ point–wise integrable, we have:
  i) $\overline{T}(v_1)$ is a random variable.
  ii) There exits $\overline{\phi}_0$ such that
  $$E|\overline{T}_1(v_1) - \overline{T}_1(v_2)| \leq CE|v_1(\overline{\phi}_0) - v_2(\overline{\phi}_0)|.$$
  iii) There is $v : \Omega \to \mathcal{D}'(\mathbb{R}^4)$ point–wise integrable such that $E|\overline{T}(v)| < \infty$.
- $T_2(\cdot)(s,x) = 1_{]a,\infty)}(s) E[\overline{T}_2(\cdot)/\mathcal{F}_a] f_2(s,x)$, where $f_2 \in C^\infty(\mathbb{R}^4)$ and $\overline{T}_2$ is such that for any $v_1, v_2 : \Omega \to \mathcal{D}'(\mathbb{R}^4)$ point–wise integrable, we have:
  i) $\overline{T}_2 : \mathcal{D}'(\Omega) \to \mathbb{R}$ is bounded.
  ii) $\overline{T}_2(v_1)$ is a random variable.
  iii) Let $\overline{\phi}_0$ be as before. Then $E|\overline{T}_2(v_1) - \overline{T}_2(v_2)| \leq CE|v_1(\overline{\phi}_0) - v_2(\overline{\phi}_0)|$ holds.

Now we are ready to state our existence and uniqueness result for the solution of equation (4.1).

THEOREM 4.1. *Assume that (H3) holds. Then equation (4.2) has a unique point–wise integrable solution.*

PROOF. We first show the existence of the solution of equation (4.2).

Fix $\phi \in \mathcal{D}(\mathbb{R}^4)$. Now by induction, we define the following family of point–wise integrable $\mathcal{D}'(\mathbb{R}^4)$–valued random variables:

$$u^{(0)}(\phi) = (4\pi)^{-1}\int_{-\infty}^{0}(-r)^{-1}\int_{\partial B(0,-r)}(T_1(v_0)\Lambda + T_2(v_0)K_g)_\phi(r,\xi)dS_\xi dr$$

and, for $n \geq 1$,

$$u^{(n)}(\phi) = (4\pi)^{-1}\int_{-\infty}^{0}(-r)^{-1}\int_{\partial B(0,-r)}(T_1(u^{(n-1)})\Lambda + T_2(u^{(n-1)})K_g)_\phi(r,\xi)dS_\xi dr.$$

So, by a standard argument and using (H3), we obtain

$$\begin{aligned}(4.3)\quad & E|u^{(n)}(\phi) - u^{(n-1)}(\phi)| \\ & \leq (M_\phi^{(1)} + M_\phi^{(2)})(M_{\phi_0}^{(1)} + M_{\phi_0}^{(2)})^{n-1}E(|u^{(0)}(\phi_0) - v_0(\phi_0)|), \quad n \geq 1.\end{aligned}$$

Consequently, there is a random linear functional $\widetilde{u}$ (i.e., $\widetilde{u}(a\phi_1 + b\phi_2) = a\widetilde{u}(\phi_1) + b\widetilde{u}(\phi_2)$ w.p.1 for each $\phi_1, \phi_2 \in \mathcal{D}(\mathbb{R}^4)$ and $a, b \in \mathbb{R}$) such that

$$(4.4)\qquad E|u^{(n)}(\phi) - \widetilde{u}(\phi)| \to 0 \qquad \text{as} \qquad n \to \infty.$$

Now we prove that $\widetilde{u}$ has a version $u : \Omega \to \mathcal{D}'(\mathbb{R}^4)$. Note that if $(\phi_n)_n \subset \mathcal{D}(\mathbb{R}^4)$ is a sequence that tends to $\phi$ as $n \to \infty$, then (4.3) implies

$$\begin{aligned}&\sup_n E|\widetilde{u}(\phi_n) - u^{(k)}(\phi_n)| \\ &\leq \left(\sup_n(M_{\phi_n}^{(1)} + M_{\phi_n}^{(2)})\right)(E|u^{(0)}(\phi_0) - v_0(\phi_0)|)\sum_{j=k}^{\infty}(M_{\phi_0}^{(1)} + M_{\phi_0}^{(2)})^j \\ &\leq C\sum_{j=k}^{\infty}(M_{\phi_0}^{(1)} + M_{\phi_0}^{(2)})^j.\end{aligned}$$

Therefore, using the regularization theorem again and (4.4), $\widetilde{u}$ has a point–wise integrable version $u : \Omega \to \mathcal{D}'(\mathbb{R}^4)$, which is a solution of equation (4.2).

Finally, if $u, v : \Omega \to \mathcal{D}'(\mathbb{R}^4)$ are two point–wise integrable solutions of equation (4.2), then by H3).ii), we get

$$E|u(\phi) - v(\phi)| \leq (M_\phi^{(1)} + M_\phi^{(2)})E|u(\phi_0) - v(\phi_0)|,$$

which, together with the fact that $M_{\phi_0}^{(1)} + M_{\phi_0}^{(2)} < 1$, implies that $u(\phi) = v(\phi)$ w.p.1. Thus (4.2) has a unique solution and the proof is complete. □

**Acknowledgement.** The authors would like to thank anonymous referee for useful comments and suggestions that improved the presentation of the paper.

## References

[D] R.C. Dalang, *Extending martingale measure stochastic integral with applications to spatially homogeneous S.P.D.E'S,* Electron. J. Probab. **4**, no. 6 (1999), pp. 1–29.

[D-F] R.C. Dalang and N.E. Frangos, *The stochastic wave equation in two spatial dimensions,* Ann. Probab. **26**, no. 1 (1998), pp. 187–212.

[E] L.C. Evans, *Partial Differential Equations,* Graduate Studies in Mathematics **19**, AMS, Providence, Rhode Island, (1991).

[K-Z] A. Karczewska and J. Zabczyk, *Stochastic PDE's with function–valued solutions,* in Infinity Dimensional Stochastic Analysis (Amsterdam, 1999), pp. 197–216, Proceedings of the Colloquium of the Royal Netherlands Academy of Arts and Sciences, North–Holland, Amsterdam, (2000).

[M-S] A. Millet and M. Sanz–Solé, *A stochastic wave equation in two space dimension: smoothness of the law,* Ann. Probab. **27**, no. 2 (1999), pp. 803–844.

[N] J. Neveu, *Processus Ponctuels,* École d'Eté de Probabilités de Saint–Flour VI–1976, Lecture Notes in Math. **598** (1977), pp. 249–445.

[P-Z1] S. Peszat and J. Zabczyk, *Nonlinear stochastic wave and heat equations,* Probab. Theory Related Fields **116** (2000), pp. 421–443.

[P-Z2] S. Peszat and J. Zabczyk, *Stochastic evolution equations with a spatially homogeneous Wiener process,* Stochastic Process. Appl. **72**, no. 2 (1997), pp. 187–204.

[S] E. Saint Loubert Bié, *Étude d'une EDPS conduite par un bruit poissonnien,* Probab. Theory Related Fields **111** (1998), pp. 287–321.

[Wa] J.B. Walsh, *An Introduction to Stochastic Partial Differential Equations,* École d'Eté de Probabilités de Saint–Flour XIV–1984, Lecture Notes in Math. **1180** (1986), pp. 265–439.

[Wi] C.H. Wilcox, *The Cauchy problem for the wave equation with distribution data: an elementary approach,* Amer. Math. Monthly **98**, no. 5 (1991), pp. 401–410.

Departamento de Matemáticas, Cinvestav del IPN, Apartado Postal 14–740, 07000 México, D.F., MEXICO

*E-mail address*: `jleon@math.cinvestav.mx`

Facultat de Matemàtiques, Gran Via 585, E–08007 Barcelona, SPAIN

*E-mail address*: `sarra@mat.ub.es`

Contemporary Mathematics
Volume **336**, 2003

# Existence of Self–Intersection Local Time of the Multitype Dawson-Watanabe Superprocess

José Alfredo López–Mimbela and José Villa

ABSTRACT. In [**D**] Dynkin introduced a notion of self-intersection local time (SILT) for a class of continuous measure-valued processes, and gave a criterion for existence of SILT in terms of superprocess multiple stochastic integrals. In this note we show that the multitype Dawson-Watanabe superprocess with symmetric $\alpha_i$-stable motions satisfies Dynkin's existence criterion provided the spatial dimension is less than $4 \min_i \alpha_i$.

## 1. Introduction and preliminary definitions

Dynkin introduced in [**D**] a definition of self-intersection local time (SILT) for a general class of continuous measure-valued processes, and gave a criterion for existence of SILT in terms of superprocess multiple stochastic integrals. In this note we prove that when the spatial dimension is small the hypothesis in Dynkin's existence criterion are satisfied by the multitype Dawson-Watanabe superprocess studied in [**G-L**], and obtain in this way a sufficient condition for existence of SILT of the Dawson-Watanabe superprocess in the multitype setting.

Let us recall that the multitype Dawson-Watanabe superprocess arises as the diffusion approximation of multitype branching particle systems in $\mathbb{R}^d$ consisting of particles of $k \geq 1$ types undergoing spatial diffusion and critical multitype branching, where the diffusions, the particle lifetimes and the branching laws depend on the types. In the model we consider here the motions of particles are symmetric $\alpha_i$-stable processes, $0 < \alpha_i \leq 2$, and the lifetimes of particles are exponentially distributed with parameters $V_i$, $i = 1, \ldots, k$. For simplicity, in this paper we restrict ourselves to the case of multitype populations which are a.s. finite, and whose branching law has a stochastic mean matrix $(m_{ij})_{1 \leq i,j \leq k}$, i.e., $m_{ij} \geq 0$ and $\sum_{j=1}^k m_{ij} = 1$, $i = 1, \ldots, k$.

Our multitype particle systems are properly represented by counting measures $\mu \in M_f(\mathbb{S})$, where $M_f(\mathbb{S})$ denotes the space of finite positive measures on $\mathbb{S} \equiv \mathbb{R}^d \times \{1, \ldots, k\}$. Here the first component of $(x, i) \in \mathbb{S}$ stands for the position and the second component for the type of an individual $\delta_{(x,i)}$. We denote by $\xi_t := (W_t, \eta_t)$,

---

2000 *Mathematics Subject Classification.* Primary: 60J55; Secondary: 60G57, 60J65, 60J27.

Research partially supported by Conacyt Grant No. 32401E.
Research partially supported by PROMEP Grant UAAGS-99-04.

$t \geq 0$, the Markov process on $\mathbb{S}$ whose type component $\eta_t$ follows a Markov chain with Q-matrix $(V_i(m_{ij} - \delta_{ij}))_{1\leq i,j\leq k}$, and whose position component $W_t$ follows an $\alpha_i$-stable motion as long as $\eta_t = i$. The generator of $\xi$ is given by

$$A\phi(x,i) = \Delta_{\alpha_i}\phi(x,i) + V_i \sum_{j=1}^{k} (m_{ij} - \delta_{ij})\, \phi(x,j), \quad (x,i) \in \mathbb{S}, \quad \phi(\cdot,i) \in \mathrm{Dom}(\Delta_{\alpha_i}),$$

where $\Delta_{\alpha_i}$ denotes the generator of the $\alpha_i$-stable process, $i = 1, \ldots, k$. Writing $\mu(\phi) \equiv \int \phi(x)\, \mu(dx)$ and denoting the semigroup associated with $A$ by $U \equiv (U_t)_{t\geq 0}$, we recall (cf. [**L**]) that

$$(U_t\phi)\,(x,i) = \mathbb{E}\mathbb{Y}_t^{(x,i)}\,(\phi)\,, \; t \geq 0,$$

where $(\mathbb{Y}_t^{(x,i)})_{0\leq t<\infty}$ stands for the particle system started from $\delta_{(x,i)}$, that is, one type $i$-individual at position $x$. In this sense, the process $\xi$ is the *expectation process* of the branching system; its transition probability equals $J_t^{(x,i)}(\cdot) := \mathbb{E}\mathbb{Y}_t^{(x,i)}(\cdot)$.

Roughly described, the renormalization of the branching particle system suitable for the diffusion limit consists in increasing the density of the population of all types, assigning a small mass and a short lifetime to every individual, and decreasing the mutation probabilities in such a way that the mutation rates remain constant; the spatial motion laws remain the same. If the initial populations (i.e., at time $t = 0$) of the renormalized branching particle systems converge weakly to $\mu \in M_f(\mathbb{S})$, and the branching mechanisms have finite second-order moments, this renormalization yields in the limit a continuous Markov process $X := \{X_t,\; t \geq 0\}$ with values in $M_f(\mathbb{S})$, having the Laplace functional

$$\mathbb{E}\left[e^{-X_t(\phi)}\right] = e^{-\mu(u_t)} \tag{1.1}$$

for any bounded, measurable $\phi : \mathbb{S} \to [0,\infty)$, where $u_t(x,i)$ is the unique solution of the nonlinear equation

$$\frac{\partial u_t(x,i)}{\partial t} = Au_t(x,i) - C_i u_t(x,i)^2, \quad u_0(x,i) = \phi(x,i),$$

and $C_i$, $i = 1, \ldots, k$ are certain positive constants determined by the lifetime parameters and the variances of the branching law (see [**G-L**]); for simplicity, we shall assume $C_i = 1$, $i = 1, \ldots, k$.

Let us now recall the construction of the superprocess stochastic integrals introduced by Dynkin in [**D**].

Let $[0,u]$ be a fixed time interval. From (1.1) one can deduce that for every $\mu \in M_f(\mathbb{S})$, any bounded, measurable functions $\phi_i : \mathbb{S} \to \mathbb{R}$, $i = 1, \ldots, n$, and positive numbers $t_1, \ldots, t_n \in [0,u]$ the random variable

$$X_{t_1}(\phi_1) \cdots X_{t_n}(\phi_n) \tag{1.2}$$

is in $L^2(P)$, where $P$ is the distribution of $X$, and that $\mathbb{E}[X_{t_1}(\phi_1) \cdots X_{t_n}(\phi_n)] = \int \left(\prod_{i=1}^{n} \varphi_i(y_i, s_i)\right) \gamma_n(dy_1, ds_1; \ldots; dy_n, ds_n)$, where

$$\varphi(x,s) := U_{t-s}\phi(x) \tag{1.3}$$

and $\gamma_n$ is the $n$-th moment measure of $X$. For our purposes we need to know the precise definition of $\gamma_n$ only for the values $n = 1, 2$, and in these cases,

$$\begin{aligned}\gamma_1(A_1 \times B_1) &= 1_{B_1}(0)\mu(A_1),\\ \gamma_2(A_1 \times B_1; A_2 \times B_2) &= 1_{B_1}(0)\mu(A_1)1_{B_2}(0)\mu(A_2)\\ &\quad +2\int 1_{B_1}(s)1_{A_1}(y)1_{B_2}(s)1_{A_2}(y)\, J_s^x(dy)\,\mu(dx)\,ds,\\ &\quad A_i \in B(\mathbb{S}),\ B_i \in B([0,u]),\ i = 1, 2.\end{aligned}$$

Let $L_n^2$ denote the smallest closed subspace of $L^2(P)$ containing all products of the form (1.2). We define $K$ as the set of functions of the form (1.3) with $\phi$ bounded and measurable, and write $\chi_n^0$ for the space of functions $\varphi : (\mathbb{S} \times [0, u])^n \to \mathbb{R}$ satisfying $(|\varphi|, |\varphi|)_n < \infty$, where

$$(\varphi, \psi)_n := \int \varphi(y_1, s_1; \dots; y_n, s_n)\psi(y_1, s_1; \dots; y_n, s_n)\, \gamma_{2n}(dy_1, ds_1; \dots; dy_n, ds_n).$$

In particular, for $n = 1$,

$$\begin{aligned}(\varphi, \psi)_1 &= \int \varphi(z_1, t_1)\psi(z_2, t_2)\, d\gamma_2\\ &= \int \varphi(z, 0)\,\mu(dz) \int \psi(z, 0)\,\mu(dz) + \int \varphi(y, s)\psi(y, s)\,\hat{\Lambda}(dy, ds),\end{aligned}$$

where $\hat{\Lambda}$ is the measure on $\mathbb{S} \times [0, u]$ given by

$$\hat{\Lambda}(C) = 2\int 1_C(y, s) J_s^x(dy)\,\mu(dx)\,ds.$$

Therefore $\varphi \in \chi_1^0$ if and only if $\hat{\Lambda}(\varphi^2) < \infty$ and $\mu(|\varphi(0, \cdot)|) < \infty$.

The stochastic integrals are defined as follows ([**D**], theorems 1.2 and 1.3):

a) For $\varphi = U_{t-s}\phi \in K$,

(1.4) $$I_1(\varphi) \equiv \int \varphi(x, s)\, dZ_{x,s} := X_t(\phi), \quad t > 0.$$

Let $\chi_1$ denote the space of equivalence classes of $\chi_1^0$ modulo $(\cdot, \cdot)_1$. There exist a unique isometry (which we again denote by) $I_1$ from $\chi_1$ to $L_1^2$ obeying (1.4). Moreover, for every $\varphi \in \chi_1^0$, the process $M_t^\varphi := I_1(\varphi) = \int \varphi(x, s)1_{s \le t}\, dZ_{x,s}$ is a martingale, and every $L_1^2$-valued martingale can be represented in this form.

b) There exists a unique mapping $I_n : \chi_n^0 \to L_n^2$ satisfying $I_n(\varphi_1 \times \cdots \times \varphi_n) = I_1(\varphi_1) \cdots I_1(\varphi_n)$ and $\mathbb{E}[I_n(\varphi)I_n(\psi)] = (\varphi, \psi)_n$ for any $\varphi_1, \dots, \varphi_n \in \chi_1^0$ and $\varphi, \psi \in \chi_n^0$. The set $I_n(\chi_n^0)$ is dense in $L_n^2$.

## 2. Existence of SILT of $X$

The self-intersection local time of $X$ can be defined heuristically by $SILT(B) = \int_B \int_{\mathbb{R}^d} \int_{\mathbb{R}^d} \delta_0(z - z')\, X_s(dz)\, X_t(dz')\, ds\, dt$, where $B \in B([0, u]^2)$. This expression is formally equivalent to

(2.1) $$SILT(B) = I_2(K_B^2) \equiv \int K_B^2(x_1, s_1; x_2, s_2)\, dZ_{x_1,s_1}\, dZ_{x_2,s_2},$$

(see [**D**] or [**F-I**]) where $K_B^2(x_1, s_1; x_2, s_2) = \int_B dt_1\, dt_2 \int_{\mathbb{S}} \prod_{i=1}^2 J_{t_i - s_i}^{x_i}(dz)$. The stochastic integral in the right-hand side of (2.1) makes sense provided $K_B^2 \in \chi_2^0$, in which case we say that the self-intersection local time of $X$ (of order two) exists. We will prove the following theorem.

THEOREM 2.1. *Let $X = \{X_t,\ t \geq 0\}$ be the multitype Dawson-Watanabe superprocess whose Laplace functional is given by (1.1), and $\Lambda = [0, u]$. Suppose $B \in B(\Lambda \times \Lambda)$ is such that $\overline{B} \cap \{(x, y) \in \Lambda \times \Lambda : x = y\} = \emptyset$. If the measures $\mu(\{i\} \times \cdot)$, $i = 1, 2$, have bounded densities with respect to d-dimensional Lebesgue measure, and $d < 4\min\{\alpha_1, \dots, \alpha_k\}$, then the self-intersection local time of $X$ exists.*

The converse result, namely, that the conditions in Theorem 2.1 are necessary for existence of SILT, remains open. The case of time sets $B$ that intersect the diagonal, which was solved by Rosen for monotype superprocesses in [**R**], also remains to be investigated.

**2.1. Proof of Theorem 2.1.** We define the measure $m$ on $(\mathbb{S}, B(\mathbb{S}))$ by $m(B \times C) = \lambda(B)\nu(C)$, $B \in B(\mathbb{R}^d)$, $C \subset \{1, \dots, k\}$, where $\nu$ is the counting measure on $\{1, \dots, k\}$ and $\lambda$ is the Lebesgue measure on $\mathbb{R}^d$. The transition kernels $J_t^{(x,h)}(\cdot)$ of the expectation process are absolutely continuous with respect to $m$, and the corresponding densities $J_t((x, h), (z, i))$ are given by [**V**]

$$J_t((x,h),(z,i)) = \int_{D_{[0,\infty)}(\{1,\dots,k\})} (q^{\alpha_1}_{L_1(t,\eta)} * \cdots * q^{\alpha_k}_{L_k(t,\eta)})(x,z) 1_{\{\eta_0 = h, \eta_t = i\}}(\eta) P_h^\eta(d\eta),$$

where $P_i^\eta$ is the distribution of the Markov chain $\eta$ starting in $i$, $L_i(t, \eta)$ is the amount of time that $\eta$ spends in type $i$ during the time interval $[0, t]$, and $\{q_r^{\alpha_i},\ r > 0\}$ are the $\alpha_i$-stable densities, $i = 1, \dots, k$. We define $J_t \equiv 0$ if $t < 0$.

In order to prove Theorem 2.1 we will show that

$$\sup_{s,y} \int_{\mathbb{S}\times\mathbb{S}} G(s, y; z) G(s, y; \zeta) H(z, \zeta)\, dz\, d\zeta < \infty, \tag{2.2}$$

where, for $0 \leq s \leq t$,

$$\begin{aligned} G(s,(y,j);(z,i)) &= \int_\Lambda J_{t-s}((y,j),(z,i))\, dt, \\ H((z,i),(\zeta,l)) &= \int_{\Lambda\times\mathbb{S}} G(s,(y,j);(z,i)) G(s,(y,j);(\zeta,l))\, m(d(y,j))\, ds. \end{aligned}$$

According to [**D**] (Theorem 1.5), under our assumptions condition (2.2) implies $K_B^2 \in \chi_2^0$.

In the remaining part of the paper we assume without loss of generality that $k = 2$ and $\alpha_1 \leq \alpha_2$. We denote by $p_t^{(i)}(x, y)$, $t > 0$, $i = 1, 2$, $x, y \in \mathbb{R}^d$, the transition densities of the position process $\{W_t,\ t \geq 0\}$, which, by the law of total probability, satisfy

$$\begin{aligned} p_t^{(i)}(x,y)\, dy &:= P[W_t \in dy | (W_0, \eta_0) = (x, i)] \\ &= \sum_{j=1}^2 P[W_t \in dy, \eta_t = j | (W_0, \eta_0) = (x, i)] \\ &= \sum_{j=1}^2 J_t((x,i),(y,j))\, dy. \end{aligned} \tag{2.3}$$

By conditioning on the time of first jump of $\eta$, and using commutativity of convolutions, it easily follows that
(2.4)

$$p_t^{(i)}(x,y) = e^{-V_i(1-m_{ii})t} q_t^{\alpha_i}(x,y) + \int_0^t \int_{\mathbb{R}^d} q_r^{\alpha_1}(x,z) q_{t-r}^{\alpha_2}(z,y)\, dz\, \theta_t^i(dr),\ i=1,2,$$

where $\theta_t^i(dr)$ is the conditional distribution of $L_i(t,\eta)$ given that a change of type occurred in the interval $(0,t]$. In order to estimate the product of stable densities inside the integral above, we need the following result from [**V**].

LEMMA 2.2. *Let* $0 \le \alpha_1 \le \alpha_2 \le 2$. *There exists a constant* $K \ge 1$ *such that* $q_t^{\alpha_2}(x) \le K q_{t^{\alpha_1/\alpha_2}}^{\alpha_1}(x)$ *for all* $t > 0$ *and* $x \in \mathbb{R}^d$. *Moreover, if* $t \ge 1$ *then* $q_t^{\alpha_2}(x) \le K t^{d(1/\alpha_1 - 1/\alpha_2)} q_t^{\alpha_1}(x) \le K t^{d/\alpha_1} q_t^{\alpha_1}(x)$, $x \in \mathbb{R}^d$.

Using (2.4) and Lemma 2.2 we deduce that for $0 < t < 1$,

$$\begin{aligned} p_t^{(i)}(x,y) &\le e^{-V_i(1-m_{ii})t} K q_{t^{\alpha_1/\alpha_i}}^{\alpha_1}(x,y) \\ &\quad + \int_0^t \int_{\mathbb{R}^d} q_r^{\alpha_1}(x,z) K q_{(t-r)^{\alpha_1/\alpha_2}}^{\alpha_1}(z,y)\, dz\, \theta_t^i(dr) \\ &\le K\left[ q_{t^{\alpha_1/\alpha_i}}^{\alpha_1}(x,y) + \int_0^t \left( q_r^{\alpha_1} * q_{(t-r)^{\alpha_1/\alpha_2}}^{\alpha_1} \right)(x,y) \theta_t^i(dr) \right] \\ &\le 2K \left[ q_{t^{\alpha_1/\alpha_i}}^{\alpha_1}(x,y) + \int_0^t q_{r+(t-r)^{\alpha_1/\alpha_2}}^{\alpha_1}(x,y) \theta_t^i(dr) \right]. \end{aligned}$$

Similarly, for $t > 1$,

$$\begin{aligned} p_t^{(i)}(x,y) &\le e^{-V_i(1-m_{ii})t} K t^{d/\alpha_1} q_t^{\alpha_1}(x,y) \\ &\quad + \int_0^t \int_{\mathbb{R}^d} \left( 1_{(0,t-1]} + 1_{(t-1,t]} \right)(r) q_r^{\alpha_1}(x,z) q_{t-r}^{\alpha_2}(z,y)\, dz\, \theta_t^i(dr) \\ &\le K t^{d/\alpha_1} q_t^{\alpha_1}(x,y) \\ &\quad + \int_0^{t-1} \int_{\mathbb{R}^d} q_r^{\alpha_1}(x,z) K (t-r)^{d/\alpha_1} q_{t-r}^{\alpha_1}(z,y)\, dz\, \theta_t^i(dr) \\ &\quad + \int_{t-1}^t \int_{\mathbb{R}^d} q_r^{\alpha_1}(x,z) K q_{(t-r)^{\alpha_1/\alpha_2}}^{\alpha_1}(z,y)\, dz\, \theta_t^i(dr) \\ &\le K \left[ t^{d/\alpha_1} q_t^{\alpha_1}(x,y) + q_t^{\alpha_1}(x,y) \int_0^{t-1} (t-r)^{d/\alpha_1} \theta_t^i(dr) \right. \\ &\quad \left. + \int_{t-1}^t q_{r+(t-r)^{\alpha_1/\alpha_2}}^{\alpha_1}(x,y) \theta_t^i(dr) \right] \\ &\le K \left[ t^{d/\alpha_1} q_t^{\alpha_1}(x,y) + t^{d/\alpha_1} q_t^{\alpha_1}(x,y) \int_0^t \theta_t^i(dr) \right. \\ &\quad \left. + \int_{t-1}^t q_{r+(t-r)^{\alpha_1/\alpha_2}}^{\alpha_1}(x,y) \theta_t^i(dr) \right] \\ &\le 2K \left[ t^{d/\alpha_1} q_t^{\alpha_1}(x,y) + \int_{t-1}^t q_{r+(t-r)^{\alpha_1/\alpha_2}}^{\alpha_1}(x,y) \theta_t^i(dr) \right]. \end{aligned}$$

Let us proceed to the proof of (2.2). In what follows we denote $(\int_a^x + \int_x^b) h(r)\, dr := \int_a^x h(r)\, dr + \int_x^b h(r)\, dr$ for any integrable function $h : [a,b] \to \mathbb{R}$ and $x \in [a,b]$.

We start by noting that, due to (2.3), $G(s,(y,j);(z,i)) = \int_s^u J_{t-s}((y,j),(z,i))\,dt \le \int_0^{u-s} p_t^{(j)}(y,z)\,dt$ for $j=1,2$. Therefore, putting $a_+ := \max\{a,0\}$, $a \in \mathbb{R}$,

$$
\begin{aligned}
&H((z,i),(\zeta,l))\\
&\quad= \sum_{j=1}^{2}\int_0^u\int_{\mathbb{R}^d} G(s,(j,y);(i,z))G(s,(j,y);(l,\zeta))\,dy\,ds\\
&\quad\le \sum_{j=1}^{2}\int_0^u\int_{\mathbb{R}^d}\left(\int_0^{u-s} p_{t_1}^{(j)}(y,z)\,dt_1\right)\left(\int_0^{u-s} p_{t_2}^{(j)}(y,\zeta)\,dt_2\right)dy\,ds\\
&\quad= \sum_{j=1}^{2}\left(\int_0^{(u-1)_+} + \int_{(u-1)_+}^{u}\right)\int_{\mathbb{R}^d}\left(\int_0^{u-s} p_{t_1}^{(j)}(y,z)\,dt_1\right)\\
&\qquad\cdot\left(\int_0^{u-s} p_{t_2}^{(j)}(y,\zeta)\,dt_2\right)dy\,ds\\
&\quad= \sum_{j=1}^{2}\int_0^{(u-1)_+}\int_{\mathbb{R}^d}\left(\int_0^1 p_{t_1}^{(j)}(y,z)\,dt_1 + \int_1^{u-s} p_{t_1}^{(j)}(y,z)\,dt_1\right)\\
&\qquad\cdot\left(\int_0^1 p_{t_2}^{(j)}(y,\zeta)\,dt_2 + \int_1^{u-s} p_{t_2}^{(j)}(y,\zeta)\,dt_2\right)dy\,ds\\
&\qquad+\sum_{j=1}^{2}\int_{(u-1)_+}^{u}\int_{\mathbb{R}^d}\left(\int_0^{u-s} p_{t_1}^{(j)}(y,z)\,dt_1\right)\left(\int_0^{u-s} p_{t_2}^{(j)}(y,\zeta)\,dt_2\right)dy\,ds.
\end{aligned}
$$

It follows that

$$
\begin{aligned}
&H((z,i),(\zeta,l))\\
&\quad\le \sum_{j=1}^{2}\Bigg(\int_0^{(u-1)_+}\int_0^1\int_0^1\int_{\mathbb{R}^d} p_{t_1}^{(j)}(w,z)p_{t_2}^{(j)}(w,\zeta)\,dw\,dt_1\,dt_2\,ds_1\\
&\qquad+\int_0^{(u-1)_+}\int_1^{u-s_1}\int_0^1\int_{\mathbb{R}^d} p_{t_1}^{(j)}(w,z)p_{t_2}^{(j)}(w,\zeta)\,dw\,dt_1\,dt_2\,ds_1\\
&\qquad+\int_0^{(u-1)_+}\int_0^1\int_1^{u-s_1}\int_{\mathbb{R}^d} p_{t_1}^{(j)}(w,z)p_{t_2}^{(j)}(w,\zeta)\,dw\,dt_1\,dt_2\,ds_1\\
&\qquad+\int_0^{(u-1)_+}\int_1^{u-s_1}\int_1^{u-s_1}\int_{\mathbb{R}^d} p_{t_1}^{(j)}(w,z)p_{t_2}^{(j)}(w,\zeta)\,dw\,dt_1\,dt_2\,ds_1\\
&\qquad+\int_{(u-1)_+}^{u}\int_0^{u-s_1}\int_0^{u-s_1}\int_{\mathbb{R}^d} p_{t_1}^{(j)}(w,z)p_{t_2}^{(j)}(w,\zeta)\,dw\,dt_1\,dt_2\,ds_1\Bigg).
\end{aligned}
$$

From the above estimates we obtain

$$
\begin{aligned}
&\int_{\mathbb{R}^d}\int_{\mathbb{R}^d} G(s,(j,y);(i,z))G(s,(j,y);(l,\zeta))H((i,z),(l,\zeta))\,dz\,d\zeta\\
&\quad\le \sum_{j=1}^{2}\int_{\mathbb{R}^d}\int_{\mathbb{R}^d}\left(\int_0^{(u-s)\wedge 1} + \int_{(u-s)\wedge 1}^{u-s}\right) p_{t_3}^{(j)}(y,z)\,dt_3
\end{aligned}
$$

$$
\begin{aligned}
&\cdot\left(\int_0^{(u-s)\wedge 1}+\int_{(u-s)\wedge 1}^{u-s}\right)p_{t_4}^{(j)}(y,\zeta)\,dt_4\\
&\cdot\left(\int_0^{(u-1)_+}\int_0^1\int_0^1\int_{\mathbb{R}^d}p_{t_1}^{(j)}(w,z)p_{t_2}^{(j)}(w,\zeta)\,dw\,dt_1\,dt_2\,ds_1\right.\\
&+\int_0^{(u-1)_+}\int_1^{u-s_1}\int_0^1\int_{\mathbb{R}^d}p_{t_1}^{(j)}(w,z)p_{t_2}^{(j)}(w,\zeta)\,dw\,dt_1\,dt_2\,ds_1\\
&+\int_0^{(u-1)_+}\int_0^1\int_1^{u-s_1}\int_{\mathbb{R}^d}p_{t_1}^{(j)}(w,z)p_{t_2}^{(j)}(w,\zeta)\,dw\,dt_1\,dt_2\,ds_1\\
&+\int_0^{(u-1)_+}\int_1^{u-s_1}\int_1^{u-s_1}\int_{\mathbb{R}^d}p_{t_1}^{(j)}(w,z)p_{t_2}^{(j)}(w,\zeta)\,dw\,dt_1\,dt_2\,ds_1\\
&\left.+\int_{(u-1)_+}^{u}\int_0^{u-s_1}\int_0^{u-s_1}\int_{\mathbb{R}^d}p_{t_1}^{(j)}(w,z)p_{t_2}^{(j)}(w,\zeta)\,dw\,dt_1\,dt_2\,ds_1\right)dz\,d\zeta\\
=\;&\sum_{j=1}^{2}\int_{\mathbb{R}^d}\int_{\mathbb{R}^d}\left(\int_0^{(u-s)\wedge 1}p_{t_3}^{(j)}(y,z)dt_3\int_0^{(u-s)\wedge 1}p_{t_4}^{(j)}(y,\zeta)\,dt_4\right.\\
&+\int_0^{(u-s)\wedge 1}p_{t_3}^{(j)}(y,z)\,dt_3\int_{(u-s)\wedge 1}^{u-s}p_{t_4}^{(j)}(y,\zeta)\,dt_4\\
&+\int_{(u-s)\wedge 1}^{u-s}p_{t_3}^{(j)}(y,z)\,dt_3\int_0^{(u-s)\wedge 1}p_{t_4}^{(j)}(y,\zeta)\,dt_4\\
&\left.+\int_{(u-s)\wedge 1}^{u-s}p_{t_3}^{(j)}(y,z)\,dt_3\int_{(u-s)\wedge 1}^{u-s}p_{t_4}^{(j)}(y,\zeta)\,dt_4\right)\\
&\cdot\left(\int_0^{(u-1)_+}\int_0^1\int_0^1\int_{\mathbb{R}^d}p_{t_1}^{(j)}(w,z)p_{t_2}^{(j)}(w,\zeta)\,dw\,dt_1\,dt_2\,ds_1\right.\\
&+\int_0^{(u-1)_+}\int_1^{u-s_1}\int_0^1\int_{\mathbb{R}^d}p_{t_1}^{(j)}(w,z)p_{t_2}^{(j)}(w,\zeta)\,dw\,dt_1\,dt_2\,ds_1\\
&+\int_0^{(u-1)_+}\int_0^1\int_1^{u-s_1}\int_{\mathbb{R}^d}p_{t_1}^{(j)}(w,z)p_{t_2}^{(j)}(w,\zeta)\,dw\,dt_1\,dt_2\,ds_1\\
&+\int_0^{(u-1)_+}\int_1^{u-s_1}\int_1^{u-s_1}\int_{\mathbb{R}^d}p_{t_1}^{(j)}(w,z)p_{t_2}^{(j)}(w,\zeta)\,dw\,dt_1\,dt_2\,ds_1\\
&\left.+\int_{(u-1)_+}^{u}\int_0^{u-s_1}\int_0^{u-s_1}\int_{\mathbb{R}^d}p_{t_1}^{(j)}(w,z)p_{t_2}^{(j)}(w,\zeta)\,dw\,dt_1\,dt_2\,ds_1\right)dz\,d\zeta\\
=\;&\sum_{j=1}^{2}\left(\int_0^{(u-1)_+}\int_0^{(u-s)\wedge 1}\int_0^{(u-s)\wedge 1}\int_0^1\int_0^1\int_{\mathbb{R}^d}\int_{\mathbb{R}^d}\int_{\mathbb{R}^d}\right.\\
&+\int_0^{(u-1)_+}\int_0^{(u-s)\wedge 1}\int_0^{(u-s)\wedge 1}\int_1^{u-s_1}\int_0^1\int_{\mathbb{R}^d}\int_{\mathbb{R}^d}\int_{\mathbb{R}^d}\\
&+\int_0^{(u-1)_+}\int_0^{(u-s)\wedge 1}\int_0^{(u-s)\wedge 1}\int_0^1\int_1^{u-s_1}\int_{\mathbb{R}^d}\int_{\mathbb{R}^d}\int_{\mathbb{R}^d}
\end{aligned}
$$

$$
\begin{aligned}
&+\int_0^{(u-1)_+}\int_0^{(u-s)\wedge 1}\int_0^{(u-s)\wedge 1}\int_1^{u-s_1}\int_1^{u-s_1}\int_{\mathbb{R}^d}\int_{\mathbb{R}^d}\int_{\mathbb{R}^d}\\
&+\int_{(u-1)_+}^{u}\int_0^{(u-s)\wedge 1}\int_0^{(u-s)\wedge 1}\int_0^{u-s_1}\int_0^{u-s_1}\int_{\mathbb{R}^d}\int_{\mathbb{R}^d}\int_{\mathbb{R}^d}\\
&+\int_0^{(u-1)_+}\int_{(u-s)\wedge 1}^{u-s}\int_0^{(u-s)\wedge 1}\int_0^{1}\int_0^{1}\int_{\mathbb{R}^d}\int_{\mathbb{R}^d}\int_{\mathbb{R}^d}\\
&+\int_0^{(u-1)_+}\int_{(u-s)\wedge 1}^{u-s}\int_0^{(u-s)\wedge 1}\int_1^{u-s_1}\int_0^{1}\int_{\mathbb{R}^d}\int_{\mathbb{R}^d}\int_{\mathbb{R}^d}\\
&+\int_0^{(u-1)_+}\int_{(u-s)\wedge 1}^{u-s}\int_0^{(u-s)\wedge 1}\int_0^{1}\int_1^{u-s_1}\int_{\mathbb{R}^d}\int_{\mathbb{R}^d}\int_{\mathbb{R}^d}\\
&+\int_0^{(u-1)_+}\int_{(u-s)\wedge 1}^{u-s}\int_0^{(u-s)\wedge 1}\int_1^{u-s_1}\int_1^{u-s_1}\int_{\mathbb{R}^d}\int_{\mathbb{R}^d}\int_{\mathbb{R}^d}\\
&+\int_{(u-1)_+}^{u}\int_{(u-s)\wedge 1}^{u-s}\int_0^{(u-s)\wedge 1}\int_0^{u-s_1}\int_0^{u-s_1}\int_{\mathbb{R}^d}\int_{\mathbb{R}^d}\int_{\mathbb{R}^d}\\
&+\int_0^{(u-1)_+}\int_0^{(u-s)\wedge 1}\int_{(u-s)\wedge 1}^{u-s}\int_0^{1}\int_0^{1}\int_{\mathbb{R}^d}\int_{\mathbb{R}^d}\int_{\mathbb{R}^d}\\
&+\int_0^{(u-1)_+}\int_0^{(u-s)\wedge 1}\int_{(u-s)\wedge 1}^{u-s}\int_1^{u-s_1}\int_0^{1}\int_{\mathbb{R}^d}\int_{\mathbb{R}^d}\int_{\mathbb{R}^d}\\
&+\int_0^{(u-1)_+}\int_0^{(u-s)\wedge 1}\int_{(u-s)\wedge 1}^{u-s}\int_0^{1}\int_1^{u-s_1}\int_{\mathbb{R}^d}\int_{\mathbb{R}^d}\int_{\mathbb{R}^d}\\
&+\int_0^{(u-1)_+}\int_0^{(u-s)\wedge 1}\int_{(u-s)\wedge 1}^{u-s}\int_1^{u-s_1}\int_1^{u-s_1}\int_{\mathbb{R}^d}\int_{\mathbb{R}^d}\int_{\mathbb{R}^d}\\
&+\int_{(u-1)_+}^{u}\int_0^{(u-s)\wedge 1}\int_{(u-s)\wedge 1}^{u-s}\int_0^{u-s_1}\int_0^{u-s_1}\int_{\mathbb{R}^d}\int_{\mathbb{R}^d}\int_{\mathbb{R}^d}\\
&+\int_0^{(u-1)_+}\int_{(u-s)\wedge 1}^{u-s}\int_{(u-s)\wedge 1}^{u-s}\int_0^{1}\int_0^{1}\int_{\mathbb{R}^d}\int_{\mathbb{R}^d}\int_{\mathbb{R}^d}\\
&+\int_0^{(u-1)_+}\int_{(u-s)\wedge 1}^{u-s}\int_{(u-s)\wedge 1}^{u-s}\int_0^{1}\int_0^{1}\int_{\mathbb{R}^d}\int_{\mathbb{R}^d}\int_{\mathbb{R}^d}\\
&+\int_0^{(u-1)_+}\int_{(u-s)\wedge 1}^{u-s}\int_{(u-s)\wedge 1}^{u-s}\int_0^{1}\int_1^{u-s_1}\int_{\mathbb{R}^d}\int_{\mathbb{R}^d}\int_{\mathbb{R}^d}\\
&+\int_0^{(u-1)_+}\int_{(u-s)\wedge 1}^{u-s}\int_{(u-s)\wedge 1}^{u-s}\int_1^{u-s_1}\int_1^{u-s_1}\int_{\mathbb{R}^d}\int_{\mathbb{R}^d}\int_{\mathbb{R}^d}\\
&+\int_{(u-1)_+}^{u}\int_{(u-s)\wedge 1}^{u-s}\int_{(u-s)\wedge 1}^{u-s}\int_0^{u-s_1}\int_0^{u-s_1}\int_{\mathbb{R}^d}\int_{\mathbb{R}^d}\int_{\mathbb{R}^d}\Bigg)
\end{aligned}
$$

$$
\begin{aligned}
& \cdot p_{t_1}^{(j)}(w,z)p_{t_2}^{(j)}(w,\zeta)p_{t_3}^{(j)}(y,z)p_{t_4}^{(j)}(y,\zeta)\,dw\,dz\,d\zeta\,dt_1\,dt_2\,dt_3\,dt_4\,ds_1 \\
=: & \sum_{j=1}^{2}\sum_{i=1}^{20} I_i(j).
\end{aligned}
$$

We will show how to estimate from above the integral $I_1(j)$; the remaining integrals can be estimated in a similar way. Applying repeatedly (2.5) renders

$$
\begin{aligned}
I_1(j) \;\le\; & 16K^4 \int_0^{(u-1)_+}\int_0^{(u-s)\wedge 1}\int_0^{(u-s)\wedge 1}\int_0^1\int_0^1\int_{\mathbb{R}^d}\int_{\mathbb{R}^d}\int_{\mathbb{R}^d} \\
& \cdot\left(q^{(1)}_{t_1^{\alpha_1/\alpha_j}}(w,z) + \int_0^{t_1} q^{(1)}_{r_1+(t_1-r_1)^{\alpha_1/\alpha_2}}(w,z)\,\theta^j_{t_1}(dr_1)\right) \\
& \cdot\left(q^{(1)}_{t_2^{\alpha_1/\alpha_j}}(w,\zeta) + \int_0^{t_2} q^{(1)}_{r_2+(t_2-r_2)^{\alpha_1/\alpha_2}}(w,\zeta)\,\theta^j_{t_2}(dr_2)\right) \\
& \cdot\left(q^{(1)}_{t_3^{\alpha_1/\alpha_j}}(y,z) + \int_0^{t_3} q^{(1)}_{r_3+(t_3-r_3)^{\alpha_1/\alpha_2}}(y,z)\,\theta^j_{t_3}(dr_3)\right) \\
& \cdot\left(q^{(1)}_{t_4^{\alpha_1/\alpha_j}}(y,\zeta) + \int_0^{t_4} q^{(1)}_{r_4+(t_4-r_4)^{\alpha_1/\alpha_2}}(y,\zeta)\,\theta^j_{t_4}(dr_4)\right) \\
& \cdot dw\,dz\,d\zeta\,dt_1\,dt_2\,dt_3\,dt_4\,ds_1.
\end{aligned}
$$

Using Chapman-Kolmogorov's equation, the scaling property of stable densities, and that $t \le t^\alpha$ for $0 < t \le 1$ and $\alpha > 0$, we obtain

$$
\begin{aligned}
I_1(j) \;\le\; & 16K^4 q_1^{(1)}(0) \int_0^{(u-1)_+}\int_0^{(u-s)\wedge 1}\int_0^{(u-s)\wedge 1} \\
& \cdot\int_0^1\int_0^1 \Big( (t_1^{\alpha_1/\alpha_j} + t_2^{\alpha_1/\alpha_j} + t_3^{\alpha_1/\alpha_j} + t_4^{\alpha_1/\alpha_j})^{-d/\alpha_1} \\
& + \int_0^{t_4} (t_1^{\alpha_1/\alpha_j} + t_2^{\alpha_1/\alpha_j} + t_3^{\alpha_1/\alpha_j} + r_4 + (t_4-r_4)^{\alpha_1/\alpha_2})^{-d/\alpha_1}\theta^j_{t_4}(dr_4) \\
& + \int_0^{t_3} (t_1^{\alpha_1/\alpha_j} + t_2^{\alpha_1/\alpha_j} + r_3 + (t_3-r_3)^{\alpha_1/\alpha_2} + t_4^{\alpha_1/\alpha_j})^{-d/\alpha_1}\theta^j_{t_3}(dr_3) \\
& + \int_0^{t_4}\int_0^{t_3} (t_1^{\alpha_1/\alpha_j} + t_2^{\alpha_1/\alpha_j} + r_3 + (t_3-r_3)^{\alpha_1/\alpha_2} + r_4 + (t_4-r_4)^{\alpha_1/\alpha_2})^{-d/\alpha_1} \\
& \cdot\theta^j_{t_3}(dr_3)\theta^j_{t_4}(dr_4) \\
& + \int_0^{t_2} (t_1^{\alpha_1/\alpha_j} + r_2 + (t_2-r_2)^{\alpha_1/\alpha_2} + t_3^{\alpha_1/\alpha_j} + t_4^{\alpha_1/\alpha_j})^{-d/\alpha_1}\theta^j_{t_2}(dr_2) \\
& + \int_0^{t_4}\int_0^{t_2} (t_1^{\alpha_1/\alpha_j} + r_2 + (t_2-r_2)^{\alpha_1/\alpha_2} + t_3^{\alpha_1/\alpha_j} + r_4 + (t_4-r_4)^{\alpha_1/\alpha_2})^{-d/\alpha_1} \\
& \cdot\theta^j_{t_2}(dr_2)\theta^j_{t_4}(dr_4) \\
& + \int_0^{t_3}\int_0^{t_2} (t_1^{\alpha_1/\alpha_j} + r_2 + (t_2-r_2)^{\alpha_1/\alpha_2} + r_3 + (t_3-r_3)^{\alpha_1/\alpha_2} + t_4^{\alpha_1/\alpha_j})^{-d/\alpha_1} \\
& \cdot\theta^j_{t_2}(dr_2)\theta^j_{t_3}(dr_3)
\end{aligned}
$$

$$
\begin{aligned}
&+\int_0^{t_4}\int_0^{t_3}\int_0^{t_2}(t_1^{\alpha_1/\alpha_j}+r_2+(t_2-r_2)^{\alpha_1/\alpha_2}+r_3+(t_3-r_3)^{\alpha_1/\alpha_2}\\
&+r_4+(t_4-r_4)^{\alpha_1/\alpha_2})^{-d/\alpha_1}\theta^j_{t_2}(dr_2)\theta^j_{t_3}(dr_3)\theta^j_{t_4}(dr_4)\\
&+\int_0^{t_1}(r_1+(t_1-r_1)^{\alpha_1/\alpha_2}+t_2^{\alpha_1/\alpha_j}+t_3^{\alpha_1/\alpha_j}+t_4^{\alpha_1/\alpha_j})^{-d/\alpha_1}\theta^j_{t_1}(dr_1)\\
&+\int_0^{t_4}\int_0^{t_1}(r_1+(t_1-r_1)^{\alpha_1/\alpha_2}+t_2^{\alpha_1/\alpha_j}+t_3^{\alpha_1/\alpha_j}+r_4+(t_4-r_4)^{\alpha_1/\alpha_2})^{-d/\alpha_1}\\
&\cdot\theta^j_{t_1}(dr_1)\theta^j_{t_4}(dr_4)\\
&+\int_0^{t_3}\int_0^{t_1}(r_1+(t_1-r_1)^{\alpha_1/\alpha_2}+t_2^{\alpha_1/\alpha_j}+r_3+(t_3-r_3)^{\alpha_1/\alpha_2}+t_4^{\alpha_1/\alpha_j})^{-d/\alpha_1}\\
&\cdot\theta^j_{t_1}(dr_1)\theta^j_{t_3}(dr_3)\\
&+\int_0^{t_4}\int_0^{t_3}\int_0^{t_1}(r_1+(t_1-r_1)^{\alpha_1/\alpha_2}+t_2^{\alpha_1/\alpha_j}+r_3+(t_3-r_3)^{\alpha_1/\alpha_2}\\
&+r_4+(t_4-r_4)^{\alpha_1/\alpha_2})^{-d/\alpha_1}\theta^j_{t_1}(dr_1)\theta^j_{t_3}(dr_3)\theta^j_{t_4}(dr_4)\\
&+\int_0^{t_2}\int_0^{t_1}(r_1+(t_1-r_1)^{\alpha_1/\alpha_2}+r_2+(t_2-r_2)^{\alpha_1/\alpha_2}+t_3^{\alpha_1/\alpha_j}+t_4^{\alpha_1/\alpha_j})^{-d/\alpha_1}\\
&\cdot\theta^j_{t_1}(dr_1)\theta^j_{t_2}(dr_2)\\
&+\int_0^{t_4}\int_0^{t_2}\int_0^{t_1}(r_1+(t_1-r_1)^{\alpha_1/\alpha_2}+r_2+(t_2-r_2)^{\alpha_1/\alpha_2}+t_3^{\alpha_1/\alpha_j}\\
&+r_4+(t_4-r_4)^{\alpha_1/\alpha_2})^{-d/\alpha_1}\theta^j_{t_1}(dr_1)\theta^j_{t_2}(dr_2)\theta^j_{t_4}(dr_4)\\
&+\int_0^{t_3}\int_0^{t_2}\int_0^{t_1}(r_1+(t_1-r_1)^{\alpha_1/\alpha_2}+r_2+(t_2-r_2)^{\alpha_1/\alpha_2}+r_3\\
&+(t_3-r_3)^{\alpha_1/\alpha_2}+t_4^{\alpha_1/\alpha_j})^{-d/\alpha_1}\theta^j_{t_1}(dr_1)\theta^j_{t_2}(dr_2)\theta^j_{t_3}(dr_3)\\
&+\int_0^{t_4}\int_0^{t_3}\int_0^{t_2}\int_0^{t_1}(r_1+(t_1-r_1)^{\alpha_1/\alpha_2}+r_2+(t_2-r_2)^{\alpha_1/\alpha_2}+r_3\\
&+(t_3-r_3)^{\alpha_1/\alpha_2}+r_4+(t_4-r_4)^{\alpha_1/\alpha_2})^{-d/\alpha_1}\\
&\cdot\theta^j_{t_1}(dr_1)\theta^j_{t_2}(dr_2)\theta^j_{t_3}(dr_3)\cdot\theta^j_{t_4}(dr_4))dt_1dt_2dt_3dt_4ds_1\\
\leq\;& 16K^4q_1^{(1)}(0)\int_0^{(u-1)_+}\int_0^{(u-s)\wedge 1}\int_0^{(u-s)\wedge 1}\int_0^1\int_0^1\Big((t_1+t_2+t_3+t_4)^{-d/\alpha_1}\\
&+\int_0^{t_4}(t_1+t_2+t_3+t_4)^{-d/\alpha_1}\theta^j_{t_4}(dr_4)+\int_0^{t_3}(t_1+t_2+t_3+t_4)^{-d/\alpha_1}\theta^j_{t_3}(dr_3)\\
&+\int_0^{t_4}\int_0^{t_3}(t_1+t_2+t_3+t_4)^{-d/\alpha_1}\theta^j_{t_3}(dr_3)\theta^j_{t_4}(dr_4)\\
&+\int_0^{t_2}(t_1+t_2+t_3+t_4)^{-d/\alpha_1}\theta^j_{t_2}(dr_2)+\int_0^{t_4}\int_0^{t_2}(t_1+t_2+t_3+t_4)^{-d/\alpha_1}
\end{aligned}
$$

$$\begin{aligned}
&\cdot\theta^j_{t_2}(dr_2)\theta^j_{t_4}(dr_4)\\
&+\int_0^{t_3}\int_0^{t_2}(t_1+t_2+t_3+t_4)^{-d/\alpha_1}\theta^j_{t_2}(dr_2)\theta^j_{t_3}(dr_3)\\
&+\int_0^{t_4}\int_0^{t_3}\int_0^{t_2}(t_1+t_2+t_3+t_4)^{-d/\alpha_1}\theta^j_{t_2}(dr_2)\theta^j_{t_3}(dr_3)\theta^j_{t_4}(dr_4)\\
&+\int_0^{t_1}(t_1+t_2+t_3+t_4)^{-d/\alpha_1}\theta^j_{t_1}(dr_1)+\int_0^{t_4}\int_0^{t_1}(t_1+t_2+t_3+t_4)^{-d/\alpha_1}\\
&\cdot\theta^j_{t_1}(dr_1)\theta^j_{t_4}(dr_4)\\
&+\int_0^{t_3}\int_0^{t_1}(t_1+t_2+t_3+t_4)^{-d/\alpha_1}\theta^j_{t_1}(dr_1)\theta^j_{t_3}(dr_3)\\
&+\int_0^{t_4}\int_0^{t_3}\int_0^{t_1}(t_1+t_2+t_3+t_4)^{-d/\alpha_1}\theta^j_{t_1}(dr_1)\theta^j_{t_3}(dr_3)\theta^j_{t_4}(dr_4)\\
&+\int_0^{t_2}\int_0^{t_1}(t_1+t_2+t_3+t_4)^{-d/\alpha_1}\theta^j_{t_1}(dr_1)\theta^j_{t_2}(dr_2)\\
&+\int_0^{t_4}\int_0^{t_2}\int_0^{t_1}(t_1+t_2+t_3+t_4)^{-d/\alpha_1}\theta^j_{t_1}(dr_1)\theta^j_{t_2}(dr_2)\theta^j_{t_4}(dr_4)\\
&+\int_0^{t_3}\int_0^{t_2}\int_0^{t_1}(t_1+t_2+t_3+t_4)^{-d/\alpha_1}\theta^j_{t_1}(dr_1)\theta^j_{t_2}(dr_2)\theta^j_{t_3}(dr_3)\\
&+\int_0^{t_4}\int_0^{t_3}\int_0^{t_2}\int_0^{t_1}(t_1+t_2+t_3+t_4)^{-d/\alpha_1}\theta^j_{t_1}(dr_1)\theta^j_{t_2}(dr_2)\theta^j_{t_3}(dr_3)\cdot\theta^j_{t_4}(dr_4))\\
\leq\;& 256K^4q_1^{(1)}(0)\int_0^{(u-1)_+}\int_0^{(u-s)\wedge 1}\int_0^{(u-s)\wedge 1}\int_0^1\int_0^1(t_1+t_2+t_3+t_4)^{-d/\alpha_1}\\
&\cdot dt_1dt_2dt_3dt_4ds_1.
\end{aligned}$$

Estimating in the same way the remaining integrals $I_i(j)$ renders

$$\begin{aligned}
&\int_{\mathbb{R}^d}\int_{\mathbb{R}^d}G(s,(j,y);(i,z))G(s,(j,y);(l,\zeta))H((i,z),(l,\zeta))\,dz\,d\zeta\\
&\leq \text{Const.}\int_0^u\int_0^{u-s}\int_0^{u-s}\int_0^{u-s_1}\int_0^{u-s_1}(t_1+t_2+t_3+t_4)^{-d/\alpha_1}\\
&\quad\cdot\,dt_1\,dt_2\,dt_3\,dt_4\,ds_1.
\end{aligned}$$

Finally, let us show that if $d<4\alpha_1$ then the right-hand side of the last inequality is finite. Indeed,

$$\begin{aligned}
&\int_0^u\int_0^{u-s}\int_0^{u-s}\int_0^{u-s_1}\int_0^{u-s_1}(t_1+t_2+t_3+t_4)^{-d/\alpha_1}\,dt_1\,dt_2\,dt_3\,dt_4\,ds_1\\
=\;&\text{Const.}\Big(4(3u-s)^{5-d/\alpha_1}+2(u-s)^{5-d/\alpha_1}+u(2u-2s)^{4-d/\alpha_1}+2u^{5-d/\alpha_1}\\
&-(2u)^{5-d/\alpha_1}-2u(u-s)^{4-d/\alpha_1}-(4u-2s)^{5-d/\alpha_1}-4(2u-s)^{5-d/\alpha_1}\\
&\cdot-(2u-2s)^{5-d/\alpha_1}\Big).
\end{aligned}$$

Hence,

$$\begin{aligned}
&\sup_{0\le s\le u}\int_0^u\int_0^{u-s}\int_0^{u-s}\int_0^{u-s_1}\int_0^{u-s_1}(t_1+t_2+t_3+t_4)^{-d/\alpha_1}\,dt_1\,dt_2\,dt_3\,dt_4\,ds_1\\
&\le \text{Const.}\left(4(3u)^{5-d/\alpha_1}+2u^{5-d/\alpha_1}+u(2u)^{4-d/\alpha_1}+2u^{5-d/\alpha_1}\right)\\
&< \infty
\end{aligned}$$

if $4-d/\alpha_1>0$. □

**Acknowledgement** The problem developed here was raised by L. Gorostiza in the Probability Seminar at CINVESTAV in the fall of 1997. The authors wish to thank an anonymous referee for his valuable comments that contributed to improve the presentation.

## References

[D] E.B. Dynkin, *Representation for functionals of superprocesses by multiple stochastic integrals, with applications to self–intersection local times,* Colloque Paul Lèvy sur les Processus Stochastiques (Palaiseau, 1987), Astérisque **157–158** (1988), pp. 147–171.

[F-I] R. E. Feldman and S. K. Iyer, *A representation for functionals of superprocesses via particle picture,* Stochastic Process. Appl. **64**, no. 2 (1996), pp. 173–186.

[G-L] L. G. Gorostiza and J. A. López-Mimbela, *The multitype measure branching process,* Adv. in Appl. Probab. **22**, no. 1 (1990), pp. 49–67.

[G-R-W] L. G. Gorostiza, S. Roelly and A. Wakolbinger, *Persistence of critical multitype particle and measure branching processes,* Probab. Theory Related Fields **92** (1992), pp. 313–335.

[L] J. A. López-Mimbela, *Fluctuation limits of multitype branching random fields,* J. Multivariate Anal. **40**, no. 1 (1992), pp. 56–83.

[R] J. Rosen, *Renormalization and limit theorems for self-intersections of superprocesses,* Ann. Probab. **20**, no. 3 (1992), pp. 1341-1368.

[V] J. Villa, *Representaciones tipo fórmula de Tanaka del tiempo local de superprocesos,* doctoral thesis. Centro de Investigación en Matemáticas, (2002).

Centro de Investigación en Matemáticas, Apartado Postal 402, 36000 Guanajuato, Mexico.

*E-mail address*: jalfredo@cimat.mx

Departamento de Matemáticas y Física, Universidad Autónoma de Aguascalientes, Mayapán 123, Colinas del Sol, Jesús María, 20900 Aguascalientes, Mexico.

*E-mail address*: villa@cimat.mx

Contemporary Mathematics
Volume **336**, 2003

# Lévy Processes in Banach Spaces: Distributional Properties and Subordination

Víctor Pérez-Abreu and Alfonso Rocha-Arteaga

ABSTRACT. The paper deals with subordination of Banach space valued Lévy processes via a random time change by a one dimensional subordinator. Distributional properties of Banach valued Lévy processes that are needed for this purpose are first derived.

## 1. Introduction

Lévy processes have received increasing attention in recent years, both from the applied and the theoretical point of view; see for example the books by Bertoin (1996) [**B**] and Sato (1999) [**Sa**] as well as the Proceedings of the two recent MaPhySto Conferences on Lévy Processes: Theory and Applications, see Barndorff-Nielsen et al (2001) [**B-M-R**].

Lévy processes taking values in separable Banach spaces have been studied by several authors. Gihman and Skorohod (1975) [**G-S**] derived the corresponding Lévy-Khintchine formula for the characteristic functional of such processes, while Dettweiler (1982) [**D**] and Albeverio and Rudiger (2002) [**A-R**] studied their associated stochastic integrals and the corresponding Lévy-Ito decomposition. The purpose of the present paper is to study distributional properties of two kind of transformations of Banach-valued Lévy processes that are still Lévy processes: linear transformations and subordination via a random time change by a one dimensional increasing Lévy process. In both cases we identified the generating triplet in an explicit manner. As a key tool for the subordination case, we first derive distributional properties that are of interest in their own, such as conditions for the existence of g-moments as well as tail and moment inequalities for Banach space valued Lévy processes.

Section 2 of this paper recalls basic facts about Banach space valued Lévy processes that are used in the sequel. In Section 3 we prove an equivalence between finiteness of g-moments of the processes and g-moments of the corresponding Lévy measures. We also derive some useful tail and moment inequalities for Banach space

2000 *Mathematics Subject Classification.* Primary 60E07, 60G51; Secondary 28C20.
*Key words and phrases.* Infinitely Divisible Law, Random Time Change.
The work of Alfonso Rocha-Arteaga was supported by a Fellowship from PROMEP grant UASIN-236.

Lévy processes which are used in Section 4 to obtain the characteristic triplet of the Banach space valued Lévy process subordinated to a one dimensional subordinator.

## 2. Preliminaries and notation

Let $B$ be a real separable Banach space with norm $\|\cdot\|$ and strong dual $B^*$. A *Lévy process* $\{X_t : t \geq 0\}$ with values in $B$ is a stochastic process defined on a probability space $(\Omega, \mathcal{F}, P)$ satisfying the following properties:
i) For any $0 \leq t_1 < t_2 < ... < t_n$, the random variables $X_{t_2} - X_{t_1}, ...., X_{t_n} - X_{t_{n-1}}$ are independent (*independent increments*).
ii) For any $s \geq 0, t \geq 0$, $X_{t+s} - X_s$ and $X_t$ have the same law (*stationary increments*).
iii) Almost surely, $X_0 = 0$.
iv) It is *stochastically continuous*, i.e., for every $\varepsilon > 0$, $P(\|X_t - X_s\| > \varepsilon) \to 0$ as $s \to t$.
v) Almost surely, it is *right-continuous* in $t \geq 0$ and has *left limits* in $t > 0$.

The structure of these processes is similar as for the finite dimensional case. Gihman and Skorohod (1975) [**G-S**] decompose the process into a sum of a non-random, discrete and stochastically continuous parts. As usual, denote $X_{s-} := \lim_{s \uparrow t} X_s$. Let $\mathcal{B}_\varepsilon$ be the ring of Borel sets of $B$ whose distance from at point 0 is at least $\varepsilon$. Define for any $A$ in $\mathcal{B}_\varepsilon$,

$$N_t(A) = \textstyle\sum_{s<t} 1_A(X_s - X_{s-}) \quad \text{and}$$
$$X_t^A = \textstyle\sum_{s<t} (X_s - X_{s-}) 1_A(X_s - X_{s-}),$$

that is, the number of jumps and the sum of jumps of the process which have occurred up to time $t$ and took place in $A$, respectively. They are well defined processes.

For any continuous linear functional $f$ on $B$, the one dimensional process $\{f(X_t)\}$ is stochastically continuous and has independent increments. This fact supplies a useful non trivial information of the Banach-valued Lévy process $\{X_t\}$ to obtain the following Lévy-Khintchine representation from Gihman and Skorohod (1975, Th. IV.5) [**G-S**].

THEOREM 2.1. *Let $\{X_t : t \geq 0\}$ be a Lévy process on a separable Banach space $B$. Then, its characteristic functional $Ee^{if(X_t)}$ has the form*

$$\text{(2.1)} \quad exp\left\{t\left(-\frac{1}{2}A(f,f) + if(\gamma) + \int \left[e^{if(x)} - 1 - if(x)1_{\{\|x\|\leq 1\}}(x)\right]\nu(dx)\right)\right\},$$

*for $f \in B^*$, where $\gamma$ is in $B$, $A(f,f)$ is a nonnegative quadratic functional in $f$, $\nu(A)$ is a finite measure in $A \in \mathcal{B}_\varepsilon$ for all $\varepsilon > 0$, such that for any continuous linear functional $f$*

$$\text{(2.2)} \qquad \int_{\|x\|\leq 1} f^2(x)\,\nu(dx) < \infty.$$

REMARK 2.2. a) For a fixed $A$ in $\mathcal{B}_\varepsilon$ with $\varepsilon > 0$, $N_t(A)$ is a Poisson process. For a fixed $t \geq 0$, $A \longmapsto N_t(A)$ is a Poisson random measure in $A \in \mathcal{B}_\varepsilon$, for all $\varepsilon > 0$, with intensity measure given by the product of Lebesgue measure on $[0,\infty)$ with $\nu$, i.e., $EN_t(A) = t\nu(A)$. Moreover, $\nu$ has a unique $\sigma$-finite extension to the $\sigma$-algebra $\mathcal{B}(B\backslash\{0\})$ containing the ring $\mathcal{B}_\varepsilon$ for all $\varepsilon > 0$ (see Albeverio and Rudiger (2002)), [**A-R**].

b) Let $\triangle_\varepsilon = \{\|x\| > \varepsilon\}$. The process $X_t^\varepsilon = X_t - X_t^{\triangle_\varepsilon}$ has moments of any order. The centered process $X_t^\varepsilon - EX_t^\varepsilon$ converges as $\varepsilon \downarrow 0$ to a continuous process with stationary e independent increments $X_t^G$. The process $X_t^G$ is Gaussian, that is, $f(X_t^G)$ is a one dimensional Gaussian process for any continuous linear functional.

c) $A(f,f)$ is a quadratic functional, i.e., $A$ is a symmetric nonnegative operator from $B^*$ into $B$ and $A(f,f)$ denotes a dual pair, that is, $(Af,f) = f(Af)$.

d) The triplet of parameters $(A, \nu, \gamma)$ in (1) is called the generating triplet of the Lévy process $\{X_t\}$.

e) This triplet is unique, recalling that for an infinite dimensional Banach space $B$, a $\sigma$-finite measure $\nu$ on $B$ with $\nu(\{0\}) = 0$ is a *Lévy measure* if the function

$$f \longmapsto \exp\{\int \left[e^{if(x)} - 1 - if(x)1_{\{\|x\|\le 1\}}(x)\right] \nu(dx)\} \qquad f \in B^*, \tag{2.3}$$

is characteristic functional of a probability measure on $B$ (see Araujo and Giné (1980) [**A-G**] and Linde (1986) [**L**]).

f) Dettweiler (1982) [**D**] has shown (see also Albeverio and Rudiger (2002) [**A-R**]) that there exist independent Lévy processes $\{X_t^0\}$ and $\{X_t^1\}$ with generating triplets $(A, \nu_0, \gamma)$ and $(0, \nu_1, 0)$ such that $\{X_t\} \overset{d}{=} \{X_t^0 + X_t^1\}$, where $\nu_0$ and $\nu_1$ are the restrictions of $\nu$ to the sets $\{\|x\| \le 1\}$ and $\{\|x\| > 1\}$ respectively.

From Theorem IV.8 of Gihman and Skorohod (1975) [**G-S**] we obtain the next result for bounded variation Banach space valued Lévy processes.

PROPOSITION 2.3. *Let $\{Z_t : t \ge 0\}$ be a $B$-valued Lévy process. Then, $\{Z_t\}$ has bounded variation on each interval $[0,t]$, with probability 1, if and only if, it has characteristic functional given by*

$$Ee^{if(Z_t)} = exp\left\{t\left(\int_B \left(e^{if(x)} - 1\right)\nu(dx) + if(\gamma)\right)\right\} \qquad f \in B^*,$$

*where $\gamma \in B$ and the Lévy measure $\nu$ satisfies*

$$\int_{0<\|x\|\le 1} \|x\|\, \nu\,(dx) < \infty. \tag{2.4}$$

*Moreover, $s(t) = var_{0\le s\le t}\, Z(s)$ is a Lévy process in $\mathbb{R}$ and*

$$Ee^{irs(t)} = exp\left\{t\left(\int_B \left(e^{ir\|x\|} - 1\right)\nu(dx) + ir\,\|\gamma\|\right)\right\} \qquad r \in \mathbb{R}.$$

## 3. Distributional Properties

In this section we prove some useful distributional properties of Lévy processes. We begin by proving that general linear transformations of Lévy processes are still Lévy processes and it is possible to describe their generating triplets. The case of linear transformations of Lévy processes on $\mathbb{R}^d$ is in Sato (1999, Th. 11.10) [**Sa**].

PROPOSITION 3.1. *Let $\{X_t : t \ge 0\}$ be a $B$-valued Lévy process with $(A, \nu, \gamma)$. Let $B_1$ be a Banach space such that the map $V \longmapsto f_V$ is an isomorphism of $B_1$ onto $B^*$. Let $T : B \to B$ and let $T' : B_1 \to B_1$ be continuous linear transformations with the property*

$$f_V(TS) = f_{T'V}(S), \tag{3.1}$$

*for every $V \in B_1$ and $S \in B$. Then $\{T(X_t) : t \geq 0\}$ is a $B$-valued Lévy process with generating triplet $(A_T, \nu_T, \gamma_T)$ given by*

$$\begin{aligned} A_T &= TAT', \\ \nu_T &= (\nu T^{-1}) \mid_{B\setminus\{0\}}, \\ \gamma_T &= T\gamma + \int Tx[1_{\{\|Tx\|\leq 1\}}(Tx) - 1_{\{\|x\|\leq 1\}}(x)]\nu(dx). \end{aligned} \tag{3.2}$$

*Here $\nu T^{-1}(C) = \nu(\{x : Tx \in C\})$ and $(\nu T^{-1}) \mid_{B\setminus\{0\}}$ denotes the restriction of the measure $\nu T^{-1}$ to $B\setminus\{0\}$ and the last integral is a Bochner integral.*

PROOF. From the continuity of $T$ and the fact that $\{X_t\}$ is a Lévy process it is clear that $\{T(X_t)\}$ is also a Lévy process. We now obtain its generating triplet. From (3.1)

$$\begin{aligned} Ee^{if_V(TX_1)} = Ee^{if_{T'V}(X_1)} &= \exp\bigg\{-\frac{1}{2}A(f_{T'V}, f_{T'V}) + if_{T'V}(\gamma) \\ &+ \int \left(e^{if_{T'V}(x)} - 1 - if_{T'V}(x)1_{\{\|x\|\leq 1\}}(x)\right)\nu(dx), \end{aligned}$$

where $A(f_{T'V}, f_{T'V}) = f_{T'V}(Af_{T'V}) = f_V(TAf_{T'V})$ (see Remark 2.2(c)). Using (3.1) the above expression becomes

$$\begin{aligned} \exp\bigg\{&-\frac{1}{2}f_V(TAf_{T'V}) + if_V(T\gamma) \\ &+ i\int f_V(Tx)\left[1_{\{\|Tx\|\leq 1\}}(Tx) - 1_{\{\|x\|\leq 1\}}(x)\right]\nu(dx) \\ &+ \int \left[e^{if_V(y)} - 1 - if_V(y)1_{\{\|y\|\leq 1\}}(y)\right]\nu T^{-1}(dy)\bigg\}. \end{aligned} \tag{3.3}$$

Next, for any $f_V \in B^*$,

$$\int (f_V^2(y) \wedge 1)\,\nu T^{-1}(dy) = \int (f_{T'V}^2(x) \wedge 1)\,\nu(dx) < \infty$$

and therefore the last integral in (3.3) is well defined. Since the infinitely divisible random variable $TX_1$ has the characteristic functional (3.3) in which the last integral is a characteristic functional, then by Remark 2.2(e) we get (3.2). □

As in the finite dimensional case (Sato (1999), Th. 25.3 [**Sa**]), for certain class of functions $g$ defined on a separable Banach space, $g$-moments of Lévy processes are intimately connected with $g$-moments of their Lévy measures. Kruglov ((1970) [**K1**], (1972) [**K2**]) proved that these are equivalent for Lévy processes on $\mathbb{R}$ and even on Hilbert spaces. For Poisson measures on Banach spaces the equivalence is due to de Acosta (1980) [**Ac**]. In Proposition 3.2 below we prove this equivalence for a general Lévy process on a separable Banach space. Let $g : B \to \mathbb{R}^+$ be a bounded function on the unit closed ball. We say that $g$ is a *submultiplicative* function if there exists a constant $a > 0$ such that $g(x + y) \leq ag(x)g(y)$ for every $x \in B$ and $y \in B$.

PROPOSITION 3.2. *Let $\{X_t : t \geq 0\}$ be a Lévy process on a separable Banach space $B$ with triplet $(A, \nu, \gamma)$ and let $g$ be a submultiplicative function. Then*

$$Eg(X_t) < \infty \quad \text{for every } t \qquad \text{if and only if} \qquad \int_{\|x\|>1} g(x)\nu(dx) < \infty.$$

PROOF. Let $b > 1$ such that $sup_{\|x\|\leq 1} |g(x)| \leq b$ and let us choose $n$ such that $n - 1 < \|x\| \leq n$, then

$$
\begin{aligned}
g(x) &= g(\frac{x}{n} + ... + \frac{x}{n}) \leq a^{n-1} g(\frac{x}{n})^n \\
&\leq a^{n-1} b^n \leq b(ab)^{n-1} = be^{c\|x\|},
\end{aligned}
\tag{3.4}
$$

where $c = \log(ab) > 0$ choosing $a > 1$.

Using the decomposition in Remark 2.2 (f) denote by $\mu$, $\mu_0$ and $\mu_1$ the distributions of $X_1$, $X_1^0$ and $X_1^1$ respectively. Assume that $Eg(X_t) < \infty$ for some $t > 0$. Then

$$Eg(X_t) = \int_B g(x)\mu^t(dx) = \int_B \int_B g(x+y)\mu_0^t(dx))\mu_1^t(dy) < \infty,$$

which implies, for some $x \in B$, that $\int_B g(x+y)\mu_1^t(dy) < \infty$ and hence

$$\sum_{n=0}^{\infty} \frac{t^n}{n!} \int_B g(x+y)\nu_1^{*n}(dy) < \infty.$$

In view of (3.4), $g(y) \leq ag(-x)g(x+y) \leq abe^{c\|x\|}g(x+y)$ and

$$\sum_{n=0}^{\infty} \frac{t^n}{n!} \int_B g(y)\nu_1^{*n}(dy) < \infty. \tag{3.5}$$

Then, for $n = 1$, $\int_B g(y)\nu_1(dy) < \infty$.

Reciprocally, assume that $\int_B g(y)\nu_1(dy) < \infty$. By submultiplicativity, we have for every $n$,

$$\int_B g(y)\nu_1^{*n}(dy) \leq a^{n-1} \left\{ \int_B g(y)\nu_1(dy) \right\}^n < \infty.$$

This implies (3.5) and therefore $Eg(X_t^1) < \infty$ for every $t > 0$. It remains to prove that $Eg(X_t^0) < \infty$ for every $t > 0$ since $Eg(X_t) \leq aE\left[g(X_t^0)\right] E\left[g(X_t^1)\right]$. Let $\{G_t^0\}, \{P_t^0\}$ and $\{D_t^0\}$ be homogeneous and independent random processes on $B$, where $\{G_t^0\}$ is Gaussian , $\{P_t^0\}$ is generalized Poisson, i.e., $P_1^0$ has the characteristic functional (2.3) with Lévy measure $\nu_0$ and $\{D_t^0\}$ is a non random function, such that $X_t^0 \stackrel{d}{=} G_t^0 + P_t^0 + D_t^0$. We only need to prove that $E\left[g(G_1^0)\right] E\left[g(P_1^0)\right] E\left[g(D_1^0)\right] < \infty$. For simplicity take $t = 1$. That $E\left[g(P_1^0)\right]$ is finite follows from Corollary 3.3 in de Acosta (1980) [**Ac**] since $\nu_0$ is concentrated in the unit ball of $B$. Now, since $G_1^0$ is a centered Gaussian random variable then $G_1^0 \stackrel{d}{=} \frac{1}{\sqrt{2}}G_1 + \frac{1}{\sqrt{2}}G_2$ where $G_1$ and $G_2$ are independent copies of $G_1^0$ (see Araujo and Giné (1980, Prop.III.6.4) [**A-G**]). Then

$$
\begin{aligned}
Eg(G_1^0) &= Eg(\frac{1}{\sqrt{2}}G_1 + \frac{1}{\sqrt{2}}G_2) \\
&\leq aE\left[g\left(\frac{1}{\sqrt{2}}G_1^0\right)\right]^2 = a\left[Eg\left(\frac{1}{(\sqrt{2})^2}G_1 + \frac{1}{(\sqrt{2})^2}G_2\right)\right]^2 \\
&\leq aa^2\left[\left\{Eg\left(\frac{1}{(\sqrt{2})^2}G_1^0\right)\right\}^2\right]^2 \leq a^7\left[Eg\left(\frac{1}{(\sqrt{2})^3}G_1^0\right)\right]^8 \\
\ldots &\leq a^{2^n-1}\left[Eg\left(\frac{1}{(\sqrt{2})^n}G_1^0\right)\right]^{2^n} \qquad \text{for } n \geq 1.
\end{aligned}
$$

We can choose $\beta > 0$ such that $Ee^{\beta\|G_1^0\|} < \infty$ which is a consequence of Fernique's theorem, see Araujo and Giné (1980, Th. III.6.5) [**A-G**]. Choose $n$ sufficiently large such that $c\frac{1}{(\sqrt{2})^n} < \beta$, then $Eg(G_1^0) < \infty$, which concludes the proof. □

REMARK 3.3. The proof of the "if" part of the above Proposition is on the same lines of the finite dimensional case of Sato (1999, Th. 25.3) [**Sa**].

For exponential and regular moments of Lévy processes we obtain the following conditions for their existence, which are independent of $t$.

COROLLARY 3.4. a) For every $f \in B^*$, $Ee^{f(X_t)} < \infty$ for all $t > 0$ if and only if $\int_{\|x\|>1} e^{f(x)}\nu(dx) < \infty$.

b) Let $\alpha > 0$. Then $Ee^{\alpha\|X_t\|} < \infty$ for all $t > 0$ if and only if $\int_{\|x\|>1} e^{\alpha\|x\|}\nu(dx) < \infty$.

c) Let $\alpha > 0$. Then $E\|X_t\|^\alpha < \infty$ for all $t > 0$ iff $\int_{\|x\|>1} \|x\|^\alpha \nu(dx) < \infty$.

PROOF. a) and b) follows from the facts that $g(x) = e^{f(x)}$ and $g(x) = e^{\alpha\|x\|}$ are submultiplicative functions.
b) Assume that $E\|X_t\|^\alpha < \infty$ for every $t > 0$, which is equivalent to the finiteness of $E[1 \vee \|X_t\|^\alpha]$ for every $t > 0$, which in turn is equivalent to $\int_{\|x\|>1} (1 \vee \|x\|^\alpha)\nu(dx) < \infty$ since $g(x) = 1 \vee \|x\|^\alpha$ is a submultiplicative function. Finally, this is equivalent to $\int_{\|x\|>1} \|x\|^\alpha \nu(dx) < \infty$. □

The following lemma on estimations of moments and tails of Lévy processes generalizes and extends, to the Banach space case, the one presented in Sato (1999, Lemma 30.3) [**Sa**] for the Euclidean case.

Let $X_t^0$ be the Lévy process in the decomposition of $X_t$ in Remark 2.2 (f). We observe by Corollary 3.4 that $E\|X_t^0\|^2 < \infty$. From now on we shall assume that $X_t^0$ is such that $E\|X_t^0\|^2 \leq \lambda t$ for some positive constant $\lambda$.

LEMMA 3.5. *Let $\{X_t : t \geq 0\}$ be a Lévy process on a separable Banach space $B$ with triplet $(A, \nu, \gamma)$ and assume that $X$ satisfies the above assumption. Then*

*a) There exist positive constants $C(\varepsilon)$, $C_1$ and $C_2$ such that, for every $t > 0$,*

$$P(\|X_t\| > \varepsilon) \le C(\varepsilon)t \qquad \textit{for } \varepsilon > 0, \tag{3.6}$$

$$E\left[\|X_t\|^2 ; \|X_t\| \le 1\right] \le C_1 t, \tag{3.7}$$

$$E\left[\|X_t\| ; \|X_t\| \le 1\right] \le C_2 t^{1/2}. \tag{3.8}$$

*b) For any continuous linear functional $f$ on $B$ there exist positive constants $C_1(f)$ and $C_2(f)$ such that, for every $t > 0$,*

$$E\left[|f(X_t)|^2 ; \|X_t\| \le 1\right] \le C_1(f)t, \tag{3.9}$$

$$|E\left[f(X_t); \|X_t\| \le 1\right]| \le C_2(f)t. \tag{3.10}$$

*c) There exists a positive constant $C_3$ such that, for every $t > 0$,*

$$\|E\left[X_t; \|X_t\| \le 1\right]\| \le C_3 t. \tag{3.11}$$

PROOF. a)

$$P(\|X_t\| > \varepsilon) = P(\|X_t^0 + X_t^1\| > \varepsilon) \le P(X_t^1 \ne 0) + P(X_t^1 = 0,\ \|X_t^0\| > \varepsilon)$$
$$\le 1 - e^{-t\nu_1(\{\|x\|>1\})} + \frac{E\|X_t^0\|^2}{\varepsilon^2} \le \left\{\nu_1(\{\|x\| > 1\}) + \frac{\lambda}{\varepsilon^2}\right\} t,$$

where we have applied Chebyshev's inequality and considered $t \le 1$. This proves (3.6). Next, note that

$$\{\|X_t^0 + X_t^1\| \le 1\} = \{\|X_t^0\| \le 1, X_t^1 = 0\} \cup \{[\|X_t\| \le 1] \cap [X_t^1 \ne 0]\}$$

then

$$E\left[\|X_t\|^2 ; \|X_t\| \le 1\right] \le E\left[\|X_t^0\|^2 1_{\{\|X_t^0\| \le 1, X_t^1 = 0\}}\right] + E\left[\|X_t\|^2 1_{\{\|X_t\| \le 1\} \cap \{X_t^1 \ne 0\}}\right]$$
$$\le E\left[\|X_t^0\|^2 1_{\{\|X_t^0\| \le 1\}}\right] + P\left(X_t^1 \ne 0\right)$$
$$\le E\|X_t^0\|^2 + P\left(X_t^1 \ne 0\right)$$

since $\|X_t\|^2 1_{\{\|X_t\| \le 1\} \cap \{X_t^1 \ne 0\}} \le 1_{\{X_t^1 \ne 0\}}$. This reduces to the above case and hence (3.7) follows. To obtain (3.8) apply Jensen's inequality to get $E\left[\|X_t\| ; \|X_t\| \le 1\right] \le \left(E\left[\|X_t\|^2 ; \|X_t\| \le 1\right]\right)^{1/2} \le C_1^{1/2} t^{1/2}$ where $C_1$ is given by (3.7).

b) Let $f$ be a continuous linear functional on $B$.
Since $E\left[|f(X_t)|^2 ; \|X_t\| \le 1\right] \le |f|^2 E\left[\|X_t\|^2 ; \|X_t\| \le 1\right]$, (3.9) follows from (3.7). To prove (3.10) we use the three elementary estimates $\left|e^{it} - 1 - it\right| \le t^2/2$, $t \in \mathbb{R}$, $|e^z - 1| \le e^{|z|} - 1$, $z \in \mathbb{C}$, and $|e^z - 1| \le \frac{7}{4}|z|$, $|z| \le 1$. Note that

$$|E\left[f(X_t); \|X_t\| \le 1\right]|$$
$$= \left| \begin{array}{c} E\left[e^{if(X_t)} - 1\right] - E\left[e^{if(X_t)} - 1; \|X_t\| > 1\right] \\ -E\left[e^{if(X_t)} - 1 - if(X_t); \|X_t\| \le 1\right] \end{array} \right|$$
$$\le \left|E\left[e^{if(X_t)} - 1\right]\right| + 2P\left(\|X_t\| > 1\right) + \frac{1}{2}E\left[f^2(X_t); \|X_t\| \le 1\right].$$

The last two terms of the sum are bounded by constant multiples of $t$ by (3.6) and (3.9) respectively. Next, $E\left|e^{if(X_t)} - 1\right| = \left|e^{t\Psi_X(f)} - 1\right|$ where

$\Psi_X(f) = if(\gamma) - \frac{1}{2}A(f,f) + \int \left[e^{if(x)} - 1 - if(x)1_{\{\|x\|\leq 1\}}(x)\right]\nu(dx)$ is given by (1). Let $z = \Psi_X(f)$. When $t \geq 1$, $E\left|e^{if(X_t)} - 1\right| \leq 2t$. If $0 \leq t \leq \frac{1}{|z|}$ then $|e^{tz} - 1| \leq \frac{7}{4}|z|\,t$. Finally, if $\frac{1}{|z|} \leq t \leq 1$, $|e^{tz} - 1| \leq e^{|tz|} - 1 \leq e^{|z|}\,|z|\,t$.

c) By (3.8) $E\left[\|X_t\|\,;\|X_t\| \leq 1\right]$ is finite for each $t$ and hence $X_t 1_{\{\|X_t\|\leq 1\}}$ is Bochner integrable (Araujo and Giné [1980] Prop. III.2.2) [**A-G**]. Let $V = X_t 1_{\{\|X_t\|\leq 1\}}$. Since $V$ is also Pettis integrable, we have

$$f(EV) = Ef(V) \qquad \text{for every } f \in B^*.$$

On the other hand, by the Hahn-Banach Theorem there exists a continuous linear functional $\tilde{f} \in B^*$ satisfying $\|EV\| = \tilde{f}(EV)$ and then $\|EV\| = E\tilde{f}(V)$. This means $\|E\left[X_t; \|X_t\| \leq 1\right]\| = E\left[\tilde{f}(X_t); \|X_t\| \leq 1\right]$. The assertion follows from (3.10) applied to $\tilde{f}$. □

## 4. Subordination

Recall that a one dimensional subordinator is an increasing Lévy process $\{Z_t; t \geq 0\}$ in $\mathbb{R}$ with generating triplet $(A, \nu, \gamma)$ satisfying

$$A = 0, \quad \nu((-\infty, 0)) \quad \text{and} \quad \gamma - \int_{(0,1]} x\nu(dx) \geq 0,$$

where $\gamma - \int_{(0,1]} x\nu(dx) = \gamma_0$, the drift of $\{Z_t\}$, and its law is specified by its Laplace transform given by

$$Ee^{-uZ_t} = \exp\{-t\Psi(u)\} \quad u \geq 0, \tag{4.1}$$

where for any $w \in \mathbb{C}$ whose $\mathrm{Re}(w) \leq 0$

$$\Psi(w) = \int_{(0,\infty)} (e^{ws} - 1)\,\nu(ds) + \gamma^0 w, \tag{4.2}$$

and

$$\int_{(0,\infty)} (1 \wedge s)\,\nu(ds) < \infty. \tag{4.3}$$

In this section we study subordination of a Banach space valued Lévy process by a real valued subordinator. That is, we study the process that is obtained as the random time change of a Banach space valued Lévy process $\{X_t\}$ by a one dimensional subordinator $\{Z_t\}$. The corresponding process $\{Y_t\}$ -called subordinated process- is also a Lévy process.

THEOREM 4.1. *Let $\{Z_t : t \geq 0\}$ be a one dimensional subordinator with Laplace transform (16) and let $\{X_t : t \geq 0\}$ be a Lévy process on a separable Banach space $B$. Assume that $\{X_t\}$ and $\{Z_t\}$ are independent and define $Y_t = X_{Z_t}$ for all $t \geq 0$. Then the process $\{Y_t : t \geq 0\}$ is a Lévy process.*

The proof of this result follows the idea of the proof of the first part of Theorem 30.1 for the Euclidean case in Sato (1999) [**Sa**].

THEOREM 4.2. *Let $\{Z_t : t \geq 0\}$ be a one dimensional subordinator with Lévy measure $\nu_Z$ and drift $\gamma_Z^0 \geq 0$. Let $\{X_t : t \geq 0\}$ be a Lévy process as in Lemma 3.5 with generating triplet $(A, \nu, \gamma)$ and let $\mu = \mathcal{L}(X_1)$. Assume that $\{X_t\}$ and $\{Z_t\}$ are independent and let $\{Y_t : t \geq 0\}$ be the subordinated process obtained from subordination of $\{X_t\}$ by $\{Z_t\}$,that is $Y_t = X_{Z_t}$. Then we have the following:*

*a) The characteristic functional of the Lévy process* $\{Y_t\}$ *is given by* $Ee^{if(Y_t)} = e^{t\Psi(\log \hat{\mu}(f))}$ *for all real-valued* $f \in B^*$.

*b) Let* $\mu_s = \mathcal{L}(X_s)$. *The generating triplet* $(A_Y, \nu_Y, \gamma_Y)$ *of* $\{Y_t\}$ *is given by*

$$A_Y = \gamma_Z^0 A, \tag{4.4}$$

$$\nu_Y(C) = \int_{(0,\infty)} \mu_s(C)\nu_Z(ds) + \gamma_Z^0 \nu(C) \qquad C \in \mathcal{B}(B\backslash\{0\}), \tag{4.5}$$

$$\gamma_Y = \int_{(0,\infty)} \int_{\|x\|\le 1} x\mu_s(dx)\nu_Z(ds) + \gamma_Z^0 \gamma, \tag{4.6}$$

*where the last integral is a Bochner integral.*

*c) If* $\gamma_Z^0 = 0$ *and* $\int_{(0,1]} s^{1/2}\nu_Z(ds) < \infty$, *then* $A_Y = 0$, $\int_{\|x\|\le 1} \|x\|\, \nu_Y(dx) < \infty$ *and* $\gamma_Y - \int_{\|x\|\le 1} x\nu_Y(dx) = 0$. *Moreover, the Banach space Lévy process* $\{Y_t\}$ *has bounded variation on each interval* $[0,t]$.

PROOF. a) Let $f$ be a real-valued continuous linear functional on $B$. We observe that $Ee^{if(X(s))} = e^{s \log \hat{\mu}(f)}$ for each $s \ge 0$. Since $\operatorname{Re}(\log \hat{\mu}(f)) \le 0$ we use (4.2) to get

$$Ee^{if(Y_t)} = E\left[\left(Ee^{if(X(s))}\right)_{s=Z_t}\right] = E\left[\left(e^{s\log \hat{\mu}(f)}\right)_{s=Z_t}\right] = e^{t\Psi(\log \hat{\mu}(f))}.$$

b) We use Lemma 3.5 in order to find the generating triplet of $\{Y_t\}$. From (4.2) we have

$$\begin{aligned} Ee^{if(Y_t)} &= e^{t\Psi(\log \hat{\mu}(f))} = \\ &\exp\left\{t\left(\int_{(0,\infty)} (e^{s\log\hat{\mu}(f)} - 1)\nu_Z(ds) + \gamma_Z^0 \log \hat{\mu}(f)\right)\right\} \\ &= \exp\left\{t\left(\int_{(0,\infty)} (\hat{\mu}_s(f) - 1)\nu_Z(ds) + \gamma_Z^0 \log \hat{\mu}(f)\right)\right\}. \end{aligned} \tag{4.7}$$

Let $g(f,x) = e^{f(x)} - 1 - if(x)1_{\{\|r\|\le 1\}}(x)$. Note that

$$\gamma_Z^0 \log \hat{\mu}(f) = -\frac{1}{2}\gamma_Z^0 (Af, f) + i\gamma_Z^0 f(\gamma) + \int_B g(f,x)\, \gamma_Z^0 \nu(dx). \tag{4.8}$$

On the other hand, by Lemma 3.5 $\int_{\|x\|\le 1} x\mu_s(dx)$ is a well defined Bochner integral. From (3.11) and (4.3) we have that $\int_{(0,\infty)} \left\|\int_{\|x\|\le 1} x\mu_s(dx)\right\| \nu_Z(ds)$ is finite and hence $\int_{(0,\infty)} \int_{\|x\|\le 1} x\mu_s(dx)\nu_Z(ds)$ is a well defined Bochner integral. Therefore

$$\int_{(0,\infty)} \int_{\|x\|\le 1} f(x)\mu_s(dx)\nu_Z(ds) = f\left(\int_{(0,\infty)} \int_{\|x\|\le 1} x\mu_s(dx)\nu_Z(ds)\right).$$

Then we have

$$\int_{(0,\infty)} (\hat{\mu}_s(f) - 1)\nu_Z(ds) = \int_{(0,\infty)} \int_B (e^{if(x)} - 1)\mu_s(dx)\nu_Z(ds)$$

$$= \int_{(0,\infty)} \int_B g(f,x)\mu_s(dx)\nu_Z(ds) + if\left(\int_{(0,\infty)} \int_{\|x\|\le 1} x\mu_s(dx)\nu_Z(ds)\right). \tag{4.9}$$

From (4.7), (4.8) and (4.9) we have

$$Ee^{if(Y_t)}$$

$$= \exp\left\{t\left[-\frac{1}{2}\gamma_Z^0 A(f,f) + if\left(\int_{(0,\infty)}\int_{\|x\|\le 1} x\mu_s(dx)\nu_Z(ds) + \gamma_Z^0\gamma\right)\right.\right.$$

$$\left.\left. + \int_B g(f,x)\left(\int_{(0,\infty)} \mu_s(\cdot)\nu_Z(ds) + \gamma_Z^0\nu(\cdot)\right)(dx)\right]\right\}. \tag{4.10}$$

Define $\nu_1(C) = \int_{(0,\infty)} \mu_s(C)\nu_Z(ds)$ for $C \in \mathcal{B}(B\backslash\{0\})$. It remains to prove that $\nu_1$ is a Lévy measure, i.e, we show that the function $f \longmapsto \exp\int_B g(f,x)\nu_1(dx)$ for all $f \in B^*$ is a characteristic functional of some Banach-valued random variable (see (2.3)). Then

$$\exp\int_B g(f,x)\,\nu_1(dx)$$

$$= \exp\left\{\int_{(0,\infty)} (Ee^{if(X(s))} - 1)\nu_Z(ds) - i\int_{(0,\infty)}\int_{\|x\|\le 1} f(x)\mu_s(dx)\nu_Z(ds)\right\}$$

$$= \exp\left\{\int_{(0,\infty)} (e^{s\log\hat{\mu}(f)} - 1)\nu_Z(ds)\right\}\exp\left\{-if\left(\int_{(0,\infty)}\int_{\|x\|\le 1} x\mu_s(dx)\nu_Z(ds)\right)\right\}.$$

The last two factors are characteristic functionals, since the first one corresponds to the characteristic functional of a subordinated process at time 1 obtained by subordination of $X_s$ by $Z_t - t\gamma_Z^0$ and the second one corresponds to the characteristic functional of a degenerated distribution. In view of (4.10) and Remark 2.2(e) we get (4.4), (4.5) and (4.6).

c) Assume that $\gamma_Z^0 = 0$ and $\int_{(0,1]} s^{1/2}\nu_Z(ds) < \infty$. Then $A_Y = 0$ by (4.4) and $\nu_Y(dx) = \int_{(0,1]} \mu_s(dx)\nu_Z(ds)$ by (4.5). We have

$$\int_{\|x\|\le 1} \|x\|\,\nu_Y(dx) = \int_{(0,\infty)}\int_{\|x\|\le 1} \|x\|\,\mu_s(dx)\nu_Z(ds) < \infty$$

by (4.3) and Lemma 3.5 relation (3.8). Now, it follows from (4.6) and (4.10) that $\gamma_Y - \int_{\|x\|\le 1} x\nu_Y(dx) = 0$ and moreover

$$Ee^{if(Y_t)} = \exp\left\{t\int_B (e^{if(x)} - 1)\nu_Y(dx)\right\}.$$

By Proposition 2.3 we conclude that $\{Y_t\}$ has bounded variation on each interval $[0,t]$. □

**Acknowledgement** The authors thank Ken-iti Sato for valuable observations on the original manuscript.

## References

[A-R] S. Albeverio and B. Rudiger, *Stochastic integrals and Lévy-Ito decomposition on separable Banach spaces*, In Mini-proceedings of 2nd MaPhySto Conference on Lévy Processes: Theory and Applications, Miscellanea 22, University of Aarhus, (2002).

[A-G] A. Araujo and E. Giné, *The Central Limit Theorem for Real and Banach Valued Random Variables*, John Wiley & Sons, New York, (1980).

[B-M-R] O.E. Barndorff-Nielsen, T. Mikosch and S.I. Resnick, *Lévy Processes: Theory and Applications*, In First MaPhySto Conference on Lévy Processes: Theory and Applications. Ed. Birkhäuser, Boston, (2001)

[B] J. Bertoin, *Lévy Processes*, Cambridge University Press, Cambridge, (1996).

[Ac] A. de Acosta, *Exponential moments of vector valued random series and triangular arrays*, Ann. Probab. **8**, (1980), pp. 381-389.

[D] E. Dettweiler, *Banach space valued processes with independent increments and stochastic integrals*, In Probability in Banach Spaces IV. Proc. Oberwolfach, Lect. Notes in Math. vol. 990, (1982), pp. 54-83 Springer Verlag, Berlin.

[G-S] I.I. Gihman and A.V. Skorohod, *The Theory of Stochastic Processes*, vol. II, Springer-Verlag, Berlin, (1975).

[K1] V.M. Kruglov, *A note on infinitely divisible distributions*, Probab.Theor. Appl. **15**, (1970), pp. 319-324.

[K2] V.M. Kruglov, *Integrals with respect to infinitely divisible distributions in a Hilbert space*, Math. Notes **11**, (1972), pp. 407-411.

[L] W. Linde, *Probability in Banach Spaces: Stable and Infinitely Divisible Distributions*, John Wiley & Sons, Berlin, (1986).

[Sa] K. Sato, *Lévy Processes and Infinitely Divisible Distributions,* Cambridge University Press, Cambridge, (1999).

Centro de Investigación en Matemáticas, Apdo Postal 402, Guanajuato, Gto. 36000, Mexico

*E-mail address*: pabreu@cimat.mx

Universidad Autónoma de Sinaloa & CIMAT, Apdo Postal 402, Guanajuato, Gto. 36000, Mexico

*E-mail address*: alfonso@cimat.mx

Contemporary Mathematics
Volume **336**, 2003

# Phase Space Path Integral Representation for the Solution of a Stochastic Schrödinger Equation

Luis A. Rincón

ABSTRACT. This paper provides a detailed proof of a path integral representation in phase space for the solution of a stochastic Schrödinger equation driven by a continuous semimartingale. We follow the approach of path integration established by Itô, Albeverio and Høegh-Krohn. The equation we consider here is given by the classical Schrödinger equation plus the stochastic term $K(t, -i\frac{\partial}{\partial x})\Psi_t(x) \circ dM_t$, where $M_t$ is a continuous semimartingale. As an application we solve the path integral for a stochastic quantum harmonic oscillator. We also give the solution of the corresponding stochastic heat equation.

## 1. Introduction

The first appearance of path integration, also known as functional integration in quantum mechanics, was due to the Nobel laureate Richard Feynman who in 1948 expressed the solution of the Schrödinger equation in terms of an ill-defined integral in configuration space [**F**]. Later in 1951, in a joint work with A. R. Hibbs, they introduced the same integral but now in phase space [**F-H**]. Integrals over a space of functions were considered earlier in 1923 by N. Wiener when he gave his mathematical formalism of Brownian motion, although these two approaches to functional integration are not quite the same. For a Wiener integral a Gaussian measure is used, whereas for the Feynman integral the problem lies in the non-existence of any corresponding measure on infinite dimensional path space.

The Feynman integral is now known as an alternative formulation of quantum mechanics, and an enormous amount of work has been devoted to finding a rigorous mathematical definition of such path integral. See references [**I**], [**K-Mc**], [**M**], [**N**] and [**Si**] to mention only a few. Several definitions have been proposed, see for instance the books [**C-D**], [**A-H**], [**A**], [**Sc**] and the references therein. One particularly successful and well established approach to Feynman path integrals is that suggested by Itô, Albeverio and Høegh-Krohn in [**A-H**]. Following this approach, T. Zastawniak proved an extension to the stochastic case of this path integral representation in configuration space [**Z**]. He considered a kind of stochastic Schrödinger

2000 *Mathematics Subject Classification.* Primary 60H15, 81S40, 81S30.
*Key words and phrases.* Stochastic Schrödinger Equation, Path Integrals, Mehler Kernel, Stochastic Heat Equation.
Financial support from CONACYT is acknowledged.

equation driven by Brownian motion and proved that a path integral representation in configuration space exists for the solution of this equation. Further extensions of path integrals on phase space have been recently done by Truman and Zastawniak [**T-Z1**], [**T-Z2**].

Theorems 4.1 and 5.1 and Corollary 5.2 are the main results in this paper. None of them is entirely new. Theorems 4.1 and 5.1 appeared first in [**T-Z1**] but using Brownian motion in the stochastic term. We here extend the same results to any continuous semimartingale but more importantly we give full details of our proofs. We actually give two different proofs of Theorem 5.1. Finally Corollary 5.2 on the exact solution of a stochastic heat equation is a rather direct and natural consequence of Theorem 5.1. This Paper is a continuation of [**R**].

## 2. Path integration in configuration space

The following is a quick overview of the path integration approach by Itô, Albeverio and Høegh-Krohn as discussed in [**A-H**]. For any fixed $t > 0$, let $\mathcal{H}_t$ be the Hilbert space of continuous paths $\gamma : [0,t] \to \mathbb{R}^d$ with $\gamma(t) = 0$ and such that each component of $\dot{\gamma}$ is square integrable in the sense of distributions. The inner product in $\mathcal{H}_t$ is defined by

$$(\gamma, \eta)_t := \int_0^t (\, \dot{\gamma}(s), \dot{\eta}(s)\,)\, ds \,,$$

where $(x,y) = x \cdot y$ is the usual Euclidean inner product in $\mathbb{R}^d$. The norm of path $\gamma$ is denoted by $|\gamma|_t$. This Hilbert space has the reproducing kernel $G : [0,t] \times [0,t] \to \mathbb{R}^{d\times d}$ given by the diagonal matrix $G(t_1, t_2) = (t - t_1 \vee t_2)I$, where $I$ is the $d \times d$ identity matrix. This means that, for fixed $t_1$, the $k$-th column path $t_2 \mapsto G_{\cdot,k}(t_1,t_2)$ belongs to $\mathcal{H}_t$ and is such that for any $\gamma \in \mathcal{H}_t$,

$$\gamma_k(t_1) = (\gamma(\cdot), G_{\cdot,k}(t_1,\cdot))_t = \sum_{j=1}^d \int_0^t \dot{\gamma}_j(s) \dot{G}_{j,k}(t_1, s)\, ds.$$

Now let $\mathcal{M}(\mathcal{H}_t)$ be the Banach algebra of bounded complex-valued measures on $\mathcal{H}_t$. The norm on this space is given by the total variation norm, that is, for $\mu \in \mathcal{M}(\mathcal{H}_t)$, we define $\|\mu\| := |\mu|(\mathcal{H}_t)$. The operations in $\mathcal{M}(\mathcal{H}_t)$ are the pointwise sum, and the product is the convolution measure $\mu * \nu\,(A) := \int \mu(A-x)\, d\nu(x) \in \mathcal{M}(\mathcal{H}_t)$. In particular, the exponential of a measure is defined as its power series expansion which is itself in $\mathcal{M}(\mathcal{H}_t)$. Finally, let $\mathcal{F}(\mathcal{H}_t)$ be the Banach algebra of Fourier transforms of measures in $\mathcal{M}(\mathcal{H}_t)$, that is $\mathcal{F}(\mathcal{H}_t) := \{\hat{\mu}(\gamma) : \mu \in \mathcal{M}(\mathcal{H}_t)\}$, where

$$\hat{\mu}(\gamma) := \int_{\mathcal{H}_t} e^{i(\gamma,\eta)_t}\, d\mu(\eta) \,.$$

The norm in $\mathcal{F}(\mathcal{H}_t)$ is defined as $\|\hat{\mu}\| := \|\mu\|$. The operations are again the pointwise sum, and the product is the pointwise multiplication $\hat{\mu} \cdot \hat{\nu} = \widehat{\mu * \nu} \in \mathcal{F}(\mathcal{H}_t)$. In particular, $\widehat{e^{\mu}}(\gamma) = e^{\hat{\mu}(\gamma)}$. The Fourier transform operation $\;\hat{}\; : \mathcal{M}(\mathcal{H}_t) \to \mathcal{F}(\mathcal{H}_t)$ defines an isomorphism between Banach algebras. We now define the Feynman path integral of any functional $f \in \mathcal{F}(\mathcal{H}_t)$ by

$$\int_{\mathcal{H}_t} e^{\frac{i}{2}|\gamma|_t^2} f(\gamma)\, d\gamma := \int_{\mathcal{H}_t} e^{-\frac{i}{2}|\gamma|_t^2}\, d\mu_f(\gamma), \tag{2.1}$$

where $\mu_f \in \mathcal{M}(\mathcal{H}_t)$ is such that $\hat{\mu}_f(\gamma) = f(\gamma)$. The integral is well defined since the mapping $\gamma \mapsto e^{-\frac{i}{2}|\gamma|_t^2}$ is a bounded continuous functional on $\mathcal{H}_t$. Then Albeverio and Høegh-Krohn show [**A-H**] that for $\psi_0$ and $V$ in $\mathcal{F}(\mathcal{H}_t)$, the solution of the Schrödinger equation

$$i\hbar\frac{\partial}{\partial t}\psi_t(x) = -\frac{\hbar^2}{2m}\triangle_x\psi_t(x) + V(t,x)\psi_t(x), \tag{2.2}$$

with initial wave function $\psi_0$, admits the path integral representation

$$\psi_t(x) = \int_{\mathcal{H}_t} e^{\frac{i}{\hbar}\left(\frac{m}{2}|\gamma|_t^2 - \int_0^t V(s,x+\gamma_s)\,ds\right)}\psi_0(x+\gamma_0)\,d\gamma. \tag{2.3}$$

## 3. Path integration in phase space

In this section we seek to extend the configuration space path integral of Itô, Albeverio and Høegh-Krohn to phase space. The Feynman's solution of the classical Schrödinger equation (2.2) as an integral in phase space can be formally written as

$$\Psi_t(x) = \int_{(q,p)} e^{i\int_0^t (p_s\dot{q}_s - \frac{1}{2}p_s^2 - V(q_s))\,ds}\Psi_0(q_0)\,d(q,p)\,. \tag{3.1}$$

We will now give a rigorous definition of phase space path integrals. As before, we take $\mathcal{H}_t$ as configuration path space. As space of momentum paths we take the Hilbert space $\mathcal{L}_t = \mathcal{L}^2[0,t]$, and we set the product $\mathcal{H}_t \times \mathcal{L}_t$ as our space of phase space paths. This space can be made into a Hilbert space by defining as its inner product the sum of the inner products of each factor. We denote by $\mathcal{M}(\mathcal{L}_t \times \mathcal{H}_t)$ the space of bounded complex-valued measures on $\mathcal{L}_t \times \mathcal{H}_t$. Then, following [**T-Z1**], for any measure $\mu \in \mathcal{M}(\mathcal{L}_t \times \mathcal{H}_t)$, we define

$$\hat{\mu}(q,p) := \int_{\mathcal{L}_t\times\mathcal{H}_t} e^{i\langle(q,p),(P,Q)\rangle}\,d\mu(P,Q)\,,$$

where

$$\langle(q,p),(P,Q)\rangle = \int_0^t P_s\dot{q}_s\,ds + \int_0^t p_s\dot{Q}_s\,ds\,.$$

We denote by $\mathcal{F}(\mathcal{L}_t \times \mathcal{H}_t)$ the space of all such Fourier transforms of measures on $\mathcal{L}_t \times \mathcal{H}_t$. Therefore, for any function $f$ of the form $f(q,p) = \hat{\mu}_f(q,p)$ with $\hat{\mu}_f \in \mathcal{F}(\mathcal{L}_t \times \mathcal{H}_t)$, we define

$$\int_{\mathcal{H}_t\times\mathcal{L}_t} e^{i\int_0^t(p_s\dot{q}_s - \frac{1}{2}p_s^2)\,ds} f(q,p)\,d(q,p) := \int_{\mathcal{L}_t\times\mathcal{H}_t} e^{-i\int_0^t(p_s\dot{q}_s + \frac{1}{2}p_s^2)\,ds}\,d\mu_f(p,q)\,.$$

The reader can find a more general definition of this type of integrals in the paper by Truman and Zastawniak [**T-Z2**]. For our purposes the above definition suffice. We next state and prove an iterative property for phase space path integrals.

PROPOSITION 3.1. *Let $0 < s < t$ be fixed. Suppose that $f, g$ in $\mathcal{F}(\mathcal{L}_t \times \mathcal{H}_t)$ are such that for any $(q,p) \in \mathcal{H}_t \times \mathcal{L}_t$, $f(q,p)$ depends only on the values of $(q,p)$ on $[0,s]$, and $g(q,p)$ depends only on the values of $(q,p)$ on $[s,t]$. Then we have the* *iterative formula*

$$\int_{\mathcal{H}_t\times\mathcal{L}_t} e^{i\int_0^t (p_s\dot q_s-\frac12 p_s^2)\,ds} f(q,p)g(q,p)\,d(q,p) =$$

$$\int_{\mathcal{H}_{[s,t]}\times\mathcal{L}_{t-s}} e^{i\int_s^t (d_u\dot c_u-\frac12 d_u^2)\,du}\, g(c,d)$$

$$\int_{\mathcal{H}_s\times\mathcal{L}_s} e^{i\int_0^s (b_u\dot a_u-\frac12 b_u^2)\,du}\, f(c_s+a,b)\,d(a,b)\,d(c,d)\,. \tag{3.2}$$

PROOF. To reduce the length of our formulae we will use the following abbreviations $\mathcal{HL}_t := \mathcal{H}_t\times\mathcal{L}_t$, $\mathcal{HL}_{[s,t]} := \mathcal{H}_{[s,t]}\times\mathcal{L}_{[s,t]}$, and corresponding meaning for $\mathcal{LH}_t$ and $\mathcal{LH}_{[s,t]}$, where $\mathcal{H}_{[s,t]}$ denotes the Hilbert space of paths defined on the interval $[s,t]$, which is a time translation of $\mathcal{H}_{t-s}$. Let us first define the measurable mapping

$$T : \mathcal{LH}_t \to (\mathcal{LH}_s)\times(\mathcal{LH}_{[s,t]})\,,$$

given by $T(p,q) := ((b,a),(d,c))$, where $a_\tau := q_\tau - q_s$ for $\tau\in[0,s]$, $b_\tau := p_\tau$ for $\tau\in[0,s]$, and $c_\tau := q_\tau$ for $\tau\in[s,t]$, $d_\tau := p_\tau$ for $\tau\in[s,t]$. We also define its inverse by

$$T^{-1}((b,a),(d,c))(\tau) := \begin{cases} (b_\tau, c_s+a_\tau) & \text{if } \tau\in[0,s]\,,\\ (d_\tau,c_\tau) & \text{if } \tau\in[s,t]\,.\end{cases}$$

We keep the order of the alphabet when writing paths in $\mathcal{H}_t\times\mathcal{L}_t$ as pairs, that is, $(a,b)$ and $(c,d)$ belong to $\mathcal{H}_t\times\mathcal{L}_t$, so $a$ and $c$ are paths in $\mathcal{H}_t$ and $b$ and $d$ are momentum paths. This, however, is not so for the pair $(q,p)$, which is a standard notation for a path in phase space. It is easy to check that

$$\int_0^t (p_u\dot q_u+\frac12 p_u^2)\,du = \int_0^s (b_u\dot a_u+\frac12 b_u^2)\,du + \int_s^t (d_u\dot c_u+\frac12 d_u^2)\,du\,.$$

Let us suppose there exist measures $\mu_f$ and $\mu_g$ on $\mathcal{H}_t\times\mathcal{H}_{[s,t]}$ such that $f(q,p) = \hat\mu_f(q,p)$, and $g(q,p)=\hat\mu_g(q,p)$. Then, defining $\mu$ as the convolution measure $\mu_f*\mu_g$, we have $\hat\mu(q,p) = f(q,p)g(q,p)$. Therefore

$$\begin{aligned}
&\int_{\mathcal{HL}_t} e^{i\int_0^t (p_s\dot q_s-\frac12 p_s^2)\,ds} f(q,p)g(q,p)\,d(q,p)\\
&= \int_{\mathcal{LH}_t} e^{-i\int_0^t (p_s\dot q_s+\frac12 p_s^2)\,ds}\, d(\mu\circ T^{-1})(T(p,q))\\
&= \int_{\mathcal{LH}_s\times\mathcal{LH}_{[s,t]}} e^{-i\int_0^s (b_u\dot a_u+\frac12 b_u^2)\,du-i\int_s^t (d_u\dot c_u+\frac12 d_u^2)\,du}\, d(\mu\circ T^{-1})((b,a),(d,c))\\
&= \int_{\mathcal{HL}_s\times\mathcal{HL}_{[s,t]}} e^{i\int_0^s (b_u\dot a_u-\frac12 b_u^2)\,du+i\int_s^t (d_u\dot c_u-\frac12 d_u^2)\,du}\\
&\qquad\qquad (\hat\mu\circ T^{-1})((b,a),(d,c))\,d((a,b),(c,d))\\
&= \int_{\mathcal{HL}_s\times\mathcal{HL}_{[s,t]}} e^{i\int_0^s (b_u\dot a_u-\frac12 b_u^2)\,du+i\int_s^t (d_u\dot c_u-\frac12 d_u^2)\,du}
\end{aligned}$$

$$[f \circ T^{-1}((b,a),(d,c))]\,[g \circ T^{-1}((b,a),(d,c))]\, d((a,b),(c,d))$$

$$= \int_{\mathcal{HL}_s \times \mathcal{HL}_{[s,t]}} e^{i\int_0^s (b_u \dot{a}_u - \frac{1}{2} b_u^2)\, du + i \int_s^t (d_u \dot{c}_u - \frac{1}{2} d_u^2)\, du}$$

$$[f(c_s + a, b))]\,[g(c,d)]\, d((a,b),(c,d))$$

$$= \int_{\mathcal{HL}_{[s,t]}} e^{i\int_s^t (d_u \dot{c}_u - \frac{1}{2} d_u^2)\, du}\, g(c,d)$$

$$\int_{\mathcal{HL}_s} e^{i\int_0^s (b_u \dot{a}_u - \frac{1}{2} b_u^2)\, du}\, f(c_s + a, b)\, d(a,b)\, d(c,d)\,.$$

□

Observe that in the formula we just proved, the momentum paths $b$ and $d$ are not related to each other as the only restriction on them is the square integrability. We now show how our phase space path integral reduces to the configuration space path integral when the function $f(q,p)$ depends only on the configuration space variable $q$. Let us assume $f(q) = \hat{\mu}_f(q)$ for some $\mu_f \in \mathcal{M}(\mathcal{H}_t)$. We define the product measure $\mu$ on $\mathcal{L}_t \times \mathcal{H}_t$ by $\mu(p,q) := \delta_{\{\dot{q}\}}(p) \times \mu_f(q)$, and the invertible transformation $S_1 : \mathcal{L}_t \times \mathcal{H}_t \to \mathcal{L}_t \times \mathcal{H}_t$ by

$$S_1(p,q) = \begin{pmatrix} 1 & 0 \\ \int_0^{\cdot} ds & 1 \end{pmatrix} \begin{pmatrix} p \\ q \end{pmatrix} = (p, q + \int_0^{\cdot} p_s\, ds)\,, \tag{3.3}$$

with inverse

$$S_1^{-1}(p,q) = \begin{pmatrix} 1 & 0 \\ -\int_0^{\cdot} ds & 1 \end{pmatrix} \begin{pmatrix} p \\ q \end{pmatrix} = (p, q - \int_0^{\cdot} p_s\, ds)\,. \tag{3.4}$$

Then $f$ can be written as the Fourier transform on phase space of the measure $\mu \circ S_1$ as we now state in our next proposition.

PROPOSITION 3.2. *Let $f \in \mathcal{F}(\mathcal{H}_t)$ with $f = \hat{\mu}_f$ for some $\mu_f \in \mathcal{M}(\mathcal{H}_t)$. Then $f \in \mathcal{F}(\mathcal{H}_t \times \mathcal{L}_t)$ with $f = \widehat{\mu \circ S_1}$, where $\mu$ and $S_1$ are as defined above.*

PROOF.

$$\begin{aligned}
\widehat{\mu \circ S_1}(q,p) &= \int_{\mathcal{L}_t \times \mathcal{H}_t} e^{i\langle (q,p),(P,Q)\rangle}\, d(\mu \circ S_1)(P,Q) \\
&= \int_{\mathcal{L}_t \times \mathcal{H}_t} e^{i\langle (q,p), S_1^{-1}(P,Q)\rangle}\, d\mu(P,Q) \\
&= \int_{\mathcal{L}_t \times \mathcal{H}_t} e^{i\langle (q,p),(P, Q - \int_0^{\cdot} P_s\, ds)\rangle}\, d(\delta_{\{\dot{Q}\}} \times \mu_f)(P,Q) \\
&= \int_{\mathcal{H}_t} e^{i\int_0^t \dot{Q}_s\, dq_s}\, d\mu_f(Q) \\
&= \int_{\mathcal{H}_t} e^{i(q,Q)_t}\, d\mu_f(Q) \\
&= \hat{\mu}_f(q) \\
&= f(q)\,.
\end{aligned}$$

□

Thus, any function $f$ in $\mathcal{F}(\mathcal{H}_t)$ can be seen as an element of $\mathcal{F}(\mathcal{L}_t \times \mathcal{H}_t)$, when we first compose with the transformation $S_1$. Finally we show the phase space path integral of such $f$, reduces to its configuration space path integral.

PROPOSITION 3.3. *Let $f$ be a function in $\mathcal{F}(\mathcal{H}_t)$. Then we have*

$$\int_{\mathcal{H}_t\times\mathcal{L}_t} e^{i\int_0^t (p_s\dot{q}_s-\frac{1}{2}p_s^2)\,ds} f(q)\,d(q,p) = \int_{\mathcal{H}_t} e^{\frac{i}{2}|q|_t^2} f(q)\,dq\,. \tag{3.5}$$

PROOF.

$$\begin{aligned}
&\int_{\mathcal{H}_t\times\mathcal{L}_t} e^{i\int_0^t (p_s\dot{q}_s-\frac{1}{2}p_s^2)\,ds} f(q)\,d(q,p)\\
&\qquad = \int_{\mathcal{L}_t\times\mathcal{H}_t} e^{-i\int_0^t (p_s\dot{q}_s+\frac{1}{2}p_s^2)\,ds}\, d(\mu\circ S_1)(p,q)\\
&\qquad = \int_{\mathcal{L}_t\times\mathcal{H}_t} e^{-i\int_0^t (p_s(\dot{q}_s-p_s)+\frac{1}{2}p_s^2)\,ds}\, d(\delta_{\{\dot{q}\}}\times\mu_f)(p,q)\\
&\qquad = \int_{\mathcal{H}_t} e^{-\frac{i}{2}|q|_t^2}\, d\mu_f(q)\\
&\qquad = \int_{\mathcal{H}_t} e^{\frac{i}{2}|q|_t^2} f(q)\,dq\,.
\end{aligned}$$

□

We will also need to know how to write functions in $\mathcal{F}(\mathcal{L}_t)$ as functions in $\mathcal{F}(\mathcal{L}_t\times\mathcal{H}_t)$. Let us assume $f(p)=\hat{\mu}_f(p)$ for some $\mu_f\in\mathcal{M}(\mathcal{L}_t)$. We define the product measure $\mu$ on $\mathcal{L}_t\times\mathcal{H}_t$ as follows $\mu(p,q):=\mu_f(p)\times\delta_{\{\int_0^{\cdot} p_s\,ds\}}(q)$, and the invertible transformation $S_2:\mathcal{L}_t\times\mathcal{H}_t\to\mathcal{L}_t\times\mathcal{H}_t$ given by

$$S_2(p,q) := \begin{pmatrix} -1 & \frac{d}{ds} \\ 0 & 1 \end{pmatrix}\begin{pmatrix} p \\ q \end{pmatrix} = (\dot{q}-p,q)\,, \tag{3.6}$$

with identical inverse $S_2^{-1}(p,q)=(\dot{q}-p,q)$. Then $f$ can be written as the Fourier transform on phase space of the measure $\mu\circ S_2$ as we next show.

PROPOSITION 3.4. *Let $f\in\mathcal{F}(\mathcal{L}_t)$ with $f=\hat{\mu}_f$ for some $\mu_f\in\mathcal{M}(\mathcal{L}_t)$. Then $f\in\mathcal{F}(\mathcal{H}_t\times\mathcal{L}_t)$ with $f=\widehat{\mu\circ S_2}$, where $\mu$ and $S_2$ are as defined above.*

PROOF. Doing similar calculations as before we have

$$\begin{aligned}
\widehat{\mu\circ S_2}(q,p) &= \int_{\mathcal{L}_t\times\mathcal{H}_t} e^{i\langle(q,p),(P,Q)\rangle}\, d(\mu\circ S_2)(P,Q)\\
&= \int_{\mathcal{L}_t\times\mathcal{H}_t} e^{i\langle(q,p),S_2^{-1}(P,Q)\rangle}\, d\mu(P,Q)\\
&= \int_{\mathcal{L}_t\times\mathcal{H}_t} e^{i\langle(q,p),(\dot{Q}-P,Q)\rangle}\, d(\mu_f\times\delta_{\{\int_0^{\cdot} P_s\,ds\}})(P,Q)\\
&= \int_{\mathcal{L}_t} e^{i\langle p,P\rangle_{\mathcal{L}_t}}\, d\mu_f(P)\\
&= \hat{\mu}_f(p)\\
&= f(p)\,.
\end{aligned}$$

□

## 4. A phase space path integral representation

We now state and prove our main result. Let $(\Omega, \mathcal{F}, \mathbb{P})$ be a complete probability space, $\{\mathcal{F}_t\}_{t\geq 0}$ a standard filtration and $M : [0,\infty)\times\Omega \to \mathbb{R}$ a continuous semimartingale. As usual, $\circ dM_t$ will denote a Stratonovich differential respect to the semimartingale $M$.

THEOREM 4.1. *Let $\Psi_0(\cdot)$, $V(t,\cdot)$, and $K(t,\cdot)$ be in $\mathcal{F}(\mathbb{R}^d)$ such that $V(t,x)$ and $K(t,x)$ are real-valued jointly continuous functions. Then the solution of the stochastic Schrödinger equation*

$$id\Psi_t(x) = \left(-\frac{1}{2}\Delta_x + V(t,x)\right)\Psi_t(x)\,dt + K(t,-i\frac{\partial}{\partial x})\Psi_t(x)\circ dM_t, \tag{4.1}$$

*with initial condition $\Psi_t(x)|_{t=0} = \Psi_0(x)$, is given by the path integral*

$$\Psi_t(x) := \int_{\mathcal{H}_t\times\mathcal{L}_t} e^{i\int_0^t (p_s\dot{q}_s - \frac{1}{2}p_s^2)\,ds}$$
$$e^{-i\int_0^t V(s,x+q_s)\,ds - i\int_0^t K(s,p_s)\,dM_s}\Psi_0(x+q_0)\,d(q,p)\ . \tag{4.2}$$

PROOF. First we prove that the functional

$$(q,p)\mapsto e^{-i\int_0^t V(s,x+q_s)\,ds - i\int_0^t K(s,p_s)\,dM_s}\Psi_0(x+q_0) \tag{4.3}$$

belongs to $\mathcal{F}(\mathcal{H}_t\times\mathcal{L}_t)$, and hence the writing of the path integral (4.2) is justified. We only need to translate measures from $\mathcal{M}(\mathcal{H}_t)$ and $\mathcal{M}(\mathcal{L}_t)$ into measures in $\mathcal{M}(\mathcal{H}_t\times\mathcal{L}_t)$. To do this we make use of our mappings $S_1$ and $S_2$ defined in last section. Thus we have

$$\begin{aligned} e^{-i\int_0^t V(s,x+q_s)\,ds}\Psi_0(x+q_0) &= \hat{\sigma}(q),\\ K(s,p_s) &= \hat{\mu_s}(p)\ , \end{aligned}$$

where $\sigma\in\mathcal{M}(\mathcal{H}_t)$, and $\mu_s\in\mathcal{M}(\mathcal{L}_t)$. Hence from Propositions 3.2 and 3.4, we have

$$\begin{aligned} e^{-i\int_0^t V(s,x+q_s)\,ds}\Psi_0(x+q_0) &= \widehat{\sigma\circ S_1}(q,p),\\ K(s,p_s) &= \widehat{\mu_s\circ S_2}(q,p)\ . \end{aligned}$$

We define the measure-valued Itô stochastic integral $\nu_t(\cdot) := \displaystyle\int_0^t \mu_s\circ S_2(\cdot)\,dM_s$. This measure is defined now on $\mathcal{L}_t\times\mathcal{H}_t$. Then $e^{-i\int_0^t K(s,p_s)\,dM_s} = \widehat{e^{-i\nu_t}}(q,p)$. Therefore

$$\begin{aligned} e^{-i\int_0^t V(s,x+q_s)\,ds - i\int_0^t K(s,p_s)\,dM_s}\Psi_0(x+q_0) &= (\widehat{\sigma\circ S_1})(\widehat{e^{-i\nu_t}})(q,p)\\ &= (\widehat{\sigma\circ S_1)*(e^{-i\nu_t}})(q,p)\ , \end{aligned}$$

and hence the writing out of the path integral (4.2) is justified. Observe the stochastic integral appearing in the path integral is an Itô integral. Before verifying that (4.2) solves equation (4.1), we first recall two preliminary results. Firstly, we observe that the Itô formula applied to the exponential function $x\mapsto e^{-ix}$ and the measure-valued diffusion process $\nu_t$ yields the formula

$$\begin{aligned} e^{-i\nu_t} &= \delta_{\{0\}} - \frac{1}{2}\int_0^t e^{-i\nu_s} * (\mu_s\circ S_2) * (\mu_s\circ S_2)\,d\langle M,M\rangle_s\\ &\qquad -i\int_0^t e^{-i\nu_s} * (\mu_s\circ S_2)\,dM_s\ . \end{aligned} \tag{4.4}$$

Here $\delta_{\{0\}}$ is the delta measure concentrated at the path identically equal to zero, $\langle M, M\rangle_s$ is the quadratic variation process of the continuous semimartingale $M$, and $*$ denotes convolution. Observe again the stochastic integral is in the Itô sense. Secondly, the solution of the non-stochastic Schrödinger equation (2.2) with $\hbar = m = 1$, initial condition $\psi_t(x)|_{t=0} = \Psi_0(x)$, and same conditions on $V$ and $\Psi_0$ as before, can be represented by the configuration space path integral

$$\left[e^{-itH}\Psi_0\right](x) = \int_{\mathcal{H}_t} e^{\frac{i}{2}|q|_t^2 - i\int_0^t V(s,x+q_s)\,ds}\,\Psi_0(x+q_0)\,dq\ ,$$

and therefore by our Proposition 3.3, we can write this solution as the phase space path integral

$$\left[e^{-itH}\Psi_0\right](x) = \int_{\mathcal{H}_t\times\mathcal{L}_t} e^{i\int_0^t (p_s\dot{q}_s - \frac{1}{2}p_s^2)\,ds - i\int_0^t V(s,x+q_s)\,ds}\,\Psi_0(x+q_0)\,d(q,p)\,,$$

where $H = -\frac{1}{2}\Delta_x + V(t,x)$. Observe we are denoting by $\Psi_t(x)$ the solution of the stochastic equation (4.1), and by $\psi_t(x)$ the solution of the non-stochastic equation (2.2). Both equations having the same initial wave function $\Psi_0$. We now prove that the path integral (4.2) solves equation (4.1). By definition and then using formula (4.4) we have

$$\begin{aligned}
\Psi_t(x) &= \int_{\mathcal{L}_t\times\mathcal{H}_t} e^{-i\int_0^t (p_s\dot{q}_s + \frac{1}{2}p_s^2)\,ds}\, d\left[(\sigma\circ S_1) * e^{-i\nu_t}\right](p,q) \\
&= \int_{\mathcal{L}_t\times\mathcal{H}_t} e^{-i\int_0^t (p_s\dot{q}_s + \frac{1}{2}p_s^2)\,ds}\, d\left[(\sigma\circ S_1) *\right. \\
&\qquad (\delta_0 - \frac{1}{2}\int_0^t e^{-i\nu_s} * (\mu_s\circ S_2) * (\mu_s\circ S_2)\, d\langle M, M\rangle_s \\
&\qquad -i\int_0^t e^{-i\nu_s} * (\mu_s\circ S_2)\, dM_s)]\ (p,q) \\
&= \int_{\mathcal{L}_t\times\mathcal{H}_t} e^{-i\int_0^t (p_s\dot{q}_s + \frac{1}{2}p_s^2)\,ds}\, d[(\sigma\circ S_1) * \delta_0](p,q) \\
&\qquad + \int_{\mathcal{L}_t\times\mathcal{H}_t} e^{-i\int_0^t (p_s\dot{q}_s + \frac{1}{2}p_s^2)\,ds}\, d\left[(\sigma\circ S_1) *\right. \\
&\qquad (-\frac{1}{2}\int_0^t e^{-i\nu_s} * (\mu_s\circ S_2) * (\mu_s\circ S_2)\, d\langle M, M\rangle_s)](p,q) \\
&\qquad + \int_{\mathcal{L}_t\times\mathcal{H}_t} e^{-i\int_0^t (p_s\dot{q}_s + \frac{1}{2}p_s^2)\,ds} \\
&\qquad\qquad d\left[(\sigma\circ S_1) * (-i\int_0^t e^{-i\nu_s} * (\mu_s\circ S_2)\, dM_s)\right](p,q) \\
&= e^{-itH}\Psi_0(x) - \frac{1}{2}\int_0^t d\langle M, M\rangle_s \int_{\mathcal{L}_t\times\mathcal{H}_t} e^{-i\int_0^t (p_s\dot{q}_s + \frac{1}{2}p_s^2)\,ds} d[(\sigma\circ S_1) * \\
&\qquad (e^{-i\nu_s} * (\mu_s\circ S_2) * (\mu_s\circ S_2))](p,q) \\
&\qquad -i\int_0^t dM_s \int_{\mathcal{L}_t\times\mathcal{H}_t} e^{-i\int_0^t (p_s\dot{q}_s + \frac{1}{2}p_s^2)\,ds} \\
&\qquad\qquad d[(\sigma\circ S_1) * (e^{-i\nu_s} * (\mu_s\circ S_2))](p,q)\,.
\end{aligned}$$

We now look at the Lebesgue integral in the last line. We will use the iterative property of Proposition (3.2) to find an expression for this integral. Thus, the referred integral is

$$\begin{aligned}
&\int_{\mathcal{L}_t\times\mathcal{H}_t} e^{-i\int_0^t (p_s\dot{q}_s+\frac{1}{2}p_s^2)\,ds} d[(\sigma\circ S_1)*(e^{-i\nu_s}*(\mu_s\circ S_2))](p,q)\\
=&\int_{\mathcal{H}_t\times\mathcal{L}_t} e^{i\int_0^t (p_s\dot{q}_s-\frac{1}{2}p_s^2)\,ds}\,(\widehat{\sigma\circ S_1})(q,p)(\widehat{e^{-i\nu_s}})(q,p)(\widehat{\mu_s\circ S_2})(q,p)\,d(q,p)\\
=&\int_{\mathcal{H}_t\times\mathcal{L}_t} e^{i\int_0^t (p_s\dot{q}_s-\frac{1}{2}p_s^2)\,ds}\,e^{-i\int_0^t V(r,x+q_r)\,dr-i\int_0^s K(r,p_r)\,dM_r}\\
&\qquad\Psi_0(x+q_0)K(s,p_s)\,d(q,p)\\
=&\int_{\mathcal{H}_{[s,t]}\times\mathcal{L}_{[s,t]}} e^{i\int_s^t (d_r\dot{c}_r-\frac{1}{2}d_r^2)\,dr}\,e^{-i\int_s^t V(r,x+c_r)\,dr}K(s,d_s)\cdot\\
&\int_{\mathcal{H}_s\times\mathcal{L}_s} e^{i\int_0^s (b_r\dot{a}_r-\frac{1}{2}b_r^2)\,dr}\,e^{-i\int_0^s V(r,x+c_s+a_r)\,dr-i\int_0^s K(r,b_s+d_r)\,dM_r}\\
&\qquad\Psi_0(x+c_s+a_0)\,d(a,b)\,d(c,d)\\
=&\int_{\mathcal{H}_{[s,t]}\times\mathcal{L}_{[s,t]}} e^{i\int_s^t (d_r\dot{c}_r-\frac{1}{2}d_r^2)\,dr}\,e^{-i\int_s^t V(r,x+c_r)\,dr}K(s,d_s)\Psi_s(x+c_s)\,d(c,d)\\
=&\; e^{-i(t-s)H}K(s,-i\frac{\partial}{\partial x})\Psi_s(x)\,.
\end{aligned}$$

The other path integral is solved similarly. We thus arrive at the expression

$$\begin{aligned}
\Psi_t(x) &= e^{-itH}\Psi_0(x)-\frac{1}{2}\int_0^t e^{-i(t-s)H}K^2(s,-i\frac{\partial}{\partial x})\Psi_s(x)\,d\langle M,M\rangle_s\\
&\quad -i\int_0^t e^{-i(t-s)H}K(s,-i\frac{\partial}{\partial x})\Psi_s(x)\,dM_s\,.
\end{aligned}$$

Multiplying through by $e^{itH}$ and differentiating respect to $t$ gives

$$\begin{aligned}
\left(e^{itH}iH\Psi_t\right)(x)+\left(e^{itH}d\Psi_t\right)(x) &= -\frac{1}{2}e^{itH}K^2(t,-i\frac{\partial}{\partial x})\Psi_t(x)\,d\langle M,M\rangle_t\\
&\quad -ie^{itH}K(t,-i\frac{\partial}{\partial x})\Psi_t(x)\,dM_t\,.
\end{aligned}$$

Hence

$$d\Psi_t(x)=-iH\Psi_t-\frac{1}{2}K^2(t,-i\frac{\partial}{\partial x})\Psi_t(x)\,d\langle M,M\rangle_t-iK(t,-i\frac{\partial}{\partial x})\Psi_t(x)\,dM_t\,,$$

but this is the Itô version of the Stratonovich equation

$$id\Psi_t(x)=H\psi_t+K(t,-i\frac{\partial}{\partial x})\Psi_t(x)\circ dM_t\,.$$

□

Applications of this result will be presented in the following section where we solve the path integral for a harmonic oscillator.

## 5. A stochastic Mehler kernel

In this section we solve the one-dimensional Schrödinger equation

$$id\Psi_t(x) = \left(-\frac{1}{2}\frac{\partial^2}{\partial x^2} + \frac{1}{2}x^2\right)\Psi_t(x)dt + \left(-i\frac{\partial}{\partial x}\right)\Psi_t(x) \circ dM_t\,,$$

where $M_t(\omega) : [0,\infty) \times \Omega \to \mathbb{R}$ is a continuous semimartingale vanishing at $t = 0$. We calculate the kernel, or propagator, using a discretisation of the path integral, and also by a change of variable in phase space. First we observe that the potential functions $V(t,x) = x^2/2$ and $K(t,x) = x$ do not belong to $\mathcal{F}(\mathbb{R})$, so we cannot use our representation theorem proved earlier. Nevertheless, the expression of the path integral *suggested* by our representation theorem will give us the correct solution of the above Schrödinger equation.

THEOREM 5.1. *The kernel $G_t(x,y)$ for the stochastic Schrödinger equation*

$$id\Psi_t(x) = \left(-\frac{1}{2}\frac{\partial^2}{\partial x^2} + \frac{1}{2}x^2\right)\Psi_t(x)dt + \left(-i\frac{\partial}{\partial x}\right)\Psi_t(x) \circ dM_t \tag{5.1}$$

*with initial condition $\Psi_0(\cdot) = \delta_y(\cdot)$, can be represented by the phase space path integral*

$$G_t(x,y) = \int_{\mathcal{H}_t\times\mathcal{L}_t} e^{iS(x+q,p)}\delta_y(x+q_0)\,d(q,p)\,,$$

*where the action $S$ is $S(x+q,p) = \int_0^t (p_s\dot{q}_s - \frac{1}{2}p_s^2 - \frac{1}{2}(x+q_s)^2)\,ds - \int_0^t p_s \circ dM_s$. This gives the kernel*

$$\begin{aligned} G_t(x,y) &= \frac{1}{\sqrt{2\pi i \sin t}} \exp\left(i\frac{(x^2+y^2)\cos t - 2xy}{2\sin t}\right) \\ &\quad \exp\left(i\int_0^t \frac{-x\cos r + y\cos(t-r)}{\sin t} \circ dM_r\right) \\ &\quad \exp\left(i\int_0^t\left[\int_0^r \frac{\cos s\ \cos(t-r)}{\sin t}\circ dM_s\right]\circ dM_r\right). \end{aligned}\tag{5.2}$$

When $M = 0$ formula (5.2) reduces to what is known as the Mehler kernel. We prove the above formula twice, first by discretising the path integral, the second method is shorter and uses a change of variable.

PROOF. (First) We here discretise the path integral for equation (5.1), and obtain solution (5.2) as a limiting process. We only give the setup of this procedure and leave the details behind as the complete process involves a great deal of calculations. Let $0 < t_1 < t_2 < \cdots < t_N < t$ be a partition of the time interval $[0,t]$ into $N+1$ subintervals such that $t_{i+1} - t_i = 2\epsilon > 0$. At each time $t_i$ we have information of the position of the particle $q_i = q(t_i)$, and at the middle of each interval we know the momenta $p_i = p(t_{i-1}+\epsilon)$. We observe that as a consequence of Heisenberg's uncertainty principle we cannot have precise information of position and momentum at the same time. Then $2(N+1)\epsilon = t$, and we have $N$ $q$-variables $q_i$, and $(N+1)$ $p$-variables $p_i$. Note also that $q_0 = y$, $q_{N+1} = q_t = x$, and no boundary constraints are imposed on the paths $p$. Thus, we have that the discretised action

$S(\bar{q},\bar{p})$ is

$$\begin{aligned} S(\bar{q},\bar{p}) &= \sum_{j=0}^{N} p_{j+1}(q_{j+1}-q_j) - \epsilon\sum_{j=0}^{N}(p_{j+1})^2 \\ &\quad -\epsilon\sum_{j=0}^{N}\left(\frac{q_j+q_{j+1}}{2}\right)^2 - \sum_{j=0}^{N} p_{j+1}\,\Delta_j M\,, \end{aligned}$$

where $\bar{q}=(q_1,q_2,\ldots,q_N)$, $\bar{p}=(p_1,p_2,\ldots,p_N,p_{N+1})$ and $\Delta_j M = M(t_{j+1})-M(t_j)$. The path integral is then *assumed* to be calculated as follows

$$G_t(x,y) = \lim_{\epsilon\to 0}\,(2\pi)^{-(N+1)}\int_{\mathbb{R}^{2N+1}} e^{iS(\bar{q},\bar{p})}\,d(\bar{q},\bar{p})\,,$$

where the limit is taken in such a way that also $N\to\infty$, and $2(N+1)\epsilon = t$ remains constant. The approximated action $S(\bar{q},\bar{p})$ can be written in matrix form as

$$S(\bar{q},\bar{p}) = \frac{1}{2}\left\{(\bar{q},\bar{p})^T\mathbb{A}(\bar{q},\bar{p}) + B^T(\bar{q},\bar{p}) + O\right\}\,,$$

where $\mathbb{A}$ is the $(2N+1)\times(2N+1)$ matrix

$$\mathbb{A} = \begin{pmatrix} A & J \\ J^T & (-2\epsilon)I \end{pmatrix},$$

with

$$A = (-\frac{\epsilon}{2})\begin{pmatrix} 2 & 1 & 0 & 0 & \cdots \\ 1 & 2 & 1 & 0 & \cdots \\ 0 & 1 & 2 & 1 & \cdots \\ 0 & 0 & 1 & 2 & \cdots \\ \vdots & \vdots & \vdots & \vdots & \end{pmatrix} \qquad J = \begin{pmatrix} 1 & -1 & 0 & 0 & 0 & \cdots \\ 0 & 1 & -1 & 0 & 0 & \cdots \\ 0 & 0 & 1 & -1 & 0 & \cdots \\ 0 & 0 & 0 & 1 & -1 & \cdots \\ \vdots & \vdots & \vdots & \vdots & & \end{pmatrix},$$

where $A$ is a $N\times N$ matrix, $J$ is a $N\times(N+1)$ matrix, and $I$ is the $(N+1)\times(N+1)$ identity matrix. The vector $B=(B_1^T,B_2^T)$ has components

$$B_1 = \begin{pmatrix} -\epsilon y \\ 0 \\ \vdots \\ 0 \\ -\epsilon x \end{pmatrix}_{N\times 1}, \qquad B_2 = \begin{pmatrix} -2\Delta_0 M - 2y \\ -2\Delta_1 M \\ \vdots \\ -2\Delta_{N-1}M \\ -2\Delta_N M + 2x \end{pmatrix}_{(N+1)\times 1},$$

and $O = -\frac{\epsilon}{2}(x^2+y^2)$. Observe that $\mathbb{A}$ is a real symmetric matrix and hence we have the formula

$$\int_{\mathbb{R}^{2N+1}} e^{\frac{i}{2}\{(\bar{q},\bar{p})^T\mathbb{A}(\bar{q},\bar{p})+B^T(\bar{q},\bar{p})+O\}}d(\bar{q},\bar{p}) = (2\pi i)^{(2N+1)/2}\mid\mathbb{A}\mid^{-1/2}e^{\{\frac{i}{2}(O-\frac{1}{4}B^T\mathbb{A}^{-1}B)\}}\,,$$

where $\mid\mathbb{A}\mid$ stands for the determinant of matrix $\mathbb{A}$. Therefore we have

$$G_t(x,y) = \lim_{\epsilon\to 0}\,(2\pi)^{-(N+1)}(2\pi i)^{(2N+1)/2}\mid\mathbb{A}\mid^{-1/2}e^{\{\frac{i}{2}(O-\frac{1}{4}B^T\mathbb{A}^{-1}B)\}}\,.$$

The main obstacle in this discretisation method is the computation of the determinant $\mid\mathbb{A}\mid$ and the inverse $\mathbb{A}^{-1}$. The former is calculated by using the following product

$$\begin{pmatrix} A & J \\ J^T & (-2\epsilon)I \end{pmatrix}\begin{pmatrix} (-2\epsilon)I & 0 \\ -J^T & I \end{pmatrix} = \begin{pmatrix} (-2\epsilon)A - JJ^T & J \\ 0 & (-2\epsilon)I \end{pmatrix},$$

from where we obtain $|\mathbb{A}| = (-2\epsilon)\,|(-2\epsilon)A - JJ^T|$, with matrix $D := (-2\epsilon)A - JJ^T$ taking the tridiagonal form

$$D = \begin{pmatrix} a & -b & 0 & 0 & \cdots \\ -b & a & -b & 0 & \cdots \\ 0 & -b & a & -b & \cdots \\ 0 & 0 & -b & a & \cdots \\ \vdots & \vdots & \vdots & \vdots & \end{pmatrix}_{N\times N},$$

with $a = 2\epsilon^2 - 2$, and $b = -(\epsilon^2 + 1)$. On the other hand, the inverse of $\mathbb{A}$ is given by

$$\mathbb{A}^{-1} = \begin{pmatrix} -2\epsilon D^{-1} & -D^{-1}J \\ -J^T D^{-1} & -\frac{1}{2\epsilon}(I + J^T D^{-1} J) \end{pmatrix},$$

and thus we have reduced the problem of the computation of $|\mathbb{A}|$ and $\mathbb{A}^{-1}$ to the calculation of the same operations for the tridiagonal matrix $D$. Expanding out the product $B^T\mathbb{A}^{-1}B$ gives

$$\begin{aligned} B^T\mathbb{A}^{-1}B = (-2\epsilon)B_1^T D^{-1} B_1 \;-\; B_1^T D^{-1} J B_2 \;-\; B_2^T J^T D^{-1} B_1 \\ -\frac{1}{2\epsilon} B_2^T (I + J^T D^{-1} J) B_2 \,. \end{aligned}$$

Hence our kernel is given by the following product of limits.

$$\begin{aligned} G_t(x,y) &= \lim_{\epsilon\to 0} (2\pi)^{-(N+1)} (2\pi i)^{(2N+1)/2} |\mathbb{A}|^{-1/2} \\ & \lim_{\epsilon\to 0} \exp\left[\frac{i}{2}(-\frac{1}{4})(-2\epsilon B_1^T D^{-1} B_1)\right] \\ & \lim_{\epsilon\to 0} \exp\left[\frac{i}{2}(-\frac{1}{4})(-B_1^T D^{-1} J B_2)\right] \\ & \lim_{\epsilon\to 0} \exp\left[\frac{i}{2}(-\frac{1}{4})(-B_2^T J^T D^{-1} B_1)\right] \\ & \lim_{\epsilon\to 0} \exp\left[\frac{i}{2}(-\frac{1}{4})(-\frac{1}{2\epsilon} B_2^T (I + J^T D^{-1} J) B_2)\right] . \end{aligned}$$

We can then show that all the above limits exist. We will omit all together the rest of the calculations as they consist basically of an enormous amount of sums and products and we cannot give all the details here. The final results are the following.

**(a)** $\lim_{\epsilon\to 0} (2\pi)^{-(N+1)} (2\pi i)^{(2N+1)/2}\, |\mathbb{A}|^{-1/2} = (2\pi i \sin t)^{-1/2}$.

**(b)** $\lim_{\epsilon\to 0} \exp\left[]\frac{i}{2}(-\frac{1}{4})(-2\epsilon B_1^T D^{-1} B_1)\right] = 1$.

**(c)** $\lim_{\epsilon\to 0} \exp\left[\frac{i}{2}(-\frac{1}{4})(-B_1^T D^{-1} J B_2)\right] = 1$.

**(d)** $\lim_{\epsilon\to 0} \exp\left[\frac{i}{2}(-\frac{1}{4})(-B_2^T J^T D^{-1} B_1)\right] = 1$.

(e) $\lim_{\epsilon\to 0} \exp\left[\frac{i}{2}(-\frac{1}{4})(-\frac{1}{2\epsilon}B_2^T(I+J^TD^{-1}J)B_2)\right] = \exp\left(i\frac{(x^2+y^2)\cos t - 2xy}{2\sin t}\right)$

$$\exp\left(i\int_0^t \frac{-x\cos r + y\cos(t-r)}{\sin t} \circ dM_r + i\int_0^t\int_0^r \frac{\cos s\ \cos(t-r)}{\sin t} \circ dM_s \circ dM_r\right).$$

Putting all these limits together gives solution (5.2). We can directly substitute in our Schrödinger equation and check that (5.2) is indeed a solution of our equation (5.1). □

PROOF. (Second) We give here a shorter but formal way to prove solution (5.2) via a change of variable. Let us assume the kernel $G_t(x,y)$ of equation (5.1) admits the phase space path integral representation

$$G_t(x,y) = \int_{\mathcal{H}_t\times\mathcal{L}_t} e^{iS(x+q,p)}\delta_y(x+q_0)\,d(q,p)\,,$$

where the action $S$ is given by

$$S(x+q,p) = \int_0^t (p_s\dot{q}_s - \frac{1}{2}p_s^2 - \frac{1}{2}|x+q_s|^2)\,ds - \int_0^t p_s \circ dM_s\,.$$

Through a translation we will reduce the above integral to the case of the classical Mehler formula. Let us define the canonical transformation $T: \mathcal{H}_t\times\mathcal{L}_t \to \mathcal{H}_t\times\mathcal{L}_t$ by

$$T(q_s,p_s) = (q_s+\dot{\xi}_s, p_s - \xi_s)\,,$$

where the path $\xi : [0,t] \to \mathbb{R}$ is the solution of the stochastic integro-differential equation

$$d\dot{\xi}_s + \xi_s\,ds = \circ\, dM_s\ , \tag{5.3}$$

for $s\in[0,t]$, subject to the boundary conditions $\xi_0 = 0$, and $\dot{\xi}_t = 0$. The stochastic equation (5.3) can be explicitly solved, its solution being

$$\xi(s) = -\frac{\sin s}{\cos t}\int_0^t \cos(u-t)\circ dM_u - \int_0^s \sin(u-s)\circ dM_u\,. \tag{5.4}$$

Then, under this translation, we have

$$\begin{aligned} S(x+q+\dot{\xi}, p-\xi) \;=\; & \int_0^t (p_s\dot{q}_s - \frac{1}{2}p_s^2 - \frac{1}{2}(x+q_s)^2)\,ds \\ & + \int_0^t (p_s\ddot{\xi}_s + p_s\xi_s)\,ds - \int_0^t p_s\circ dM_s \\ & + \int_0^t (-\xi\dot{q}_s - q_s\dot{\xi}_s)\,ds \\ & + \int_0^t (-\xi_s\ddot{\xi}_s - \frac{1}{2}\xi_s^2 - x\dot{\xi}_s^2 - \frac{1}{2}\dot{\xi}_s^2)\,ds + \int_0^t \xi_s\circ dM_s\,. \end{aligned}$$

The first line of the above expression we will keep inside the path integral. The second line vanishes by using the differential equation for $\xi_s$ multiplied by $p_s$ and integrated. The third line also vanishes by using the integration by parts formula and the boundary conditions for $\xi_s$. The fourth and last line we will place outside the path integral. This then gives us the expression

$$\begin{aligned} G_t(x,y) &= e^{i\int_0^t(-\xi_s\ddot{\xi}_s-\frac{1}{2}\xi_s^2-x\dot{\xi}_s^2-\frac{1}{2}\dot{\xi}_s^2)\,ds+i\int_0^t\xi_s\circ dM_s} \\ &\quad \int_{\mathcal{H}_t\times\mathcal{L}_t} e^{i\int_0^t(p_s\dot{q}_s-\frac{1}{2}p_s^2-\frac{1}{2}(x+q_s)^2)\,ds}\delta_{y-\dot{\xi}_0}(x+q_0)\,d(q,p)\ , \end{aligned}$$

where the last path integral is solved by the Mehler kernel formula

$$(2\pi i\sin t)^{-1/2}\exp\left\{i\frac{[x^2+(y-\dot{\xi}_0)^2]\cos t-2x(y-\dot{\xi}_0)}{2\sin t}\right\}.$$

Comparing the above solution with (5.2), we see that the coefficients of the terms $x^2$, $y^2$, and $-2xy$ in the exponent, are the right ones. We then need to verify the coefficients corresponding to the terms $y$, $x$, and the independent term, to completely verify our solution. For this we need from (5.4) the equation

$$\dot{\xi}_s=-\frac{\cos s}{\cos t}\int_0^t\cos(u-t)\circ dM_u+\int_0^s\sin(u-s)\,dM_u\ ,$$

and hence

$$\dot{\xi}_0=-\frac{1}{\cos t}\int_0^t\cos(u-t)\circ dM_u\ .$$

With these formulae we can easily prove the following identities.

For the coefficient of $y$,

$$-\dot{\xi}_0\frac{\cos t}{\sin t}=\frac{1}{\sin t}\int_0^t\cos(u-t)\circ dM_r\ .$$

For the coefficient of $x$,

$$\dot{\xi}_0\frac{1}{\sin t}-\int_0^t\dot{\xi}_s\,ds=-\frac{1}{\sin t}\int_0^t\cos r\circ dM_r\ .$$

For the independent term,

$$\begin{aligned} &\dot{\xi}_0^2\frac{\cos t}{2\sin t}+\int_0^t(-\xi_s\ddot{\xi}_s-\frac{1}{2}\xi_s^2-\frac{1}{2}\dot{\xi}_s^2)\,ds+\int_0^t\xi_s\circ dM_s \\ &=\int_0^t\left[\int_0^r\frac{\cos s\ \cos(t-r)}{\sin t}\circ dM_s\right]\circ dM_r\ . \end{aligned}$$

Again, the proofs of the above three identities are straightforward but require some lengthy computations. After this we can finally reconstruct our solution (5.2) once more. □

When the constants $\hbar$ and $m$ are incorporated in equation (5.1), it is computationally easier to consider that the potentials are also multiplied by $m$, and thus we consider the slightly more general equation

$$(5.5)\qquad i\hbar d\Psi_t(x)=\left(-\frac{\hbar^2}{2m}\frac{\partial^2}{\partial x^2}+\frac{m}{2}w^2x^2\right)\Psi_t(x)dt+m\left(-i\frac{\partial}{\partial x}\right)\Psi_t(x)\circ dM_t,$$

which has propagator

$$
\begin{aligned}
G_t(x,y) \quad &= \quad \sqrt{\frac{mw}{2\pi i\hbar \sin wt}} \exp\left(\frac{imw}{2\hbar}\frac{(x^2+y^2)\cos wt - 2xy}{\sin wt}\right) \\
&\exp\left(\frac{im}{\hbar}\int_0^t \frac{-x\cos wr + y\cos w(t-r)}{\sin wt}\circ dM_r\right) \\
&\exp\left(\frac{im}{\hbar w}\int_0^t \left[\int_0^r \frac{\cos ws\ \cos w(t-r)}{\sin wt}\circ dM_s\right]\circ dM_r\right). \qquad (5.6)
\end{aligned}
$$

This formula is easily derived from the one we obtained when we considered the simplifying case $\hbar = m = 1$. Finally we consider the associated heat equation.

COROLLARY 5.2. *The kernel of the stochastic heat equation*

$$
du_t(x) = \left(\frac{\mu}{2}\frac{\partial^2}{\partial x^2} - \frac{1}{2\mu}w^2x^2\right)u_t(x)dt - \frac{1}{\mu}\left(-i\frac{\partial}{\partial x}\right)u_t(x)\circ dM_t. \qquad (5.7)
$$

*is given by*

$$
\begin{aligned}
G_t(x,y) \quad &= \quad \sqrt{\frac{w}{2\pi\mu \sinh wt}} \exp\left(-\frac{w}{2\mu}\frac{(x^2+y^2)\cosh wt - 2xy}{\sinh wt}\right) \\
&\exp\left(\frac{1}{\mu}\int_0^t \frac{-x\cosh wr + y\cosh w(t-r)}{\sinh wt}\circ dM_r\right) \\
&\exp\left(\frac{1}{\mu w}\int_0^t \left[\int_0^r \frac{\cosh ws\ \cosh w(t-r)}{\sinh wt}\circ dM_s\right]\circ dM_r\right). \qquad (5.8)
\end{aligned}
$$

PROOF. From (5.5), making the change of time to purely imaginary time, $t \mapsto -it$ and setting $\mu = \hbar/m$, we obtain the associated stochastic heat equation (5.7). Using the formulae $\sinh t = i\sin(-it)$ and $\cosh t = \cos(-it)$ we can easily transform solution (5.6) into (5.8). This turns out to be an explicit expression for the solution of (5.7) as can be checked directly. □

## 6. Conclusion

We have given a detailed proof of a path integral representation for the solution of a stochastic Schrödinger equation driven by any continuous semimartingale. With the aid of this result we have been able to confirm, twice, a formula derived previously by A. Truman and T. Zastawniak [**T-Z1**], [**T-Z2**], concernig a stochastic version of the Mehler kernel. The discretization method we used in our first proof only works because the potentials involved are at most quadratic. It is certainly not the best method to get our result. The second proof is only a change of variable and thus it is a quicker road to the same destiny. As a consequence, via a complex time continuity argument we have also given the explicit solution to the corresponding initial value heat equation.

**Acknowledgments.** It is a pleasure to thank Prof. A. Truman of the University of Wales Swansea for suggesting the topic of research and for very helpful discussions. I would also like to thank the anonymous referee for the careful reading of the manuscript and the valuable comments.

## References


[A-H] S.A. Albeverio and R.J. Høegh-Krohn, *Mathematical Theory of Feynman Path Integrals.*, Lect. Notes Math., Vol. **523**, Springer-Verlag, Berlin, (1976).

[A] A.M. Arthurs (Ed.), *Functional Integration and its Applications.*, Clarendon Press, Oxford, (1975).

[C-D] M. Chaichian and A. Demichev, *Path Integrals in Physics Vol. 1: Stochastic Processes and Quantum Mechanics.*, Institute of Physics Publishing, (2001).

[F] R.P. Feynman, *Space-time Approach to Non-Relativistic Quantum Mechanics.*, Rev. Mod. Phys. **20** (1948), pp. 367-387.

[F-H] R.P. Feynman and A.R. Hibbs, *Quantum Mechanics and Path Integrals.*, McGraw Hill, New York, (1965).

[I] K. Itô *Generalized Uniform Complex Measures in the Hilbertian Metric Space with Their Application to the Feynman Path Integral.*, Proceedings of the Fifth Berkeley Symposium on Mathematical Statistics and Probability Vol II, Univ. of California Press, Berkeley, (1967), pp. 145–161.

[K-Mc] J.B. Keller and D.W. McLaughlin, *The Feynman Integral.*, Amer. Math. Month., **82**, (1975), pp. 451–465.

[M] C. Morette, *On the Definition and Approximation of Feynman's Path Integrals.*, Phys. Rev., **81**, (1951), pp. 848–852.

[N] E. Nelson, *Feynman Integrals and the Schrödinger Equation.*, J. Math. Phys., **5**, (1964), pp. 332–343.

[R] L.A. Rincon, *Path Integral Representation for the Solution of a Stochastic Schrödinger Equation Driven by a Semimartingale*, Manuscript submitted to the Journal of Interdisciplinary Mathematics, (2002).

[Sc] L.S. Schulman, *Techniques and Applications of Path Integration.*, John Wiley & Sons, New York, (1981).

[Si] B. Simon, *Functional Integration and Quantum Physics.*, Academic Press, New York, (1979).

[T-Z1] A. Truman and T. Zastawniak *Stochastic PDE's of Schrödinger Type and Stochastic Mehler Kernels: A Path Integral Approach.*, Proceedings of the Seminar on Stochastic Analysis, Random Fields and Applications, 1997. Editor Marco Dozzi, (1998).

[T-Z2] A. Truman and T. Zastawniak, *Stochastic Mehler kernels Via oscillatory Path Integrals.*, J. Korean Math. Soc. **38**, No. 2, (2001), pp. 469–483.

[Z] T. J. Zastawniak, *Fresnel Type Path Integral for the Stochastic Schrödinger Equation*, Lett. Math. Phys. **41**, No. 1, (1997), pp. 93–99.



Departamento de Matemáticas, Facultad de Ciencias UNAM, Circuito Exterior de CU, 04510 México DF.

*E-mail address*: lars@fciencias.unam.mx

Contemporary Mathematics
Volume **336**, 2003

# A Note on Covariance Characterization of some Generalized Gaussian Random Fields

Anna Talarczyk

ABSTRACT. Conditions characterizing some space–homogeneous Gaussian random fields, such as conditions which imply that an $\mathcal{S}'(\mathbb{R}^d)$ process is in fact function–valued or conditions on existence of self–intersection local time, are often given in terms of the spectral measure. In this paper we show how to express these conditions in terms of the spatial correlation.

## 1. Introduction

We will consider processes or random variables taking values in the space of tempered distributions (denoted by $\mathcal{S}'(\mathbb{R}^d)$), conjugated to the Schwartz space $\mathcal{S}(\mathbb{R}^d)$ of smooth rapidly decreasing functions.

A generalized space–homogeneous Gaussian random field $Z$ is a Gaussian random variable with values in $\mathcal{S}'(\mathbb{R}^d)$ such that

$$E\langle Z,\varphi\rangle\langle Z,\psi\rangle = \langle \Gamma, \varphi * \psi_{(s)}\rangle, \qquad \varphi,\psi\in\mathcal{S}(\mathbb{R}^d), \tag{1.1}$$

where $\psi_{(s)}(x) = \psi(-x)$, $*$ denotes convolution, $\Gamma$ is a positive definite tempered distribution. $\Gamma$ is called *space correlation.* In particular if $\Gamma$ is a bounded function then $Z$ is function–valued and $EZ(x)Z(y) - \Gamma(x-y)$.

(1.1) can be written in terms of the *spectral measure* $\mu = \mathcal{F}^{-1}\Gamma$ ($\mathcal{F}$ is the Fourier transform)

$$E\langle Z,\varphi\rangle\langle Z,\psi\rangle = \int_{\mathbb{R}^d} \mathcal{F}\varphi(x)\overline{\mathcal{F}\psi(x)}\mu(dx), \qquad \varphi,\psi\in\mathcal{S}(\mathbb{R}^d). \tag{1.2}$$

$\mu$ is a non-negative tempered measure.

In many examples [**K-Z**], [**K-Z1**], [**P-Z**], [**S-S**], [**L**], [**B-G**] there appear processes which are solutions of stochastic equations driven by a space homogeneous Wiener process $W$, i.e. a centered Gaussian process in $\mathcal{S}'(\mathbb{R}^d)$ such that

$$E\langle W_t,\varphi\rangle\langle W_u,\psi\rangle = t\wedge u\langle \Gamma, \varphi * \psi_{(s)}\rangle, \qquad \varphi,\psi\in\mathcal{S}(\mathbb{R}^d),$$

---

2000 *Mathematics Subject Classification.* Primary 60G20, 60G15, 60H15.

*Key words and phrases.* Function–valued Solutions, Self–intersection Local Time, Tempered Distributions, Gaussian Random Fields.

Research supported in part by KBN grant 2P03A02622, Poland.

The characterization of these processes is often given in terms of the spectral measure $\mu$.

Consider for example the equation

$$dX = \Delta_\alpha X dt + dW_t, \tag{1.3}$$

with zero initial condition, where $\Delta_\alpha$ is the fractional Laplacian $\Delta_\alpha = -(-\Delta)^{\alpha/2}$, $0 < \alpha \leq 2$. It was shown in [**K-Z1**] that the equation (1.3) has a function–valued solution if and only if

$$\int_{\mathbb{R}^d} \frac{1}{1+|x|^\alpha} \mu(dx) < \infty. \tag{1.4}$$

The same condition is necessary and sufficient for existence of function–valued solutions of the corresponding wave equation. Conditions of similar form were obtained in [**P-Z**], [**L**] and [**S-S**], where the authors considered existence of function–valued solutions of heat and wave equations and their Hölder continuity.

In [**B-G**] a condition for the existence of self–intersection local time of order 2 for the solution of (1.3) was also given in terms of the spectral measure $\mu$:

$$\int_{\mathbb{R}^d} |\varphi(x+y)| \frac{(\mu+\mu_0)(dx)}{1+|x|^{2\alpha}} \frac{(\mu+\mu_0)(dy)}{1+|y|^{2\alpha}} < \infty, \quad \text{for all } \varphi \in \mathcal{S}(\mathbb{R}^d), \tag{1.5}$$

where $\mu_0$ is the spectral measure of the generalized space–homogeneous random field $X_0$. Equivalently, by Lemma 6.2 in [**B-G**], $|\varphi(x+y)|$ can be replaced by $|\mathcal{F}\varphi(x+y)|^2$ in formula (1.5). In the same paper, the authors also dealt with the self–intersection local time of order 2 of the Wiener process itself. They gave the condition for the existence in terms of $\mu$ and also showed that if $\Gamma$ has a nonnegative density, this condition is equivalent to the fact that the density is square tempered.

In this paper we show how to express (1.4) and (1.5) with help of the space correlation $\Gamma$.

For (1.4), Karczewska and Zabczyk in [**K-Z1**] stated that if $\Gamma$ is a nonnegative tempered measure and $\mu = \mathcal{F}^{-1}\Gamma$ then

$$\int_{\mathbb{R}^d} \frac{1}{1+|x|^\alpha} \mu(dx) = \int_{\mathbb{R}^d} G_\alpha(x)\Gamma(dx), \tag{1.6}$$

where $G_\alpha(x) = \int_0^\infty e^{-t} p_t^\alpha(x) dt$ and $p_t^\alpha$ denotes the transition density of a symmetric $\alpha$-stable process. (1.6) can be considered as a generalized Plancherel equality, but this requires a proof since $1/(1+|x|^\alpha)$ does not belong to $\mathcal{S}(\mathbb{R}^d)$. Equality (1.6) was proved in [**K-Z**] only for $\alpha = 2$. For $0 < \alpha < 2$ the equation (1.6) was formulated in [**K-Z1**], but the suggestion of the proof turned out to be incomplete.

Another approach to write (1.4) with the help of $\Gamma$ was taken in [**S-S**], but there was an error, which was pointed out in [**L**] p.42.

In the current paper we show that (1.6) is indeed true for any $0 < \alpha \leq 2$ (Proposition 2.1). Next, in Theorems 2.2 and 2.3 we express the condition (1.5) for existence of self–intersection local time in terms of $\Gamma$, under the assumption that $\Gamma$ is a nonnegative measure.

Since there are some known estimates for $G_\alpha$, these results allow us to reduce the problem of existence of function–valued solutions or existence of self–intersection local time to proper integrability of $\Gamma$.

## 2. Results

Recall that $\mathcal{F}$ denotes the Fourier transform, i.e.

$$\mathcal{F}\psi(x) = \int_{\mathbb{R}^d} e^{ix\cdot y}\psi(y)dy,$$
$$\mathcal{F}^{-1}\psi(x) = \frac{1}{(2\pi)^d}\int_{\mathbb{R}^d} e^{-ix\cdot y}\psi(y)dy, \qquad \psi \in \mathcal{S}(\mathbb{R}^d).$$

In what follows $\mu$ and $\Gamma$ always stand for nonnegative tempered measures such that $\mu = \mathcal{F}^{-1}\Gamma$, i.e.

$$\langle \mu, \psi\rangle = \left\langle \Gamma, \mathcal{F}^{-1}\psi\right\rangle, \qquad \text{for all } \psi \in \mathcal{S}(\mathbb{R}^d). \tag{2.1}$$

For $0 < \alpha \le 2$ let $p_t^\alpha$ denote the $\alpha$-stable density and set

$$G_\alpha(x) = \int_0^\infty e^{-t}p_t^\alpha(x)dt.$$

PROPOSITION 2.1. *If $\mu$ and $\Gamma$ are nonnegative tempered measures such that $\mu = \mathcal{F}^{-1}\Gamma$ and $0 < \alpha \le 2$ then*

$$\int_{\mathbb{R}^d} \frac{1}{1+|x|^\alpha}\mu(dx) = \int_{\mathbb{R}^d} G_\alpha(x)\Gamma(dx). \tag{2.2}$$

The equality (2.2) is intuitively clear since $\mathcal{F}G_\alpha$ is well defined and $\mathcal{F}G_\alpha = 1/(1+|x|^\alpha)$, but we cannot apply directly (2.1) since $G_\alpha \notin \mathcal{S}(\mathbb{R}^d)$.

PROOF. Let $f \in \mathcal{S}(\mathbb{R}^d)$ be symmetric, nonnegative, with support contained in the ball $\{|x| \le 1\}$ and $\int_{\mathbb{R}^d} f(x)dx = 1$. Moreover, assume that $\mathcal{F}f \ge 0$ and $\mathcal{F}^{-1}f \ge 0$ (to find such a function it is enough to take $f = g * g$, where $g \in \mathcal{S}(\mathbb{R}^d)$ is symmetric, nonnegative, $\int_{\mathbb{R}^d} g(x)dx = 1$ and the support of $g$ is contained in $\{|x| \le 1/2\}$).

Denote

$$h(x) = \frac{1}{1+|x|^\alpha}, \qquad g_s(x) = e^{-s|x|^\alpha} \qquad s > 0$$
$$f_\varepsilon(x) = \frac{1}{\varepsilon^d} f\left(\frac{x}{\varepsilon}\right), \qquad 0 < \varepsilon < 1. \tag{2.3}$$

We will approximate the integrand on the left hand side of (2.2) by $f_\varepsilon * (g_s h)$ and pass with $s$ and $\varepsilon$ to 0. Note that the function $f_\varepsilon * (g_s h)$ belongs to $\mathcal{S}(\mathbb{R}^d)$.

It is easy to see that

$$\mathcal{F}^{-1}\left[f_\varepsilon * (g_s h)\right](x) = (\mathcal{F}f)(\varepsilon x)(p_s^\alpha * G_\alpha)(x). \tag{2.4}$$

Since $f_\varepsilon * (g_s h) \in \mathcal{S}(\mathbb{R}^d)$, using (2.4) and the assumption $\mu = \mathcal{F}^{-1}\Gamma$, we can write

$$\int_{\mathbb{R}^d} [f_\varepsilon * (g_s h)](x)\mu(dx) = \int_{\mathbb{R}^d} \mathcal{F}f(\varepsilon x)(p_s^\alpha * G_\alpha)(x)\Gamma(dx). \tag{2.5}$$

Notice first that

$$e^{-s}p_s^\alpha * G_\alpha(x) = \int_0^\infty e^{-(s+t)}p_{s+t}^\alpha(x)dt$$
$$= \int_s^\infty p_t^\alpha(x)dt \quad \nearrow G_\alpha(x) \qquad \text{as } s \to 0. \tag{2.6}$$

Also,

$$f_\varepsilon * (g_s h)(x) \quad \nearrow f_\varepsilon * h(x) \qquad \text{as } s \to 0. \tag{2.7}$$

Thus, by (2.5) and the monotone convergence theorem, we have

$$\int_{\mathbb{R}^d} f_\varepsilon * h(x)\mu(dx) = \int_{\mathbb{R}^d} \mathcal{F}f(\varepsilon x)G_\alpha(x)\Gamma(dx). \tag{2.8}$$

Now we pass to 0 with $\varepsilon$ in (2.8). Using the obvious estimate

$$1 + |x|^\alpha \leq 2^\alpha(1 + |x-y|^\alpha)(1 + |y|^\alpha)$$

and the fact that $f_\varepsilon(x) = 0$ for $|x| > \varepsilon$, we have

$$\begin{aligned} f_\varepsilon * h(x) &= \int_{\mathbb{R}^d} \frac{1}{1 + |x-y|^\alpha} f_\varepsilon(y)dy \\ &\leq 2^\alpha \int_{\mathbb{R}^d} \frac{1 + |y|^\alpha}{1 + |x|^\alpha} f_\varepsilon(y)dy \\ &= 2^\alpha h(x) \int_{|y|\leq\varepsilon} (1 + |y|^\alpha) f_\varepsilon(y)dy \\ &\leq 2^\alpha(1 + \varepsilon^\alpha)h(x) \int_{\mathbb{R}^d} f_\varepsilon(y)dy \end{aligned}$$

and, consequently, we obtain

$$f_\varepsilon * h(x) \leq Const\ h(x). \qquad \text{for } 0 < \varepsilon < 1. \tag{2.9}$$

Assume first that $\int_{\mathbb{R}^d} h(x)\mu(dx) < \infty$. Then, by (2.9) and the dominated convergence theorem,

$$\lim_{\varepsilon\to 0} \int_{\mathbb{R}^d} (f_\varepsilon * h)(x)\mu(dx) = \int_{\mathbb{R}^d} h(x)\mu(dx). \tag{2.10}$$

Also, using the fact that $\lim_{\varepsilon\to 0} \mathcal{F}f(\varepsilon x) = 1$, Fatou's Lemma, and (2.8), we have

$$\begin{aligned} \int_{\mathbb{R}^d} G_\alpha(x)\Gamma(dx) &\leq \liminf_{\varepsilon\to 0} \int_{\mathbb{R}^d} \mathcal{F}f(\varepsilon x)G_\alpha(x)\Gamma(dx) \\ &= \liminf_{\varepsilon\to 0} \int_{\mathbb{R}^d} (f_\varepsilon * h)(x)\mu(dx) \\ &= \int_{\mathbb{R}^d} h(x)\mu(dx) < \infty. \end{aligned}$$

Then, by the dominated convergence theorem,

$$\lim_{\varepsilon\to 0} \int_{\mathbb{R}^d} \mathcal{F}f(\varepsilon x)G_\alpha\Gamma(dx) = \int_{\mathbb{R}^d} G_\alpha(x)\Gamma(dx). \tag{2.11}$$

Hence, by (2.8), (2.10) and (2.11), we obtain (2.2) if $\int_{\mathbb{R}^d} h(x)\mu(dx)$ is finite.

Now, assuming that $\int_{\mathbb{R}^d} G_\alpha(x)\Gamma(dx) < \infty$, and using Fatou's Lemma similarly as above, we see that $\int_{\mathbb{R}^d} h(x)\mu(dx)$ is also finite and the equality in (2.2) holds. This finishes the proof of the Proposition. □

The next result is associated with the self–intersection local time of an Ornstein–Uhlenbeck process with values in $\mathcal{S}'(\mathbb{R}^d)$.

THEOREM 2.2. *Let $\mu$ and $\Gamma$ be symmetric nonnegative and tempered measures such that $\mu = \mathcal{F}^{-1}\Gamma$ and let $0 < \alpha \leq 2$.*

(a) *If* $\varphi \geq 0$, $\varphi \in \mathcal{S}(\mathbb{R}^d)$ *then*

$$\int_{\mathbb{R}^{2d}} |\mathcal{F}\varphi(x+y)|^2 \frac{1}{1+|x|^\alpha} \frac{1}{1+|y|^\alpha} \mu(dx)\mu(dy) = \int_{\mathbb{R}^{3d}} \varphi * \varphi_{(s)}(v) G_\alpha(z-v) G_\alpha(w-v) dv \Gamma(dx)\Gamma(dy). \tag{2.12}$$

(b) *Condition*

$$\int_{\mathbb{R}^{2d}} |\mathcal{F}\varphi(x+y)|^2 \frac{1}{1+|x|^\alpha} \frac{1}{1+|y|^\alpha} \mu(dx)\mu(dy) < \infty \qquad \textit{for all } \varphi \in \mathcal{S}(\mathbb{R}^d) \tag{2.13}$$

*is satisfied if and only if for all* $\varphi \geq 0$, $\varphi \in \mathcal{S}(\mathbb{R}^d)$

$$\int_{\mathbb{R}^{3d}} \varphi * \varphi_{(s)}(v) G_\alpha(z-v) G_\alpha(w-v) dv \Gamma(dx)\Gamma(dy) < \infty. \tag{2.14}$$

*If either* (2.13) *or* (2.14) *holds then* (2.12) *is satisfied for all* $\varphi \in \mathcal{S}(\mathbb{R}^d)$.

PROOF. (a): Let

$$\Phi_{\varepsilon,f,s}(x,y) := |\mathcal{F}\varphi(x+y)|^2 (f_\varepsilon * (g_s h))(x)(f_\varepsilon * (g_s h))(y),$$

where $f_\varepsilon$, $g_s$ and $h$ are defined by (2.3). It is easy to see that

$$\mathcal{F}_{2d}^{-1}\Phi_{\varepsilon,f,s}(x,y) = \int_{\mathbb{R}^d} \varphi * \varphi_s(v) \mathcal{F}^{-1}(f_\varepsilon * (g_s h))(x-v) \mathcal{F}^{-1}(f_\varepsilon * (g_s h))(y-v) dv,$$

where $\mathcal{F}_{2d}$ denotes the Fourier transform of a function from $\mathcal{S}(\mathbb{R}^{2d})$. By the assumption $\mu = \mathcal{F}^{-1}\Gamma$ we have

$$\int_{\mathbb{R}^{2d}} |\mathcal{F}\varphi(x+y)|^2 (f_\varepsilon * (g_s h))(x)(f_\varepsilon * (g_s h))(y)\mu(dx)\mu(dy) = \int_{\mathbb{R}^{3d}} \varphi * \varphi_s(v) \mathcal{F}^{-1}(f_\varepsilon * (g_s h))(x-v) \mathcal{F}^{-1}(f_\varepsilon * (g_s h))(y-v) dv \Gamma(dx)\Gamma(dy) \tag{2.15}$$

We use (2.4) and we proceed similarly as in the proof of Proposition 2.1. Using (2.4), (2.6), (2.7) and the monotone convergence theorem, we pass to zero in $s$. Next, passing to 0 in $\varepsilon$, using the dominated convergence theorem and Fatou's Lemma, we obtain (2.12).

(b): By (a), if (2.13) is satisfied then (2.14) also holds. Now assume that (2.14) holds and fix $\varphi \in \mathcal{S}(\mathbb{R}^d)$. By Lemma 6.2 in [**B-G**] there exists $\psi \in \mathcal{S}(\mathbb{R}^d)$ such that $|\varphi| \leq \psi$. Hence, by part (a) of the theorem, we have that

$$\int_{\mathbb{R}^{3d}} |\varphi| * |\varphi_{(s)}|(v) G_\alpha(z-v) G_\alpha(w-v) dv \Gamma(dx)\Gamma(dy) < \infty. \tag{2.16}$$

Using (2.15), (2.16) and Fatou's Lemma we obtain (2.13).

It remains to prove (2.12) for any $\varphi \in \mathcal{S}(\mathbb{R}^d)$ under condition (2.13) or (2.16). We start again with (2.15). The only difference with the proof of part (a) is that now, passing to 0 with $s$, we cannot apply directly the monotone convergence theorem to the right hand side of (2.15). This problem is easily solved by splitting the set of integration into two subsets $\{v : \varphi * \varphi_{(s)}(v) \geq 0\}$ and $\{v : \varphi * \varphi_{(s)}(v) < 0\}$. □

Below we show how to deal with powers higher than 2 in $1/(1+|x|^\alpha)$ in formula (2.15).

THEOREM 2.3. *Let* $\mu$ *and* $\Gamma$ *be symmetric, nonnegative and tempered measures such that* $\mu = \mathcal{F}^{-1}\Gamma$ *and let* $0 < \alpha \leq 2$. *If* $\varphi \geq 0$, $\varphi \in \mathcal{S}(\mathbb{R}^d)$ *then*

$$\int_{\mathbb{R}^{2d}} |\mathcal{F}\varphi(x+y)|^2 \frac{1}{(1+|x|^\alpha)^2} \frac{1}{(1+|y|^\alpha)^2} \mu(dx)\mu(dy)$$
$$= \int_{\mathbb{R}^{3d}} \varphi * \varphi_{(s)}(v) G_\alpha * G_\alpha(z-v) G_\alpha * G_\alpha(w-v) dv \Gamma(dx)\Gamma(dy).$$

The proof of this theorem is similar to the previous one and therefore we omit it. One can also formulate an analogue of part (b) of Theorem 2.2.

**Acknowledgement.** The author is grateful to Prof. J. Zabczyk for suggesting the problem.

## References

[B-G] T. Bojdecki, L.G. Gorostiza, *Self-intersection local time for $\mathcal{S}'(\mathbb{R}^d)$-Wiener process and related Ornstein-Uhlenbeck processes*, Infinite Dimen. Anal. Quant. Probab. Related Topics **2** (1999), pp. 569-615.

[K-Z] A. Karczewska, J. Zabczyk, *Stochastic PDE's with function–valued solutions*, Infinite Dimensional Stochastic Analysis, Proceedings of the Colloquium of Royal Netherlands Academy of Arts and Sciences, Amsterdam, (1999), pp. 197-216.

[K-Z1] A. Karczewska, J. Zabczyk, *Regularity of solutions to stochastic Volterra equations*, Rend. Mat. Acc. Lincei **9**, no. 11, (2000), pp. 141-154.

[L] O. Lévêque, *Hyperbolic stochastic partial differential equations driven by boundary noises*, PhD thesis, École Polytechnique Fédérale de Lausanne (2001).

[P-Z] S. Peszat, J. Zabczyk, *Nonlinear stochastic wave and heat equations*, Probab. Theory Relat. Fields **116** (2000), pp. 421-443.

[S-S] M. Sanz-Solé, M. Sarrà, *Path properties of a class of Gaussian processes with applications to SPDE's*, Proceedings of the Canadian Mathematical Society **28** (2000), pp. 303-316.

INSTITUTE OF MATHEMATICS, UNIVERSITY OF WARSAW, UL. BANACHA 2, 02-097 WARSZAWA, POLAND

*E-mail address*: annatal@mimuw.edu.pl

Contemporary Mathematics
Volume **336**, 2003

# On Two-Parameter Stieltjes Integrals for Functions in Besov-Liouville Spaces and Stochastic Integrals

Constantin Tudor

ABSTRACT. The classical Lebesgue-Stieltjes integral $\int_{a_1}^{b_1}\int_{a_2}^{b_2} fdg$ is extended to a large class of functions $f, g$ of unbounded variation. Fractional integrals and derivatives of functions of two variables are used as main toool. In particular if $f, g$ are Hölder, with the sum of orders biger than 1, then the convergence of the corresponding Riemann-Stieltjes sums to the integral in obtained. If $g$ is a two-parameter Wiener process and $f$ is an adapted process with the paths in some Besov-Liouville space $\Lambda_{\alpha_1,\alpha_2,,p}$, then the Itô integral and the generalized Stieltjes integral agree if $f \in \Lambda_{\alpha_1,\alpha_2,,2}$ for some $\alpha_i > \frac{1}{2}$ and the Itô integral is limit in probability of generalized Stieltjes integrals if $f \in \Lambda_{\alpha_1,\alpha_2,,2}$ for all $0 < \alpha_i < \frac{1}{2}$.

## 1. Introduction

For functions of one variable the Stieltjes integral was extended to functions with unbounded variation by using $p$-variation or fractional integrals and derivatives (see Dudley and Norvaisa [**D-N**], Feyel and De la Pradelle [**F-P**], Kondurar [**Ko**], Young [**Yo**], Zähle [**Z**]). In particular this is the case of the fractional Brownian motion as a driving process.

Such extension make possible the study of the corresponding deterministic and stochastic equations (see Klingenhöfer and M. Zähle [**K-Z**], Kubilius [**Ku1**], [**Ku2**], Lyons [**Ly1**], [**Ly2**], Mikosch and Norvaisa [**M-N**], Nualart and Răşcanu [**N-R**], Ruzmaikina [**R**], Zähle [**Z**]).

In the present paper we consider the case of functions of two variables (extension to many variables is straithforward). By using fractional integrals and derivatives for functions of two variables we extend the Zähle definition for integrands and integrators in Besov-Liouville spaces. We show that in standard cases the integral $\int_{a_1}^{b_1}\int_{a_2}^{b_2} f(s,t)dg(s,t)$ agrees with the usual Stieltjes integral. In particular if $f$ is $(\alpha_1, \alpha_2)$-Hölder and $g$ is $(\beta_1, \beta_2)$-Hölder with $\alpha_i + \beta_i > 1$, we obtain the convergence of the Riemann-Stieltjes sums to the integral. Since the Brownian motion and the fractional Brownian motion have Hölder paths we can define the stochastic integral

2000 *Mathematics Subject Classification.* Primary: 60H05; Secondary: 26A33, 26B99.

*Key words and phrases.* Stieltjes integral, Besov-Liouville space, Itô integral, stochastic integral.

by using the above approach. In particular such an extension makes possible the study of the corresponding deterministic and stochastic equations.

For adapted integrands with the paths in Besov-Liouville spaces we show the relationship between the generalized Stieltjes integral and the Itô integral.

## 2. Preliminaries

Let $f \in L^1(a,b)$ and $0 < \alpha < 1$. Then *left-sided* and *right-sided Riemann-Liouville fractional integrals of order* $\alpha$ of $f$ are defined for almost all $x \in (a,b)$ by

$$I_{a+}^{\alpha} f(x) = \frac{1}{\Gamma(\alpha)} \int_a^x \frac{f(t)}{(x-t)^{1-\alpha}} dt, \tag{2.1}$$

$$I_{b-}^{\alpha} f(x) = \frac{(-1)^{-\alpha}}{\Gamma(\alpha)} \int_x^b \frac{f(t)}{(t-x)^{1-\alpha}} dt. \tag{2.2}$$

REMARK 2.1. (see [**S-K-M**]) ($\mathrm{i}_1$) If $1 < p < \frac{1}{\alpha}$, $\frac{1}{q} = \frac{1}{p} - \alpha$, then the integral operators $I_{a+}^{\alpha}$, $I_{b-}^{\alpha}$ are bounded from $L^p(a,b)$ into $L^q(a,b)$ and moreover

$$I_{a+}^{\alpha}\left(L^p(a,b)\right) = I_{b-}^{\alpha}\left(L^p(a,b)\right).$$

($\mathrm{i}_2$) (*Semigroup property*). For every $\alpha, \beta$,

$$I_{a+}^{\alpha} I_{a+}^{\beta} = I_{a+}^{\alpha+\beta},\ I_{b-}^{\alpha} I_{b-}^{\beta} = I_{b-}^{\alpha+\beta}.$$

($\mathrm{i}_3$) If $f \in I_{a+}^{\alpha}\left(L^p(a,b)\right)$ (resp. $f \in I_{b-}^{\alpha}\left(L^p(a,b)\right)$ then the function $\varphi \in L^p(a,b)$ such that $I_{a+}^{\alpha}\varphi = f$ (resp. $I_{b-}^{\alpha}\varphi = f$ ) is unique and we denote it by $D_{a+}^{\alpha} f$ (resp. $D_{b-}^{\alpha} f$). The function $D_{a+}^{\alpha} f$ (resp. $D_{b-}^{\alpha} f$) is called *the left-sided* (resp. *right-sided*) *Riemann-Liouville fractional derivative of order* $\alpha$ *of* $f$ and we have the *Weyl representations*

$$D_{a+}^{\alpha} f(x) = \frac{1}{\Gamma(1-\alpha)} \left( \frac{f(x)}{(x-a)^{\alpha}} + \alpha \int_a^x \frac{f(x)-f(t)}{(x-t)^{1+\alpha}} dt \right) 1_{(a,b)}(x), \tag{2.3}$$

(resp.,

$$D_{b-}^{\alpha} f(x) = \frac{(-1)^{\alpha}}{\Gamma(1-\alpha)} \left( \frac{f(x)}{(b-x)^{\alpha}} + \alpha \int_x^b \frac{f(x)-f(t)}{(t-x)^{1+\alpha}} dt \right) 1_{(a,b)}(x)) \tag{2.4}$$

(the above Weyl representations are defined for almost all $x \in (a,b)$ and the convergence of the integrals at the singularity $t = x$ holds pointwise for almost all $x \in (a,b)$ if $p = 1$ and moreover in $L^p$ if $1 < p < \infty$).

($\mathrm{i}_4$) The following inversion formulas are satisfied:

$$I_{a+}^{\alpha} D_{a+}^{\alpha} f = f,\ \forall f \in I_{a+}^{\alpha}\left(L^p(a,b)\right),$$

$$D_{a+}^{\alpha} I_{a+}^{\alpha} f = f,\ \forall f \in L^1(a,b),$$

and similar formulas are true for $\left(I_{b-}^{\alpha}, D_{b-}^{\alpha}\right)$.

Now we pass to the case of two variables. Let $T = [a_1,b_1] \times [a_2,b_2]$, $\alpha = (\alpha_1,\alpha_2)$, $\alpha_i \in (0,1)$ and $p = (p_1,p_2)$, $p_i \geq 1$.

For a rectangle $D = (s_1,t_1] \times (s_2,t_2] \subset T$ we define the increment on $D$ of the function $f : T \to R$ by

$$f(D) = f(t_1,t_2) - f(t_1,s_2) - f(s_1,t_2) + f(s_1,s_2).$$

We consider the mixed norm

$$\|f\|_p = \left\{ \int_{a_2}^{b_2} \left[ \int_{a_1}^{b_1} |f(t_1,t_2)|^{p_1} \, dt_1 \right]^{\frac{p_2}{p_1}} dt_2 \right\}^{\frac{1}{p_2}},$$

and we define the mixed $L^p$-space by

$$L^p(T) = \left\{ f : T \longrightarrow R : \|f\|_p < \infty \right\}.$$

It is clear that if $p_1 = p_2$ then $L^p(T) = L^{p_1}(T)$.

For a function $f : T \to R$ we define *the left-sided* (resp. *right-sided*) *Riemann-Liouville fractional integral of order* $\alpha$ by

$$\left( I^{\alpha}_{(a_1,a_2)+} f \right)(x_1,x_2) = \left( I^{\alpha_1}_{a_1+} I^{\alpha_2}_{a_2+} f \right)(x_1,x_2) \tag{2.5}$$

$$= \frac{1}{\Gamma(\alpha_1)\Gamma(\alpha_2)} \int_{a_1}^{x_1} \int_{a_2}^{x_2} \frac{f(t_1,t_2)}{(x_1-t_1)^{1-\alpha_1}(x_2-t_2)^{1-\alpha_2}} dt_1 dt_2, (x_1,x_2) \in T,$$

(resp.

$$\left( I^{\alpha}_{(b_1,b_2)-} f \right)(x_1,x_2) = \left( I^{\alpha_1}_{b_1-} I^{\alpha_2}_{b_2-} f \right)(x_1,x_2) \tag{2.6}$$

$$= \frac{(-1)^{-(\alpha_1+\alpha_2)}}{\Gamma(\alpha_1)\Gamma(\alpha_2)} \int_{x_1}^{b_1} \int_{x_2}^{b_2} \frac{f(t_1,t_2)}{(t_1-x_1)^{1-\alpha_1}(t_2-x_2)^{1-\alpha_2}} dt_1 dt_2, (x_1,x_2) \in T).$$

REMARK 2.2. (see [**S-K-M**]) (a) If $1 < p_i < \frac{1}{\alpha_i}$, $\frac{1}{q_i} = \frac{1}{p_i} - \alpha_i$, then the integral operators $I^{\alpha}_{(a_1,a_2)+}$, $I^{\alpha}_{(b_1,b_2)-}$ are bounded from $L^p(T)$ into $L^q(T)$ and moreover

$$I^{\alpha}_{(a_1,a_2)+}\left(L^p(T)\right) = I^{\alpha}_{(b_1,b_2)-}\left(L^p(T)\right).$$

(b) (*Semigroup property*). For every $\alpha, \beta$,

$$I^{\alpha}_{(a_1,a_2)+} I^{\beta}_{(a_1,a_2)+} = I^{\alpha+\beta}_{(a_1,a_2)+}, \; I^{\alpha}_{(b_1,b_2)-} I^{\beta}_{(b_1,b_2)-} = I^{\alpha+\beta}_{(b_1,b_2)-}. \tag{2.7}$$

(c) If $f \in I^{\alpha}_{(a_1,a_2)+}\left(L^p(T)\right)$ (resp. $f \in I^{\alpha}_{(b_1,b_2)-}\left(L^p(T)\right)$) then the function $\varphi \in L^p(T)$ such that $I^{\alpha}_{(a_1,a_2)+}\varphi = f$ (resp. $I^{\alpha}_{(b_1,b_2)-}\varphi = f$ ) is unique and we denote it by $D^{\alpha}_{(a_1,a_2)+}f$ (resp. $D^{\alpha}_{(b_1,b_2)-}f$). The function $D^{\alpha}_{(a_1,a_2)+}f$ (resp. $D^{\alpha}_{(b_1,b_2)-}f$) is called *the left-sided* (resp. *right-sided*) *Riemann-Liouville fractional derivative of order* $\alpha$ *of* $f$.

By using arguments as in Section 24.7 of [**S-K-M**] for two variables and infinite interval, we obtain that

$$D^{\alpha}_{(a_1,a_2)+}f = D^{\alpha_1}_{a_1+} D^{\alpha_2}_{a_2+} f, D^{\alpha}_{(b_1,b_2)-}f = D^{\alpha_1}_{b_1-} D^{\alpha_2}_{b_2-} f, \tag{2.8}$$

and, taking into account (2.3) and (2.4), we obtain the *Weyl representations*

$$D^{\alpha}_{(a_1,a_2)+} f(x_1,x_2) \tag{2.9}$$

$$= \frac{1}{\Gamma(1-\alpha_1)\Gamma(1-\alpha_2)} \left\{ \frac{f(x_1,x_2)}{(x_1-a_1)^{\alpha_1}(x_2-a_2)^{\alpha_2}} \right.$$

$$+ \frac{\alpha_2}{(x_1-a_1)^{\alpha_1}} \int_{a_2}^{x_2} \frac{f(x_1,x_2) - f(x_1,t_2)}{(x_2-t_2)^{1+\alpha_2}} dt_2$$

$$+\frac{\alpha_1}{(x_2-a_2)^{\alpha_2}}\int_{a_1}^{x_1}\frac{f(x_1,x_2)-f(t_1,x_2)}{(x_1-t_1)^{1+\alpha_1}}dt_1$$
$$\left.+\alpha_1\alpha_2\int_{a_1}^{x_1}\int_{a_2}^{x_2}\frac{f((t_1,x_1]\times(t_2,x_2])}{(x_1-t_1)^{1+\alpha_1}(x_2-t_2)^{1+\alpha_2}}dt_1dt_2\right\},$$

(resp.

$$D^{\alpha}_{(b_1,b_2)-}f(x_1,x_2) \tag{2.10}$$
$$=\frac{(-1)^{\alpha_1+\alpha_2}}{\Gamma(1-\alpha_1)\Gamma(1-\alpha_2)}\left\{\frac{f(x_1,x_2)}{(b_1-x_1)^{\alpha_1}(b_2-x_2)^{\alpha_2}}\right.$$
$$+\frac{\alpha_2}{(b_1-x_1)^{\alpha_1}}\int_{x_2}^{b_2}\frac{f(x_1,x_2)-f(x_1,t_2)}{(t_2-x_2)^{1+\alpha_2}}dt_2$$
$$+\frac{\alpha_1}{(b_2-x_2)^{\alpha_2}}\int_{x_1}^{b_1}\frac{f(x_1,x_2)-f(t_1,x_2)}{(t_1-x_1)^{1+\alpha_1}}dt_1$$
$$\left.+\alpha_1\alpha_2\int_{x_1}^{b_1}\int_{x_2}^{b_2}\frac{f((t_1,x_1]\times(t_2,x_2])}{(t_1-x_1)^{1+\alpha_1}(t_2-x_2)^{1+\alpha_2}}dt_1dt_2\right\}).$$

The convergence of the integrands in (2.9), (2.10) is understood as limit in $L^p$, as $\varepsilon_i\to 0$, of the corresponding approximations obtained by replacing $\int_{a_i}^{x_i}$(resp. $\int_{x_i}^{b_i}$) by $\int_{a_i}^{x_i-\varepsilon_i}$(resp. $\int_{x_i+\varepsilon_i}^{b_i}$) and $\int_{a_1}^{x_1}\int_{a_2}^{x_2}$(resp. $\int_{x_1}^{b_1}\int_{x_2}^{b_2}$) with $\int_{a_1}^{x_1-\varepsilon_1}\int_{a_2}^{x_2-\varepsilon_2}$(resp. $\int_{x_1+\varepsilon_1}^{b_1}\int_{x_2+\varepsilon_2}^{b_2}$).

(d) (*Semigroup property*). For every $\alpha,\beta$,

$$D^{\alpha}_{(a_1,a_2)+}D^{\beta}_{(a_1,a_2)+}=D^{\alpha+\beta}_{(a_1,a_2)+},\ D^{\alpha}_{(b_1,b_2)-}D^{\beta}_{(b_1,b_2)-}=D^{\alpha+\beta}_{(b_1,b_2)-}. \tag{2.11}$$

(e) (*Integration by parts*). If $f\in L^p(T)$, $g\in L^q(T)$, $\frac{1}{p_i}+\frac{1}{q_i}\le 1+\alpha_i$, then

$$\int_T f(t)I^{\alpha}_{(a_1,a_2)+}g(t)dt=(-1)^{\alpha_1+\alpha_2}\int_T g(t)I^{\alpha}_{(b_1,b_2)-}f(t)dt, \tag{2.12}$$

$$(-1)^{\alpha_1+\alpha_2}\int_T\left(D^{\alpha}_{(a_1,a_2)+}f(t)\right)g(t)dt=\int_T f(t)D^{\alpha}_{(b_1,b_2)-}g(t)dt. \tag{2.13}$$

For $\alpha,p$ satisfying the conditions of Remark 2.2-(a), the space $\Lambda_{\alpha,p}=I^{\alpha}_{(a_1,a_2)+}(L^p(T))=I^{\alpha}_{(b_1,b_2)-}(L^p(T))$ is called the *Liouville space* (or the *Besov space*) and it becomes a separable Banach space with respect to the norm

$$\left\|I^{\alpha}_{(a_1,a_2)+}f\right\|_{\alpha,p}=\|f\|_p\,.$$

We denote by $H_{[a_i,b_i],\alpha_i}$ the space of all $\alpha_i$-Hölder functions on $[a_i,b_i]$ and

$$\|f\|_{[a_i,b_i],\alpha_i}=\sup_{u\neq v,a_i\le u,v\le b_i}\frac{|f(u)-f(v)|}{|u-v|^{\alpha_i}}.$$

Also, we denote by $H_{T,\alpha_1,\alpha_2}$ the space of all $(\alpha_1,\alpha_2)$-Hölder functions on $T$, i.e., $f\in H_{T,\alpha_1,\alpha_2}$ if $f$ is continuous,

$$\|f(a_1,.)\|_{[a_2,b_2],\alpha_2}<\infty,\ \|f(.,a_2)\|_{[a_1,b_1],\alpha_1}<\infty,$$

and

$$\|f\|_{T,\alpha_1,\alpha_2}=\sup_{u_i\neq v_i}\frac{|f((u_1,v_1]\times(u_2,v_2])|}{|u_1-v_1|^{\alpha_1}|u_2-v_2|^{\alpha_2}}<\infty.$$

REMARK 2.3. Assume that $0<\beta_1<\alpha_1, 0<\beta_2<\alpha_2$ and $p_1, p_2>1$. Then we have

$$H_{T,\alpha_1,\alpha_2} \subset \Lambda_{\beta_1,\beta_2,p} \text{ if } \beta_i p_i < 1.$$

## 3. Generalized Stieltjes integral

For functions of one variable an extension of the Stieltjes integral is given in [**Z**] in the following way. Let $u, v : (a,b) \longrightarrow R$ and define

$$u_{a+}(x) = 1_{(a,b)}(x)\left[u(x) - u(a+)\right], \; v_{b-}(x) = 1_{(a,b)}(x)\left[v(x) - v(b-)\right].$$

If $u, v$ satisfy the *Hypothesis* (FGS), i.e.,

$$u_{a+} \in I_{a+}^{\alpha}(L^p(a,b)), \; v_{b-} \in I_{b-}^{1-\alpha}(L^q(a,b)),$$

for some $\frac{1}{p}+\frac{1}{q} \le 1$, $0<\alpha<1$, then the *Stieltjes integral* of $u$ with respect to $v$ is defined by

$$\int_a^b u dv = u(a+)\left[v(b-) - v(a+)\right] \tag{3.1}$$
$$+(-1)^{\alpha} \int_a^b D_{a+}^{\alpha} u_{a+}(x) D_{b-}^{1-\alpha} u_{b-}(x) dx.$$

If $f, g : T \longrightarrow R$ then we introduce the following functions (assuming that all left and right limits which appear exist): for $t_i \in (a_i, b_i)$,

$$f_{a_1+}(t_1, a_2+) = \left[f(t_1,a_2+) - f(a_1+,a_2+)\right] 1_{(a_1,b_1)}(t_1),$$
$$f_{(a_1,a_2)+}(t_1,t_2)$$
$$= \left[f(t_1,t_2) - f(t_1,a_2+) - f(a_1+,t_2) + f(a_1+,a_2+)\right] 1_{(a_1,b_1)\times(a_2,b_2)}(t_1,t_2),$$
$$g_{(b_1,b_2)-}(t_1,t_2)$$
$$= \left[g(t_1,t_2) - g(t_1,b_2-) - g(b_1-,t_2) + g(b_1-,b_2-)\right] 1_{(a_1,b_1)\times(a_2,b_2)}(t_1,t_2),$$

(the other functions $f_{a_1+}(a_1+,t_2)$, $g_{b_1-}(t_1,b_2-)$, $g_{b_1-}(b_1-,t_2)$ are defined in a similar manner).

Next we extend the Stieltjes to functions of two variables.

Let $\alpha = (\alpha_1, \alpha_2)$, $\alpha_i \in (0,1)$, $p = (p_1, p_2)$, $q = (q_1, q_2)$ be such that $\frac{1}{p_i}+\frac{1}{q_i} \le 1$ and let $f, g : T \longrightarrow R$.

We introduce the following hypothesis:

*Hypothesis* (FGS):

$f_{(a_1,a_2)+} \in I_{(a_1,a_2)+}^{\alpha}(L^p(T)), g_{(b_1,b_2)-} \in I_{(b_1,b_2)-}^{1-\alpha}(L^q(T))$,
$f_{a_1+}(.,a_2+) \in I_{a_1+}^{\alpha_1}(L^{p_1}(a_1,b_1))$,
$g_{b_1-}(.,b_2-), g_{b_1-}(.,a_2+) \in I_{a_1+}^{1-\alpha_1}(L^{q_1}(a_1,b_1))$,
$f_{a_2+}(a_1+,.) \in I_{a_2+}^{\alpha_2}(L^{p_2}(a_2,b_2))$,
$g_{b_2-}(b_1-,.), g_{b_2-}(a_1+,.) \in I_{b_2-}^{1-\alpha_2}(L^{q_2}(a_2,b_2))$.

Standard properties of the Lebesgue-Stieltjes integral for smooth functions suggest the following extension of the Stieltjes integral.

DEFINITION 3.1. *Assume (FGS) is satisfied. Then we define the forward generalized Stieltjes integral of $f$ with respect to $g$ by*

$$\int_T f(t_1,t_2) d^- g(t_1,t_2) \tag{3.2}$$
$$= f(a_1+,a_2+) g_{(b_1,b_2)-}(a_1+,a_2+)$$

$$+(-1)^{\alpha_1}\int_{a_1}^{b_1} D^{\alpha_1}_{a_1+}f_{a_1+}(t_1,a_2+)\left[D^{1-\alpha_1}_{b_1-}g_{b_1-}(t_1,b_2-)-D^{1-\alpha_1}_{b_1-}g_{b_1-}(t_1,a_2+)\right]dt_1$$

$$+(-1)^{\alpha_2}\int_{a_2}^{b_2} D^{\alpha_2}_{a_2+}f_{a_2+}(a_1+,t_2)\left[D^{1-\alpha_2}_{b_2-}g_{b_2-}(b_1-,t_2)-D^{1-\alpha_2}_{b_2-}g_{b_2-}(a_1+,t_2)\right]dt_2$$

$$+(-1)^{\alpha_1+\alpha_2}\int_T D^{\alpha}_{(a_1,a_2)+}f_{(a_1,a_2)+}(t_1,t_2)D^{1-\alpha}_{(b_1,b_2)-}g_{(b_1,b_2)-}(t_1,t_2)dt_1t_2.$$

Proposition 3.2. *The definition (3.2) is correct, i.e., independent of the choice of* $\alpha_i$.

Proof. First assume that $\alpha_1=\alpha_2+\beta,\ \beta>0$. Then, by using the semigroup property (2.11) and the integration by parts (2.13), we have

$$\int_T D^{\alpha}_{(a_1,a_2)+}f_{(a_1,a_2)+}(t_1,t_2)D^{1-\alpha}_{(b_1,b_2)-}g_{(b_1,b_2)-}(t_1,t_2)dt_1t_2$$

$$=\int_T D^{(\beta,0)}_{(a_1,a_2)+}D^{\alpha}_{(a_1,a_2)+}f_{(a_1,a_2)+}(t_1,t_2)D^{1-\alpha}_{(b_1,b_2)-}g_{(b_1,b_2)-}(t_1,t_2)dt_1t_2$$

$$=\int_T D^{(\alpha_2,\alpha_2)}_{(a_1,a_2)+}f_{(a_1,a_2)+}(t_1,t_2)D^{(\beta,0)}_{(b_1,b_2)-}D^{1-\alpha}_{(b_1,b_2)-}g_{(b_1,b_2)-}(t_1,t_2)dt_1t_2$$

$$=\int_T D^{(\alpha_2,\alpha_2)}_{(a_1,a_2)+}f_{(a_1,a_2)+}(t_1,t_2)D^{(1-\alpha_1+\beta,1-\alpha_2)}_{(b_1,b_2)-}g_{(b_1,b_2)-}(t_1,t_2)dt_1t_2$$

$$=\int_T D^{(\alpha_2,\alpha_2)}_{(a_1,a_2)+}f_{(a_1,a_2)+}(t_1,t_2)D^{(1-\alpha_2,1-\alpha_2)}_{(b_1,b_2)-}g_{(b_1,b_2)-}(t_1,t_2)dt_1t_2.$$

Therefore it is enough to prove the assertion for $\alpha_1=\alpha_2$, in which case we proceed as above.

For the other terms in (3.2) the proof is given in [**Z**]. □

Proposition 3.3. (1) *If* $f\in C^2$ *(resp.* $g\in C^2$ *) and* $g$ *(resp.* $f$*) is Riemann integrable and satisfies (FGS), then* $\int_T fd^-g$ *exists and agrees with the Stieltjes integral* $(S)\int_T fdg$ .

(2) *If* $f$ *is continuous and* $g$ *is with bounded variation and* $g$ *satisfies (FGS), then* $\int_T fd^-g$ *exists and agrees with the Stieltjes integral* $(S)\int_T fdg$ .

(3) *If* $T_1=[c_1,d_1]\times[c_2,d_2]\subset T$ *then*

$$\int_{T_1} fd^-g=\int_T 1_{T_1}fd^-g,$$

*if for both integrals the Hypothesis (FGS) is fulfilled.*

Proof. (1) It is straightforward using the integration by parts.

The assertion (2)(resp. (3)) is a consequence of (1) by approximating $f$ (resp. $g$) with the corresponding mollifier(regularization) and then passing to the limit by the dominated convergence theorem. □

Let $\alpha'=(\alpha'_1,\alpha'_2)$, $\alpha'_i\in(0,1)$, $p'=(p'_1,p'_2)$, $q'=(q'_1,q'_2)$ be such that $\frac{1}{p'_i}+\frac{1}{q'_i}\le 1$ and let $f,g:T\longrightarrow R$ .

We introduce the following hypothesis:

*Hypothesis* (BGS):

$f_{(b_1,b_2)-} \in I^{\alpha'}_{(b_1,b_2)-}(L^{p'}(T)), g_{(a_1,a_2)+} \in I^{1-\alpha'}_{(a_1,a_2)+}(L^{q'}(T)),$
$f_{b_1-}(.,b_2-) \in I^{\alpha'_1}_{b_1-}(L^{p'_1}(a_1,b_1)),$
$g_{a_1+}(.,b_2-), g_{a_1+}(.,a_2+) \in I^{1-\alpha'_1}_{a_1+}(L^{q'_1}(a_1,b_1)),$
$f_{b_2-}(b_1-,.) \in I^{\alpha'_2}_{b_2-}(L^{p'_2}(a_2,b_2)),$
$g_{a_2+}(b_1-,.), g_{a_2+}(a_1+,.) \in I^{1-\alpha'_2}_{a_2+}(L^{q'_2}(a_2,b_2)).$

DEFINITION 3.4. *Assume (BGS) is satisfied. Then we define the backward generalized Stieltjes integral of $f$ with respect to $g$ by*

$$\int_T f(t_1,t_2)d^+g(t_1,t_2) = f(b_1-,b_2-)g_{(a_1,a_2)+}(b_1-,b_2-) \tag{3.3}$$

$$+(-1)^{-(\alpha'_1+\alpha'_2)}\int_T D^{(\alpha'_1,\alpha'_2)}_{(b_1,b_2)-}f_{(b_1,b_2)-}(t_1,t_2)D^{(1-\alpha'_1,1-\alpha'_2)}_{(a_1,a_2)+}g_{(a_1,a_2)+}(t_1,t_2)dt_1t_2$$

$$+(-1)^{-\alpha'_1}\int_{a_1}^{b_1} D^{\alpha'_1}_{b_1-}f_{b_1-}(t_1,b_2-)\left[D^{1-\alpha'_1}_{a_1+}g_{a_1+}(t_1,b_2-) - D^{1-\alpha'_1}_{a_1+}g_{a_1+}(t_1,a_2+)\right]dt_1$$

$$+(-1)^{-\alpha'_2}\int_{a_2}^{b_2} D^{\alpha'_2}_{b_2-}f_{b_2-}(b_1-,t_2)\left[D^{1-\alpha'_2}_{a_2+}g_{a_2+}(b_1-,t_2) - D^{1-\alpha'_2}_{a_2+}g_{a_2+}(a_1+,t_2)\right]dt_2.$$

REMARK 3.5. By similar arguments as for the forward Stieltjes integral, the previous integral is well defined.

PROPOSITION 3.6. *Assume $f,g$ satisfy (FSG) and (BSG). Then*

$$\int_T fd^-g = \int_T fd^+g. \tag{3.4}$$

*Moreover assume that:*

$$(g(.,a_2+),f(.,a_2+)),(g(.,b_2-),f(.,b_2-)),$$

$$(g(a_1+,.),f(a_1+,.)),(g(b_1-,.),f(b_1-,.)),$$

*satisfy (FGS).*

*Then the following integration by parts formula hold:*

$$\int_T fd^-g = (fg)_{(a_1,a_2)+}(b_1-,b_2-) \tag{3.5}$$

$$+\int_{a_1}^{b_1} g(t_1,a_2+)df(t_1,a_2+) - \int_{a_1}^{b_1} g(t_1,b_2-)df(t_1,b_2-)$$

$$+\int_{a_2}^{b_2} g(a_1+,t_2)df(a_1+,t_2) - \int_{a_2}^{b_2} g(b_1-,t_2)df(b_1-,t_2) + \int_T gd^-f.$$

PROOF. If $f,g$ are smooth then both integrals in (3.4) coincide with the classical Stieltjes integral $(S)\int_a^b fdg$ and (3.5) is the integration by parts formula for functions in two variables (see [**Ye**]). Then approximate $f,g$ with the corresponding mollifiers and then pass to the limit. □

## 4. Riemann-Stieltjes sums and Hölder continuous functions

In this section we see that for Hölder functions, the generalized Stieltjes integral is obtained as limit of Riemann-Stieltjes sums.

LEMMA 4.1. *Assume that*

$$f(t_1,t_2) = 1_{(c_1,d_1]\times(c_2,d_2]}(t_1,t_2),\ a_i < c_i < d_i < b_i,$$

*and $g$ is $(\lambda_1,\lambda_2)$-Hölder. Then*

$$\int_T fdg = g\left((c_1,d_1]\times(c_2,d_2]\right). \tag{4.1}$$

PROOF. Choose $\alpha_1 = \alpha_2 = 1-\varepsilon,\ 0<\varepsilon<\min(\lambda_1,\lambda_1)$. Since we have

$$D^{(1-\varepsilon,1-\varepsilon)}_{(a_1,a_2)+}1_{(c_1,d_1]\times(c_2,d_2]}(t_1,t_2) = D^{1-\varepsilon}_{a_1+}1_{(c_1,d_1]}(t_1)D^{1-\varepsilon}_{a_2+}1_{(c_2,d_2]}(t_2),$$

$$D^{\alpha}_{a+}1_{(c,d]}(u) = \frac{1}{\Gamma(1-\alpha)}\left[1_{(c,b)}(u)\frac{1}{(u-c)^{\alpha}} - 1_{(d,b)}(u)\frac{1}{(u-d)^{\alpha}}\right],$$

if $a<c<d<b$, it is easily seen that

$$(-1)^{2(1-\varepsilon)}\int_T D^{(1-\varepsilon,1-\varepsilon)}_{(a_1,a_2)+}f_{(a_1,a_2)+}(t_1,t_2)D^{(\varepsilon,\varepsilon)}_{(b_1,b_2)-}g_{(b_1,b_2)-}(t_1,t_2)dt_1dt_2$$

$$= g\left((c_1,d_1]\times(c_2,d_2]\right).$$

The other terms in the definition of the generalized Stieltjes integral $\int_T fdg$ vanish. □

The proof of Theorem 4.1.4 from [**Z**] implies the following useful result.

LEMMA 4.2. *Let $f\in H_{[a,b],\lambda}$ and $0<\alpha<\lambda$. For $\Delta_1>0$ and a partition*

$$\mathcal{P}(\Delta_1): a = s_0 < s_1 < ... < s_n = b,$$

*with $max_i(s_{i+1}-s_i)\le\Delta_1$, we define*

$$f_{\mathcal{P}(\Delta_1)}(x) = \sum_{i=0}^{n-1} f(s_i)1_{(s_i,s_{i+1}]}(x).$$

*Then*

$$\left\|\int_a^{\cdot}\frac{f_{\mathcal{P}(\Delta_1)}(.)-f(.)-\left[f_{\mathcal{P}(\Delta_1)}(y)-f(y)\right]}{(.-y)^{\alpha+1}}dy\right\|_1 \tag{4.2}$$

$$\le C(\lambda,\alpha)(b-a)\left\|f\right\|_{[a,b],\lambda}\Delta_1^{\lambda-\alpha}.$$

REMARK 4.3. Recall the following property: If $f\in H_{[a,b],\lambda}$ and $0<\alpha<\lambda$, then the function

$$g(.) = \frac{f(.)-f(a)}{(.-a)^{\alpha}} \in H_{[a,b],\lambda-\alpha}, \tag{4.3}$$

and

$$\left\|g\right\|_{[a,b],\lambda-\alpha} \le C\left\|f\right\|_{[a,b],\lambda}, \tag{4.4}$$

where $C$ does not depend on $f$ (see for example [**M**, p.14] or prove it directly).

LEMMA 4.4. *If $f \in H_{T,\lambda}$ and $0 < \alpha_i < \lambda_i$ . Then*

$$\left\| D^{\alpha}_{(a_1,a_2)+} f_{(a_1,a_2)+} \right\|_{\infty} \leq C \left\| f \right\|_{T,\lambda}, \tag{4.5}$$

$$\left\| D^{\alpha}_{(b_1,b_2)-} f_{(b_1,b_2)-} \right\|_{\infty} \leq C \left\| f \right\|_{T,\lambda}, \tag{4.6}$$

*where $C = C(a_i, b_i, \alpha_i, \lambda) < \infty$.*

PROOF. We prove only (4.5) (for (4.6) the proof is similar). We use the Weyl representation (2.9). By using the Lipschitz property it follows that

$$\left| \frac{f_{(a_1,a_2)+}(x_1,x_2)}{(x_1-a_1)^{\alpha_1}(x_2-a_2)^{\alpha_2}} \right| \leq \left\| f \right\|_{T,\lambda} (b_1-a_1)^{\lambda_1-\alpha_1} (b_2-a_2)^{\lambda_2-\alpha_2},$$

$$\begin{aligned}
&\left| \frac{1}{(x_1-a_1)^{\alpha_1}} \int_{a_2}^{x_2} \frac{f_{(a_1,a_2)+}(x_1,x_2) - f_{(a_1,a_2)+}(x_1,t_2)}{(x_2-t_2)^{1+\alpha_2}} dt_2 \right| \\
&\leq \frac{1}{(x_1-a_1)^{\alpha_1}} \int_{a_2}^{x_2} \frac{f\left((a_1,x_1] \times (a_2,x_2]\right)}{(x_2-t_2)^{1+\alpha_2}} dt_2 \\
&\leq \frac{\left\| f \right\|_{T,\lambda}}{\lambda_2-\alpha_2} (x_1-a_1)^{\lambda_1-\alpha_1} (x_2-a_2)^{\lambda_2-\alpha_2} \\
&\leq \frac{\left\| f \right\|_{T,\lambda}}{\lambda_2-\alpha_2} (b_1-a_1)^{\lambda_1-\alpha_1} (b_2-a_2)^{\lambda_2-\alpha_2}.
\end{aligned}$$

Similarly we get

$$\begin{aligned}
&\left| \frac{1}{(x_2-a_2)^{\alpha_2}} \int_{a_1}^{x_1} \frac{f_{(a_1,a_2)+}(x_1,x_2) - f_{(a_1,a_2)+}(t_1,x_2)}{(x_1-t_1)^{1+\alpha_1}} dt_1 \right| \\
&\leq \frac{\left\| f \right\|_{T,\lambda}}{\lambda_1-\alpha_1} (b_1-a_1)^{\lambda_1-\alpha_1} (b_2-a_2)^{\lambda_2-\alpha_2},
\end{aligned}$$

$$\begin{aligned}
&\left| \int_{a_1}^{x_1} \int_{a_2}^{x_2} \frac{f\left((t_1,x_1] \times (t_2,x_2]\right)}{(x_1-t_1)^{1+\alpha_1}(x_2-t_2)^{1+\alpha_2}} dt_1 dt_2 \right| \\
&\leq \frac{\left\| f \right\|_{T,\lambda}}{(\lambda_1-\alpha_1)(\lambda_2-\alpha_2)} (b_1-a_1)^{\lambda_1-\alpha_1} (b_2-a_2)^{\lambda_2-\alpha_2}.
\end{aligned}$$

□

Next for $\Delta > 0$ we consider partitions of the form

$$\mathcal{P}_\Delta : a_1 = s_0 < s_1 < \ldots < s_{n(1)} = b_1,\ a_2 = t_0 < t_1 < \ldots < t_{n(2)} = b_2,$$

such that

$$\max_i (s_{i+1}-s_i) + \max_j (t_{j+1}-t_j) \leq \Delta.$$

For a function $f : T \longrightarrow R$ we define

$$f_{\mathcal{P}(\Delta)}(x_1,x_2) = \sum_{i=0}^{n(1)-1} \sum_{j=0}^{n(2)-1} f(s_i,t_j) 1_{(s_i,s_{i+1}] \times (t_j,t_{j+1}]}(x_1,x_2). \tag{4.7}$$

PROPOSITION 4.5. *If* $f \in H_{T,\lambda}$, $f(.,b_1) = f(a_1,.) = 0$ *and* $0 < \alpha_i < \lambda_i$ , *then*

$$\lim_{\Delta \to 0} \sup_{\mathcal{P}(\Delta)} \left\| f_{\mathcal{P}(\Delta)} - f \right\|_{L^\infty(T)} = 0, \tag{4.8}$$

$$\lim_{\Delta \to 0} \sup_{\mathcal{P}(\Delta)} \left\| D^{\alpha}_{(a_1,a_2)+} \left[ \left( f_{\mathcal{P}(\Delta)} \right)_{(a_1,a_2)+} - f_{(a_1,a_2)+} \right] \right\|_{L^1(T)} = 0, \tag{4.9}$$

$$\lim_{\Delta \to 0} \sup_{\mathcal{P}(\Delta)} \left\| D^{\alpha}_{(b_1,b_2)-} \left[ \left( f_{\mathcal{P}(\Delta)} \right)_{(b_1,b_2)-} - f_{(b_1,b_2)-} \right] \right\|_{L^1(T)} = 0. \tag{4.10}$$

PROOF. The statement (4.8) is obvious. We only prove (4.9) (the conclusion (4.10) is similar). Next we have

$$\begin{aligned} &\left| D^{(\alpha_1,\alpha_2)}_{(a_1,a_2)+} \left[ \left( f_{\mathcal{P}(\Delta)} \right)_{(a_1,a_2)+} - f_{(a_1,a_2)+} \right] (x_1,x_2) \right| \\ &\quad \le \frac{1}{\Gamma(1-\alpha_1)\Gamma(1-\alpha_2)} \sum_{i=1}^{4} J_i(x_1,x_2), \end{aligned} \tag{4.11}$$

$$J_1(x_1,x_2) = \left| \frac{f_{\mathcal{P}(\Delta)}(x_1,x_2) - f(x_1,x_2)}{(x_1-a_1)^{\alpha_1}(x_2-a_2)^{\alpha_2}} \right|, \tag{4.12}$$

$$J_2(x_1,x_2) = \frac{\alpha_2}{(x_1-a_1)^{\alpha_1}}$$

$$\times \left| \int_{a_2}^{x_2} \frac{f_{\mathcal{P}(\Delta)}(x_1,x_2) - f_{\mathcal{P}(\Delta)}(x_1,u_2) - (f(x_1,x_2) - f(x_1,u_2))}{(x_2-u_2)^{1+\alpha_2}} du_2 \right|,$$

$$J_3(x_1,x_2) = \frac{\alpha_1}{(x_2-a_2)^{\alpha_2}}$$

$$\times \left| \int_{a_1}^{x_1} \frac{f_{\mathcal{P}(\Delta)}(x_1,x_2) - f_{\mathcal{P}(\Delta)}(u_1,x_2) - (f(x_1,x_2) - f(u_1,x_2))}{(x_1-u_1)^{1+\alpha_1}} du_1 \right|,$$

$$J_4(x_1,x_2) = \alpha_1\alpha_2 \left| \int_{a_1}^{x_1} \int_{a_2}^{x_2} \frac{\left( f_{\mathcal{P}(\Delta)} - f \right) ((u_1,x_1] \times (u_2,x_2])}{(x_1-u_1)^{1+\alpha_1}(x_2-u_2)^{1+\alpha_2}} du_1 du_2 \right|.$$

Since for $(x_1,x_2) \in (s_i,s_{i+1}] \times (t_j,t_{j+1}]$ we have

$$\begin{aligned} &|f(s_i,t_j) - f(x_1,x_2)| \\ &\le \|f\|_{T,\lambda} \left[ (b_1-a_1)^{\lambda_1} + (b_2-a_2)^{\lambda_2} \right] \left[ (x_1-s_i)^{\lambda_1} + (x_2-t_j)^{\lambda_2} \right] \\ &\quad \le \|f\|_{T,\lambda} \left[ (b_1-a_1)^{\lambda_1} + (b_2-a_2)^{\lambda_2} \right] \left( \Delta^{\lambda_1} + \Delta^{\lambda_2} \right), \end{aligned}$$

we deduce that

$$\begin{aligned} &\|J_1\|_{L^1(T)} \\ &\le \|f\|_{T,\lambda} \left[ (b_1-a_1)^{\lambda_1} + (b_2-a_2)^{\lambda_2} \right] \left( \Delta^{\lambda_1} + \Delta^{\lambda_2} \right) \\ &\times \sum_{i=0}^{n(1)-1} \sum_{j=0}^{n(2)-1} \int_{s_i}^{s_{i+1}} \int_{t_j}^{t_{j+1}} \frac{1}{(x_1-a_1)^{\alpha_1}(x_2-a_2)^{\alpha_2}} dx_1 dx_2 \\ &= C \left( \Delta^{\lambda_1} + \Delta^{\lambda_2} \right). \end{aligned} \tag{4.13}$$

Next we define

$$f_{\mathcal{P}_1(\Delta)}(x_1,x_2)=\sum_{i=0}^{n(1)-1} f(s_i,x_2)1_{(s_i,s_{i+1}]}(x_1),$$

$$f_{\mathcal{P}_2(\Delta)}(x_1,x_2)=\sum_{j=0}^{n(2)-1} f(x_1,t_j)1_{(t_j,t_{j+1}]}(x_2).$$

It is easily seen that

($\mathrm{i}_1$) For every $x_i$,

$$f_{\mathcal{P}(\Delta)}(x_1,x_2)=\left(f_{\mathcal{P}_1(\Delta)}(x_1,.)\right)_{\Delta_2}(x_2). \tag{4.14}$$

($\mathrm{i}_2$) For every $0\le u_1\le x_1$, $x_2\in[a_2,b_2]$, (here the fact that $f(.,b_1)=f(a_1,.)=0$ is used),

$$\begin{aligned} f_{\mathcal{P}(\Delta)}&\left((u_1,x_1]\times(a_2,x_2]\right) \\ &=\left(f_{\mathcal{P}_1(\Delta)}\left((u_1,x_1]\times(a_2,.]\right)\right)_{\mathcal{P}_2(\Delta)}(x_2), \\ x_2&\longrightarrow f_{\mathcal{P}_1(\Delta)}\left((u_1,x_1]\times(a_2,x_2]\right)\in H_{[a_2,b_2],\lambda_2}, \end{aligned} \tag{4.15}$$

$$\left\|f_{\mathcal{P}_1(\Delta)}\left((u_1,x_1]\times(a_2,.]\right)\right\|_{[a_2,b_2],\lambda_2}\le\|f\|_{T,\lambda}(x_1-u_1)^{\lambda_1}. \tag{4.16}$$

It is clear that

$$J_4(x_1,x_2)\le\alpha_1\alpha_2\left[J_{41}(x_1,x_2)+J_{42}(x_1,x_2)\right], \tag{4.17}$$

$$J_{41}(x_1,x_2)=\left|\int_{a_1}^{x_1}\int_{a_2}^{x_2}\frac{\left[f_{\mathcal{P}(\Delta)}-f_{\mathcal{P}_1(\Delta)}\right]\left((u_1,x_1]\times(u_2,x_2]\right)}{(x_1-u_1)^{1+\alpha_1}(x_2-u_2)^{1+\alpha_2}}du_1du_2\right|, \tag{4.18}$$

$$J_{42}(x_1,x_2)=\left|\int_{a_1}^{x_1}\int_{a_2}^{x_2}\frac{\left[f_{\mathcal{P}_1(\Delta)}-f\right]\left((u_1,x_1]\times(u_2,x_2]\right)}{(x_1-u_1)^{1+\alpha_1}(x_2-u_2)^{1+\alpha_2}}du_1du_2\right|. \tag{4.19}$$

Next $C$ is a constant which may change from time to time. By (4.15), (4.16) and Lemma 4.2 we have

$$\int_{a_2}^{x_2}\frac{\left[f_{\mathcal{P}(\Delta)}-f_{\mathcal{P}_1(\Delta)}\right]\left((u_1,x_1]\times(u_2,x_2]\right)}{(x_2-u_2)^{1+\alpha_2}}du_2 \tag{4.20}$$

$$=\int_{a_2}^{x_2}\frac{\left(f_{\mathcal{P}_1(\Delta)}\left((u_1,x_1]\times(u_2,.]\right)\right)_{\mathcal{P}_2(\Delta)}(x_2)-f_{\mathcal{P}_1(\Delta)}\left((u_1,x_1]\times(u_2,x_2]\right)}{(x_2-u_2)^{1+\alpha_2}}du_2,$$

$$\begin{aligned}&\left\|\int_{a_2}^{x_2}\frac{\left(f_{\mathcal{P}_1(\Delta)}\left((u_1,x_1]\times(u_2,.]\right)\right)_{\mathcal{P}_2(\Delta)}(x_2)-f_{\mathcal{P}_1(\Delta)}\left((u_1,x_1]\times(u_2,x_2]\right)}{(x_2-u_2)^{1+\alpha_2}}du_2\right\|_{L^1([a_2,b_2])}\\ &\le C\left\|f_{\mathcal{P}_1(\Delta)}\left((u_1,x_1]\times(a_2,.]\right)\right\|_{[a_2,b_2],\lambda_2}\Delta^{\lambda_2-\alpha_2}\le C(x_1-u_1)^{\lambda_1}\Delta^{\lambda_2-\alpha_2}.\end{aligned}$$

Since for $x_1\in[a_1,b_1]$, $a_2\le u_2\le x_2$ we have

$$x_1\longrightarrow f\left((a_1,x_1]\times(u_2,x_2]\right)\in H_{[a_1,b_1],\lambda_1},$$

$$\left\|f\left((a_1,.]\times(u_2,x_2]\right)\right\|_{[a_1,b_1],\lambda_1}\le\|f\|_{T,\lambda}(x_2-u_2)^{\lambda_2},$$

by Lemma 4.2 we obtain
(4.21)
$$\int_{a_1}^{b_1}\left|\int_{a_1}^{x_1}\frac{\left[f_{\mathcal{P}_1(\Delta)}-f\right]((u_1,x_1]\times(u_2,x_2])}{(x_1-u_1)^{1+\alpha_1}}du_1\right|dx_1\le C\,(x_2-u_2)^{\lambda_2}\,\Delta^{\lambda_1-\alpha_1}.$$

Now from (4.17)-(4.21) we obtain easily that

$$\|J_4\|_{L^1(T)}\le C\left(\Delta^{\lambda_1-\alpha_1}+\Delta^{\lambda_2-\alpha_2}\right). \tag{4.22}$$

In a similar manner we obtain that

$$\|J_2\|_{L^1(T)}\le C\left(\Delta^{\lambda_1-\alpha_1}+\Delta^{\lambda_2-\alpha_2}\right), \tag{4.23}$$

$$\|J_3\|_{L^1(T)}\le C\left(\Delta^{\lambda_1-\alpha_1}+\Delta^{\lambda_2-\alpha_2}\right). \tag{4.24}$$

Finally from (4.13), (4.22)-(4.24) we obtain that

$$\left\|D^{(\alpha_1,\alpha_2)}_{(a_1,a_2)+}\left[\left(f_{\mathcal{P}(\Delta)}\right)_{(a_1,a_2)+}-f_{(a_1,a_2)+}\right]\right\|_{L^1(T)}$$
$$\le C\left(\Delta^{\lambda_1}+\Delta^{\lambda_2}+\Delta^{\lambda_1-\alpha_1}+\Delta^{\lambda_2-\alpha_2}\right),$$

which implies the desired conclusion. □

THEOREM 4.6. *Assume that $f\in H_{T,\lambda}$ and $g\in H_{T,\mu}$ with $\lambda_i+\mu_i>1$. Then the generalized Stieltjes integral $\int_T fdg$ exists and moreover*

$$\lim_{\Delta\to 0}\sup_{\mathcal{P}_\Delta}\left|\sum_{i=0}^{n(1)-1}\sum_{j=0}^{n(2)-1}f(s_i,t_j)g\left((s_i,s_{i+1}]\times(t_j,t_{j+1}]\right)-\int_T fdg\right|=0. \tag{4.25}$$

PROOF. Choose $1-\mu_i<\alpha_i<\lambda_i$. As a consequence of the Hölder property of $f,g$ it follows that the Hypothesis (FGS) is fulfiled for $\alpha=(\alpha_1,\alpha_2)$ (see also Lemma 4.4). By Lemma 4.1 we have

$$\sum_{i=0}^{n(1)-1}\sum_{j=0}^{n(2)-1}f(s_i,t_j)g\left((s_i,s_{i+1}]\times(t_j,t_{j+1}]\right)=\int_T f_{\mathcal{P}_\Delta}dg.$$

By considering if necessary the new function

$$f_1(u_1,u_2)=f\left((a_1,u_1]\times(a_2,u_2]\right),$$

which is in $H_{T,\lambda}$ and satisfies $f_1(a_1,.)=f_1(.,b_1)=0$, it follows that we can assume $f(a_1,.)=f(.,b_1)=0$. Then, Lemma 4.4, Proposition 4.5 and Theorem 4.2.1 of [**Z**] imply that (4.25) is satisfied. □

## 5. A pathwise stochastic integral

The generalized Stieltjes integral gives the possibility of defining a pathwise stochastic integral for two-parameter integrands and driving processes with trajectories in Besov-Liouville spaces.

Next we illustrate this in the case of so called fractional Brownian sheet (or two-parameter fractional Brownian motion).

Let $\left\{W_t^h\right\}_{t\in T}$ be a two-parameter fractional Brownian motion with the Hurst parameter $h=(h_1,h_2)$, $h_i\in(0,1)$. Recall that $W^h$ is a continuous centered Gaussian process, vanishing on the axes, with the covariance

$$R_h^{(2)}(s,t)=R_{h_1}^{(1)}(s_1,t_1)R_{h_2}^{(1)}(s_2,t_2),$$

if $s = (s_1, s_2), t = (t_1, t_2)$, where for $\gamma \in (0,1)$,

$$R^{(1)}_\gamma(u,v) = \frac{1}{2}\left(u^{2\gamma} + v^{2\gamma} - |u-v|^{2\gamma}\right), \quad u, v \geq 0.$$

Clear that if $h_i = \frac{1}{2}$ then $W^{(\frac{1}{2},\frac{1}{2})}$ is the two-parameter Brownian motion.

From the Kolmogorov criterium (see [**F-P**]) it follows that $W^h$ has $\beta$-Hölder continuous paths for every $0 < \beta_i < h_i$.

Then Theorem 4.6 allows to define a pathwise stochastic integral of the form $\int_T f dW^h$ for a process $f$ such that $f_{(0,0)+} \in I^\alpha_{(0,0)+}(L^1(T))$ a.s. (in particular if $f \in H_{T,\alpha}$ a.s.), where $\alpha_i + h_i > 1$.

In the case of the Brownian sheet we have the following relationship between the pathwise stochastic integral previously defined and the Itô integral denoted by $(I)\int_T f dW^{\frac{1}{2}}$ (the proof follows along the same arguments as Theorem 5.2.1 and Corollary 5.2.2 of [**Z**] and we omit the details).

THEOREM 5.1. (1) *If $f \in \Lambda_{\alpha,2}$ a.s., for some $\alpha_i > \frac{1}{2}$, $f$ adapted, then*

$$\int_T f dW^{\frac{1}{2}} = (I)\int_T f dW^{\frac{1}{2}}. \tag{5.1}$$

(2) *If $f \in \Lambda_{\alpha,2}$ a.s. for all $0 < \alpha_i < \frac{1}{2}$, $f$ adapted, then*

$$\lim_{\varepsilon \to 0} \int_T I^{(\varepsilon,\varepsilon)}_{(a_1,a_2)+} f dW^{\frac{1}{2}} = (I)\int_T f dW^{\frac{1}{2}}. \tag{5.2}$$

## References


[A-L-P] A. Ayache, S. Leger and M. Pontier, *Drap brownian fractionnaire,* Potential Anal. **17**, no. 1 (2002), pp. 31–43.

[D-N] R.M. Dudley and R. Norvaiša, *Product integrals, Young integrals and p-variation,* in Differentiability of Six Operators on Nonsmooth Functions and $p$-Variation. With the collaboration of Jinghua Qian. Lecture Notes in Math. **1703** (1999), Springer-Verlag, Berlin.

[F-P] D. Feyel and A. de La Pradelle, *On fractional Brownian processes.* Potential Anal. **10** (1999), pp. 273-288.

[K-Z] F. Klingenhöfer and M. Zähle, *Ordinary differential equations with fractal noise,* Proc. Amer. Math. Soc. **127** (1999), pp. 1021–1028.

[Ko] V. Kondurar, *Sur l'intégrale de Stieljes,* Rec. Math. Moscou, n. Ser. **2** (1937), pp. 361–366.

[Ku1] K. Kubilius, *The existence and uniqueness of the solution of the integral equation driven by a bounded p-variation function,* Proc. of the Lithuanian Math. Soc.. vol. **III**. Tehnika (1999), pp. 136–142.

[Ku2] K. Kubilius, *The existence and uniqueness of the solution of the integral equation driven by fractional Brownian motion,* Lithuanian Math. J. **40** (2000), (special issue), pp. 104–110.

[L-P] S. Leger and M. Pontier, *Drap brownian fractionnaire,* C. R. Acad. Sci. Paris Sér. I Math. **329** (1999), pp. 893–898.

[Ly1] T. Lyons, *Differential equations driven by rough signal. I. An extension of an inequality of L.C. Young,* Math. Res. Lett. **1** (1994), pp. 451–464.

[Ly2] T. Lyons, *The interpretation and solution of ordinary differential equations driven by rough signals,* Stochastic Analysis. Proceedings of the Summer Research Institute on Stochastic Analysis, eds. M. C. Cranston et al., held at Cornell University, Ithaca, NY, USA, July 11–30, 1993. Providence, RI: American Mathematical Society. Proc. Symp. Pure Math. **57** (1995), pp. 115–128.

[M-N] T. Mikosch and R. Norvaisa, *Stochastic integral equations without probability,* Bernoulli **6** (2000), pp. 401–434.

[M] N.I. Muskhelishvili, *Singular Integral Equations,* Second Edition, P. Noordhoff N.V. Groningen–Holland, (1953).

[N-R] D. Nualart and A. Răşcanu, *Differential equations driven by fractional Brownian motion,* Collect. Math. **53** (2002), pp. 55–81.

[R] A. A. Ruzmaikina, *Stieltjes integrals of Hölder continuous functions with applications to fractional Brownian motion,* J. Statist. Phys. **100**, nos. 5–6, pp. 1049–1069.

[S-K-M] S. G. Samko, A. A. Kilbas and O.I. Marichev, *Fractional Integrals and Derivatives. Theory and applications,* Edited and with a foreword by S. M. Nikol'skiĭ. Gordon and Breach Science Publishers, (1993).

[Ye] J. Yeh, *Cameron-Martin translation theorems in the Wiener space of functions of two-variables,* Trans. Amer. Math. Soc. **107** (1963), pp. 409–420.

[Yo] L.C. Young, *An inequality of the Hölder type connected with Stieltjes integration,* Acta Math. **67** (1936), pp. 251–282.

[Z] M. Zähle, *Integration with respect to fractal functions and stochastic calculus I,* Probab. Theory Related Fields **111** (1998), pp. 333–374.

University of Bucharest and CIMAT, Department of Mathematics, 70109 Bucharest, Romania.

*E-mail address*: ctudor@pro.math.unibuc.ro, tudor@cimat.mx